NASA's Voyager Missions

Exploring the Outer Solar System and Beyond

Springer
London
Berlin
Heidelberg
New York
Hong Kong
Milan
Paris
Tokyo

Ben Evans with David M Harland

NASA's Voyager Missions

Exploring the Outer Solar System and Beyond

Ben Evans	David M Harland
Space Writer	Space Historian
Atherstone	Kelvinbridge
Warwickshire	Glasgow
UK	UK

SPRINGER–PRAXIS BOOKS IN ASTRONOMY AND SPACE SCIENCES
SUBJECT *ADVISORY EDITOR*: John Mason B.Sc., M.Sc., Ph.D.

ISBN 1-85233-745-1 Springer-Verlag London Berlin Heidelberg New York Hong Kong Milan Paris Tokyo

British Library Cataloguing-in-Publication Data
Evans, Ben, 1976
NASA's Voyager missions: exploring the outer solar system and beyond. – (Springer-Praxis books in astronomy and space sciences)
1. Voyager Project 2. Outer planets 3. Solar system 4. Outer space – Exploration
I. Title II. Harland, David M. (David Michael), 1955–
629.4′354

ISBN 1-85233-745-1

Library of Congress Cataloging-in-Publication Data
Evans, Ben, 1976–
NASA's Voyager missions: exploring the outer solar system and beyond / Ben Evans with David M. Harland.
p. cm.
Includes bibliographical references and index.
ISBN 1-85233-745-1 (alk. paper)
1. Outer space – Exploration. 2. Voyager Project. 3. United States, National Aeronautics and Space Administration. I. Harland, David M. (David Michael), 1955– II. Title.

QB500.262.E83 2003
919.9′204–dc22 2003058513

Project Copy Editor: Alex Whyte
Cover design: Jim Wilkie
Typesetting: BookEns Ltd, Royston, Herts., UK

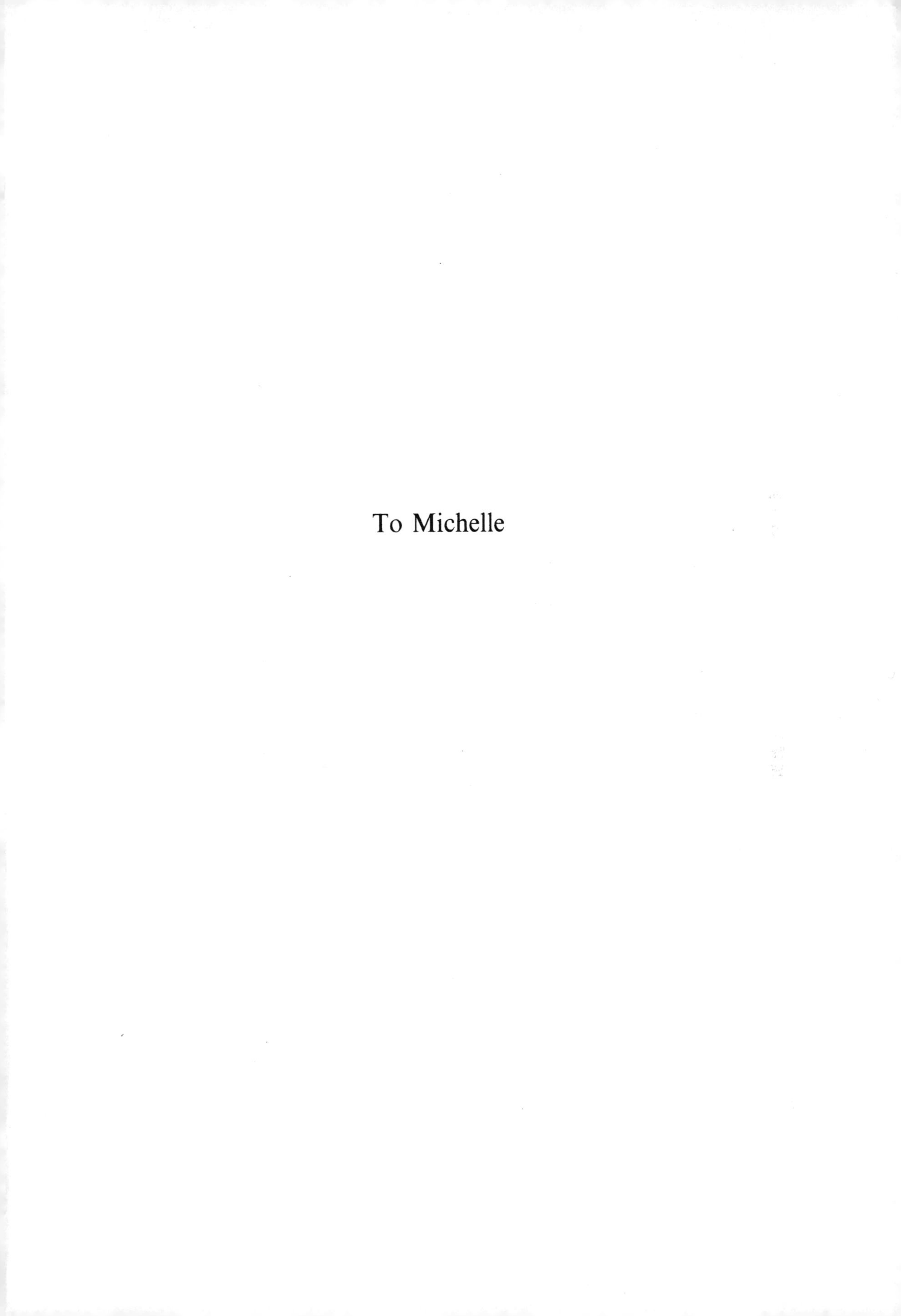

To Michelle

Contents

List of illustrations

Colour section: between pages 156–157

Author's preface

A quarter of a century ago, perhaps the most ambitious and far-sighted adventure in human history got underway when a pair of identical spacecraft were launched on a journey to unlock the secrets of the outer Solar System. Dubbed Voyagers 1 and 2, they were originally destined only to visit the giant planets Jupiter and Saturn – a remarkable achievement in its own right – to investigate their turbulent atmospheres, their rings, their multitudes of attendant moons and their celestial neighbourhood.

If their mission had ended after departing the realm of Saturn, they would easily have provided more information about these two giant worlds than had been gathered over several hundred years of ground-based telescopic observations. Indeed, as Voyager Imaging Team Leader, Brad Smith, said shortly after Voyager 1 flew through the Saturnian system: "We learned more about Saturn in one week than in the entire span of human history".

Yet for one of the Voyagers, the mission would also involve close encounters with the distant giants Uranus and Neptune: two worlds which were, until quite recently, on the fringes of not only the Solar System, but also on the fringes of human knowledge. A decade and a half have now elapsed since Voyager 2 left Neptune behind in the summer of 1989, and both it and its sister ship are presently about 10 billion km from Earth, heading for the as-yet-elusive 'heliopause' – the region of space in which the 1.5 million km/h solar wind slows down to a quarter of its previous speed. This region, which may be reached by the Voyagers in as little as a few years from now, is thought by many scientists to represent the waning of the Sun's magnetic influence.

This book aims to introduce the reader to the history of the discovery and observation of Jupiter, Saturn, Uranus and Neptune, providing context into which the Voyager missions can be neatly slotted. It will discuss the development of procedures for sending spacecraft deep into the Solar System, which ultimately led to the construction of the Voyager spacecraft, and relate the discoveries made at each of the giant planets in turn. However, in view of the tremendous amount of new data that has been gained in recent years – by Galileo at Jupiter, by the Hubble Space Telescope in orbit around the Earth and by ground-based observatories using advanced optical systems, to name but a few – some important aspects of the story will be brought up to date.

The book was conceived on what may be the eve of the Voyagers' most exciting discovery: the first direct observations of conditions in the Milky Way Galaxy beyond the Sun's magnetic influence. It is intended primarily as a means of commemorating what may be one of the few space science missions remembered clearly a thousand years from now – the missions which enabled humanity to explore the outer Solar System and set sail for the stars.

Ben Evans
Atherstone, June 2003

Acknowledgements

Writing a book has been my dream for more years than I can remember and I must thank a number of people for helping to make it come true. I must thank my parents, Marilyn and Tim Evans, who have fostered and supported my love of space exploration, astronomy and writing from childhood. They bought me my first telescope, took me to astronomy and space clubs and encouraged me to take my first steps into publishing. Other friends who have helped me immeasurably along the way have been Andy Salmon, Mike Bryce, Dave Shayler and others from the Midlands Spaceflight Society, who have patiently endured and answered my endless questions and taken the time to read and publish my articles in their journal, *Capcom*. Thanks are also due to Sandie Dearn for her support.

My grateful appreciation must go to Clive Horwood of Praxis for giving me this opportunity. Clive has been eternally enthusiastic and supportive of this project since February 2002, when he received a rather sheepish e-mail request from someone who wanted to "write a book about Voyager". David Harland's truly encyclopedic knowledge of spaceflight history has been invaluable over the past year. Whenever I sought a reference, paper, article or picture, no matter how obscure, he had it readily to hand. I am also grateful to David for reading an early draft of the book and for producing the illustrations.

A number of scientists involved, directly or indirectly, with Voyager, have very kindly offered their help and guidance over the last year. Thanks are due in particular to Ellis Miner, Bob Brown, Andy Ingersoll, Gary Flandro and Michael Kaiser. As a non-scientist, I deeply appreciate the support and assistance of individuals whose knowledge and understanding of technical or scientific concepts greatly surpasses my own. Any glaring errors or omissions that remain in the text are down to me.

Finally, I must thank Michelle Chawner, who has encouraged and pushed me during the frequent occasions when I never thought this project would work. She has endured, without complaint, more 'lost' weekends than I dare admit, which I have spent typing on the PC from dawn until dusk. I would like to dedicate this, my first book, to Michelle, for her love and support and for playing an enormous role in making all this possible.

1

Wanderers

DIVINE ORIGINS

To the ancients, they were just five wandering stars. Nightly, from time immemorial, astronomers, astrologers, priests and wise men the world over watched, marvelled and wondered at their strange dance through the heavens. They looked like stars – tiny specks of light in an ocean of unfathomable darkness – yet their regular, predictable journeys across the sky set them apart from the rest. Some moved faster than others and did not seem to 'twinkle' like ordinary stars. Small wonder, then, that the ancients named these five distant wanderers after some of their most important deities.

Today, we know them as Mercury, Venus, Mars, Jupiter and Saturn; names handed down through the generations from ancient Roman gods. We also know them not as 'stars', but 'planets', from a Greek word that literally means 'wanderer'. Mercury, the closest planet to the Sun, speeds around its parent star in under three months and perhaps for this reason was named after the fleet-footed messenger of the gods. Venus, a beautiful gem of the morning and evening sky, owes her name to the goddess of love and looks. Blood-red Mars' mythological namesake was the harbinger of warfare.

That leaves only far-off Jupiter and Saturn. The former has always been one of the brightest objects in Earth's skies, and clearly so impressed the Romans that they named it after their chief god. With hindsight, this allusion could not have been more appropriate, for the celestial Jupiter is also the largest of nine planets now known to circle the Sun. The Romans, of course, did not know or probably even suspect this. They had a 'geocentric' conception of the Universe in which Earth lay at the centre and everything else – Sun, Moon, planets and stars – revolved around it.

To the Romans, Jupiter was the supreme god, whose most important role was guardianship of the state, its laws and the dispensation of justice. However, most of the names of deities and mythological heroes or monsters handed down to us by the Romans have their roots in much older Greek tales. Hence, Mercury to the Greeks was Hermes, Venus became the lovely Aphrodite and Mars' counterpart was Ares. Jupiter and Saturn are no exception. The former was simply a Latinised version of Zeus, king of the Greek pantheon, while the latter was a Roman variant of his father, Cronus.

The Greek poet Hesiod first related the story of Zeus and Cronus to us, sometime in the eighth century BC. One of his most important epic works, *Theogony*, is a genealogy of the gods and not only introduces us to Zeus and Cronus, but also to a host of other colourful characters who will lend their names to the worlds we will encounter in this book. According to Hesiod, it began when the primordial sky god Uranus and his wife Gaia (Mother Earth) conceived a dozen immensely strong demigods called the Titans, one of whom was Cronus.

Unfortunately, the Titans were not the happy couple's only offspring. They were soon joined by three new arrivals known as Hecatonchires – hideous, enormously powerful monsters, each endowed with 50 heads and a hundred hands – whose appearance so disgusted their father that he imprisoned them in Tartarus, an abyssal cave deep within the bowels of Earth. In desperation, Gaia appealed to Cronus to oppose Uranus' tyrannical behaviour and prevent him from siring further children. Cronus dutifully attacked his father and castrated him with a sickle; and the falling blood droplets created numerous other races of goddesses and monsters.

Yet even this was not the end of Hesiod's story. The harrowing episode seems to have left Cronus so paranoid that one of his own children might rise up against him that he swore to swallow each of his offspring as soon as they were born. True to his word, the first five children he fathered by his wife Rhea – including the sea god Poseidon, later called Neptune by the Romans – quickly ended up in Cronus' bottomless stomach, unharmed but powerless to attack him.

(*Left*) Jupiter, the chief god of the Roman pantheon. (*Right*) A wild-haired and wide-eyed Cronus devours his children, as painted by Francisco de Goya (1746–1828).

Understandably, neither Rhea nor Gaia was particularly ecstatic at the prospect of losing more children. As a result, when her sixth child, Zeus, was born, Rhea wrapped a stone in the baby's swaddling clothes and handed it to her husband. Cronus, for all his divine attributes, was apparently unable to tell the difference and gobbled up the stone, thinking it to be his newborn son. Zeus, meanwhile, was quietly whisked away to Crete, where he was raised by the goat Amalthea. Upon reaching adulthood, he returned to his father's domain and forced him to regurgitate his five brothers and sisters.

In the resultant war between Zeus and Cronus' dynasty, the latter were eventually deposed. Zeus was then able to assume the throne as king of the gods. Despite his bad press, however, Cronus' reign is regarded by Hesiod as a golden age and it was perhaps with this in mind that the Romans later compared him to their own god of agriculture, Saturn. He is also often portrayed in classical art as quite elderly – the famed Bringer of Old Age – and the association between him and the planet Saturn may arise from the latter's slow progress across the heavens.

Zeus' subsequent adventures and sexual escapades are as legendary as the god himself, but a few are worthy of note as the young men, maidens and offspring involved have given their names to many of the moons that presently orbit the planets Jupiter and Saturn. One of Zeus' liaisons was with a maiden named Dione, which, according to some legends, led to the conception of the goddess Aphrodite. Other notches on the divine bedpost included the Spartan queen Leda, the Phoenician princess Europa, the Trojan prince Ganymede, the nymph Callisto and Io, daughter of the river god Inachus.

Still more names have arisen from Cronus' siblings. Titan is the name of Saturn's largest moon and will become an object of particular scrutiny in January 2005 when the European Huygens spacecraft attempts to land there. Of the other Titans besides Cronus, the names Rhea, Iapetus, Tethys, Hyperion and Phoebe also lend themselves to moons of Saturn. Hesiod thus provides us with an introduction to the background of not only Jupiter and Saturn, but also Uranus and Neptune, two other planets lying even further from the Sun. These four mysterious worlds are the focus of this book.

THE EARTH MOVES!

Although both Uranus and Neptune are several times larger than Earth, they were much too distant and faint to be seen without a telescope. For millennia, probably longer than writing had been in existence, educated people accepted that the Solar System had only five planets, the outermost of which was Saturn. Opinion as to the precise layout of the planets, admittedly, has changed quite considerably. For more than a thousand years, since the teachings of Claudius Ptolemaeus in the second century AD, it was assumed that Earth was at the centre of the Universe and everything else revolved around it.

Ptolemaeus, or 'Ptolemy' as he is more popularly known, was a Greek from Alexandria who devised a mathematical framework to explain the Earth-centred

Claudius Ptolemaeus (c. 100–170 AD), the Greek Alexandrian whose notion of a geocentric Universe persisted for more than 1,400 years.

model of the Universe originally espoused by the great thinker Aristotle more than four centuries earlier. He was concerned that Aristotle's theory was unable to reproduce or predict certain observed phenomena, such as the periodic and temporary retrograde motion of Mars. His solution was to imagine each planet orbiting Earth in a circular path, called a 'deferent'. However, according to Ptolemy, our planet was slightly offset from the centre of these deferents.

In his work *Almagest*, he argued that planets appeared to speed up and slow down because they did not revolve around Earth at a uniform rate. Instead, he identified a point – called an 'equant' – that was equal and opposite Earth from the centre of the deferent. If an observer were to stand there, it would be possible to see the planet orbiting at a uniform rate. Ptolemy's incredibly complicated model also helped to explain the retrograde motion of the planets: he decided that they must also travel through smaller orbits, called 'epicycles', along the deferent.

This explained why planets like Mars, which ordinarily travelled eastwards across the sky, occasionally slowed down, 'stopped' and for a short time began to move in a

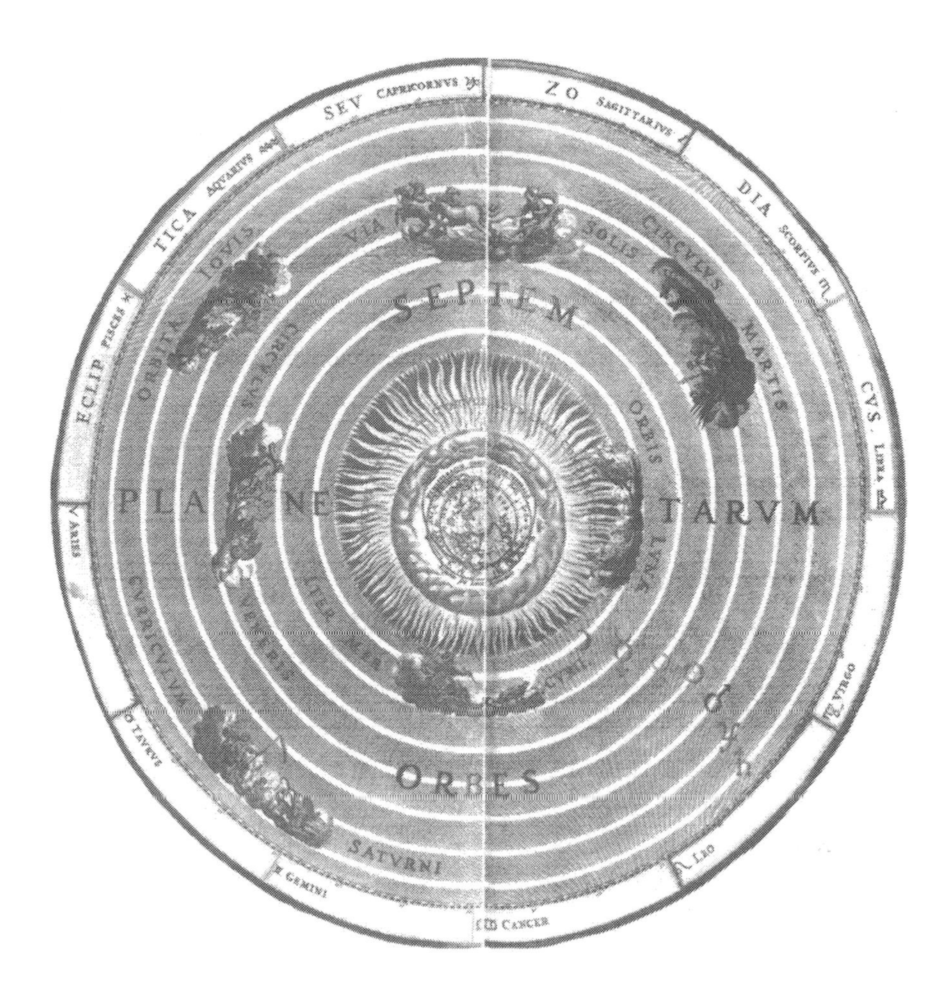

The Earth-centred Ptolemaic view of the Universe, in which the Sun, Moon, planets and stars circled our world, endured in scientific and theological circles from Roman times until the Renaissance.

westerly direction. During the time the planet was at the furthest point in its epicycle, according to Ptolemy, it would seem to a terrestrial observer that it was moving eastwards. However, by the time it reached the nearest point on the epicycle to Earth, it would apparently be travelling in a westerly – or 'retrograde' – path. Ptolemy's model remained current in scientific circles for more than 1,400 years.

The Ptolemaic view of a geocentric Universe was also, by the sixteenth century, well established in theological circles and arguments to the contrary were likely to bring charges of heresy or even the dreaded Inquisition. Nevertheless, in November 1543, Polish astronomer Nicolaus Copernicus – wisely, on his deathbed – published his 'heliocentric' theory, in which he stated that Earth and the other planets actually circled a central, stationary Sun. Although he continued to have the planets orbiting

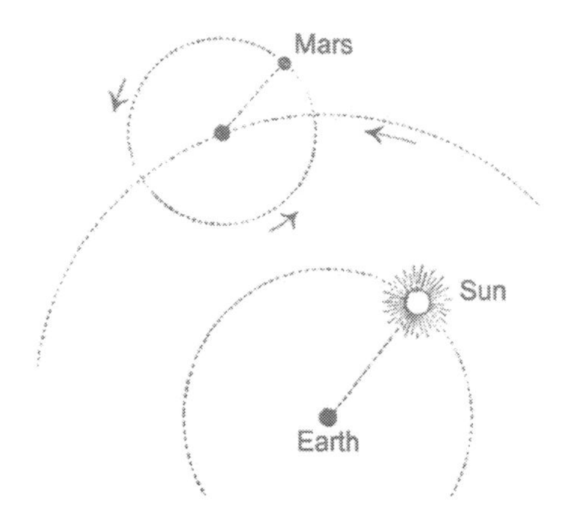

This diagram shows how Ptolemy envisaged epicycles to operate.

Nicolaus Copernicus (1473–1543), who realised that the planets orbit the Sun.

the Sun on Ptolemaic epicycles, the Copernican model nevertheless helped to explain the true mechanics of their mysterious retrograde motions.

A Universe with the Sun very close to its centre, he reasoned, meant that planets further from their parent star would take longer to circle it and thus move much more slowly than closer ones. As 'faster' planets like Venus overtake Earth in their orbits, their motion against background stars, as seen by a terrestrial observer, appears to change. Moreover, in his own commentaries, written as early as 1514, Copernicus acknowledged that a heliocentric Universe and moving Earth also helped to account for other observed phenomena, such as the daily motion of the Sun and stars.

Despite his immense contribution to modern cosmology, Copernicus was not the first person to suggest the idea of a heliocentric Universe. Nearly two thousand years earlier, in the third century BC, the astronomer Aristarchus of Samos had pondered

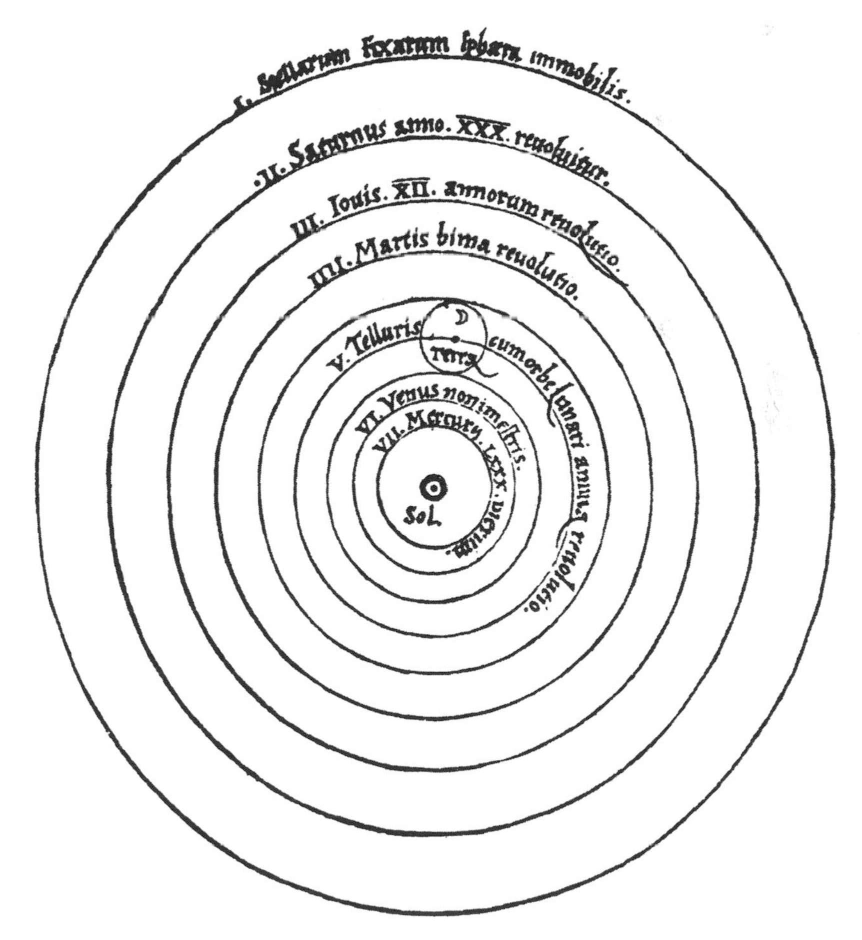

By the mid-sixteenth century the Earth had, quite literally, moved for Nicolaus Copernicus. Although considered heretical for more than a century after his death, his Sun-centred model of the Universe gradually received scientific support.

a similar possibility. Our main source for this is Greek mathematician Archimedes, who related Aristarchus' ideas in his work *The Sand-Reckoner* and criticised them as mathematically meaningless. Whatever its merits, Aristarchus' original work is now lost and certainly did not appear to have fostered much support even in ancient times.

SUPPORT FROM KEPLER

Unlike Aristarchus, however, support slowly materialised for Copernicus and one courageous soul who established himself as an early proponent of a heliocentric Universe was German astronomer Johannes Kepler, who later empirically discovered three laws of planetary motion. Although introduced to geocentric astronomy at university, he was not content simply to calculate the planets' positions and ignore whether or not the models corresponded with actual observed phenomena in the heavens.

Kepler seems to have embraced Copernicanism almost from the outset and, in his first foray into heliocentric astronomy, he attempted to calculate the relative size of

Johannes Kepler (1571–1630), an early proponent of Copernicanism and discoverer of the three laws of planetary motion.

each planet's orbital path. He realised that if the annual component of a planet's motion was a reflection of Earth's own annual motion, it should be possible to calculate through direct observation the size of each planet's orbital path around the Sun. In an incredibly complex 'cosmological model', published in 1596, Kepler showed that there were enormous distances between the planets.

He suggested that if an imaginary sphere were drawn to touch the inside of Saturn's orbital path, and if a cube were inscribed within it and a second sphere inscribed within that cube, then the innermost sphere would very precisely trace the orbital path of Jupiter. Furthermore, if a regular tetrahedron was inscribed inside this sphere and another, yet smaller, sphere was squeezed into the tetrahedron, it would trace the orbit of Mars. Kepler repeated the process three more times, using a regular dodecahedron, icosahedron and octahedron to trace the paths of Earth, Venus and Mercury.

Remarkably, despite its cumbersome appearance, the model not only agreed with the relative sizes of the planets' orbital paths predicted by Copernicus – yielding an error margin no greater than 10 per cent – but even today it is an astonishingly accurate mathematical yardstick. In Kepler's mind, it vindicated Copernicanism with observational evidence. In his work *Mysteries of the Cosmos*, he noted that the key advantage of the heliocentric theory over its geocentric counterpart was its greater ability to explain observed phenomena such as why Venus and Mercury were never far from the Sun.

Despite the obvious advantages of the heliocentric theory over Ptolemaic thought, Copernicus had still felt obliged to incorporate epicycles into the motions of his planets, partly because he assumed that all heavenly bodies should orbit the Sun along perfectly circular paths. By the early seventeenth century, however, Kepler had a new position as imperial mathematician to the Holy Roman Emperor. He was thus able to draw on a vast collection of observational material, which led him to conclude that Mars' orbit, at least, was not circular, but slightly elliptical, with the Sun in one of its foci.

This important discovery was subsequently extended to all the planets and became the basis of Kepler's First Law. Taking into account the elliptical nature of these orbits, and the off-centre position of the Sun within them, he realised that a planet would move faster when it is closest to its parent star (at 'perihelion') and slower when it is further away (at 'aphelion'). This Second Law was followed, in 1619, by a Third Law, which found that the length of time a planet takes to travel around the Sun increases rapidly with the radius of its orbit.

THE GENIUS OF GALILEO

Kepler's laws marked a watershed in astronomical thinking by establishing for the first time the crucial role of the Sun in regulating the motions of the planets in their meandering orbits. Potentially more explosive, however, was the work of the great Italian scientist Galileo Galilei, who in 1598 privately stated in a letter to Kepler that he too supported Copernicanism. In his role as a mathematics professor at the

Galileo Galilei (1564–1642).

University of Padua, Galileo gave a series of public lectures as early as 1604 in which he argued that a newly found star ('Kepler's Star', now known to be a supernova) could not be close to Earth.

Galileo's genius was multi-faceted; he initially read medicine, but his real love was mathematics and natural philosophy. In May 1609, he heard about a new spyglass that Dutchman Hans Lippershey had unveiled in Venice and succeeded in building his own version of the first telescope, which he called a 'perspicillum'. Within a year, Galileo published his book *Starry Messenger*, claiming to have seen mountains on the Moon, tiny stars comprising the Milky Way and four small moons in orbit around Jupiter.

The latter seemed to change their relative positions from night to night, and although the notion of moons in orbit around another planet did not conclusively prove Copernicus' heliocentric theory, it certainly offered strong evidence in favour of it. In fact, Galileo's long-standing German rival, astronomer Simon Marius, may have seen them at an earlier date with his Dutch-built telescope, but does not appear to have recognised their true nature. As a result, Galileo was given credit for being first to publish his observations, although it was Marius who gave them the names that we know today.

By 1612, Galileo had not only ascertained the orbital parameters of the moons – now known, in order of increasing distance from Jupiter, as Io, Europa, Ganymede and Callisto – but made detailed studies of other planets. He noticed that Venus displayed similar illumination phases to the Moon and suspected that it must orbit the Sun rather than Earth. This provided an important new piece of evidence in support of Copernicanism, because Ptolemaic astronomers were unable to reproduce such phases through geocentric arguments. Instead, they had to concede that Venus orbited the Sun, which in turn circled Earth.

At length, Galileo's research attracted the attention of the Inquisition in Rome. According to the Church, Copernicanism contradicted long-held Scriptural beliefs that Earth was at the centre of God's Universe and therefore deeply personal and relevant to humanity. However, in a 1616 letter to Grand Duchess Christina of Lorraine, Galileo stated that in his mind Copernican theory was not simply a calculating tool but a physical reality. He revealed that he vigorously opposed Ptolemaic doctrines and argued that mathematical and astronomical observations cast severe doubt on its authenticity. Galileo was subtly, but firmly, warned neither to defend nor to uphold Copernicanism.

Nevertheless, perhaps in light of perceived support from Pope Urban VIII, in 1632 Galileo published a *Dialogue* between proponents of the Ptolemaic and Copernican models, which concluded in favour of the latter. Its merits were apparently considered irrelevant by Inquisition officials and he soon found himself accused of heresy and summoned to Rome. In view of his failing health, he was condemned to house arrest for the remainder of his life and forced to renounce his belief that the Sun lay at the centre of the Universe. He died in 1642, aged 77.

Found guilty of advocating Copernicanism, in 1633 Galileo Galilei was summoned to face the Inquisition in Rome. The dramatic scene is captured in this 1857 painting by Cristiano Banti.

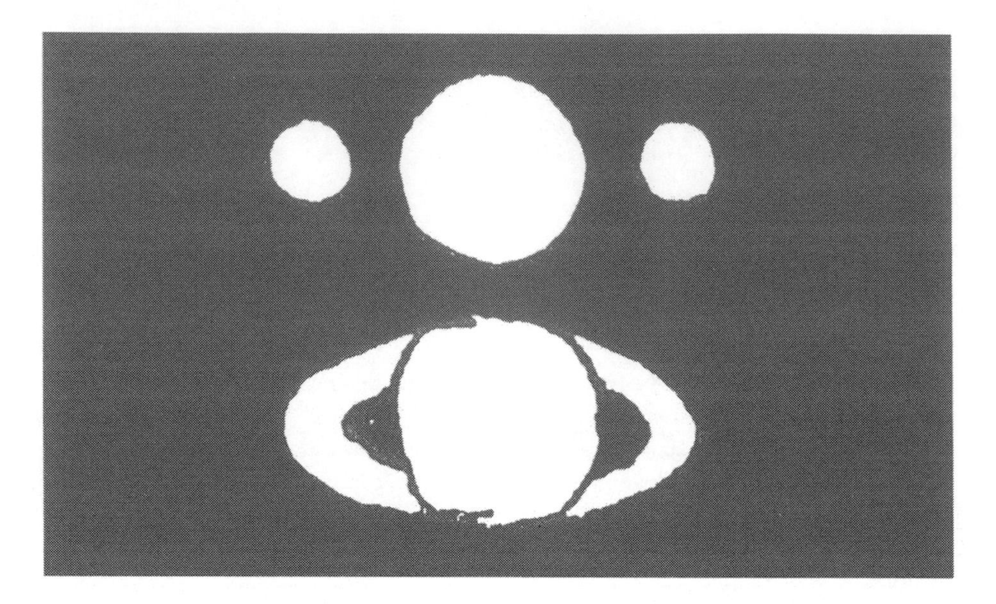

Galileo was never quite sure what caused the strange oval appearance of Saturn. His sketches hint at great arms, handles or enormous attendant moons, but it was not until after his death that its true nature as a ringed world became clear.

Although he is chiefly remembered in astronomical circles for his discovery of the four large Jovian moons, Galileo also undertook the first telescopic observations of Saturn. He did not recognise the ringed planet for what it is, but was nevertheless the first person to identify something unusual about its appearance. His sketches variously show an oval, a world with two large attendant moons or a pair of great arms sweeping out to each side. Clearly, his primitive telescopes were simply not powerful enough to resolve their true disk-like nature. He pondered the mystery for the rest of his life.

Saturn has been known since prehistoric times, although not until Galileo's day and the arrival of the telescope were these 'appendages' – or 'ansae' – first seen in detail. In fact, it was not until 1655 that Dutch astronomer Christiaan Huygens, who discovered Saturn's large moon, Titan (and after whom the spacecraft presently heading for that moon is named), worked out that they are in fact disk-like in nature. To astronomers watching the planet before the invention of the telescope, Saturn appeared as little more than a distant point of light.

The ancient Assyrians – notorious for their cruelty to vanquished enemies – provide us with the oldest extant literary reference to Saturn, which they called the 'Star of Ninib', around 700 BC. However, they derived their wealth of astronomical knowledge from the Babylonians who, as early as 3000 BC, may have recognised major constellations and possibly the five distant wanderers too. By 400 BC, the Greeks were using the name Cronus for the far-off object and ultimately the Romans called it Saturn.

Not until Galileo's time was it possible to undertake detailed studies. Although he eventually formulated a theory that arms or handles stretched away from Saturn, at one stage he actually thought he was seeing a 'triple planet', in which two attendants circled a large central body. His ideas changed when these attendants vanished. "I do not know what to say in a case so surprising, so unlooked for and so novel," he confessed. Today, of course, we know that his peculiar observations were due to the apparent disappearance of Saturn's rings as his view of the ring plane changed.

HUYGENS AND SATURN'S RINGS

Undoubtedly, the puzzle of Saturn's appendages troubled Galileo until the day he died. Thirteen years later, Huygens made a critical breakthrough by suggesting that a 'ring' of material was present around the planet. Using a telescope several times more powerful than Galileo's perspicillum, he speculated that around Saturn lay "a thin, flat ring, nowhere touching and inclined to the ecliptic". In fact, the rings are inclined 27 degrees with respect to Saturn's orbit, which explains why they periodically vanish from view and their orientation changes as viewed from Earth.

Some other ideas circulating at the time came from Polish astronomer Johannes Hevelius, who argued that the rings were actually a pair of crescents somehow attached to an oval central body, and English architect Christopher Wren visualised

Christiaan Huygens (1629–1695), who correctly inferred that a disk of material encircling Saturn was responsible for its peculiar appearance.

a corona around Saturn that was so thin as to be almost invisible when seen edge-on from Earth. By 1659, however, when Huygens published his work *The Saturnian System*, their true nature as a continuous, unconnected ring seemed more plausible. He explained that every 14–15 years, Earth appears to pass 'through' the plane of Saturn's rings, which helped to explain the planet's changing appearance over time.

Although little other definite information was available, this at least gave astronomers some idea of what they were seeing. Nevertheless, there was still controversy over the precise composition and dynamics of the rings. How did they remain stable over time? Huygens and most other scientists believed that they were solid structures, although one interesting argument from Frenchman Jean Chapelain in 1660 proposed that they might be made of hundreds of tiny moons circling Saturn. The Franco-Italian astronomer Giovanni Cassini also doubted that the rings were solid, but such opinions were largely overlooked for another two centuries.

Subsequent research, however, led many observers to conclude that, whatever the nature of their composition, there seemed to be more than one ring around Saturn. As early as 1664, Italian optician Giuseppe Campani noticed that the inner half of the ring was much brighter than the outer half. Unfortunately, he did not realise that this was due to the presence of two separate rings. In fact, it was Cassini, who, in 1675, spotted the largest of several gaps in the rings. The two rings thus detected were called 'A' (innermost) and 'B' (outermost).

THE STRUCTURE OF THE RINGS

Further work yielded tantalising clues that more divisions might exist. In June 1780, German-born English astronomer William Herschel saw dark linear markings near the innermost edge of the B ring and in December 1825 English metrologist, Captain Henry Kater, spotted what he thought were three gaps around the southern face of the A ring. Kater noticed that the central gap – a broad, low-contrast feature known today as the Encke Minima – seemed to be the widest and darkest of the three and concluded that he had "little doubt that which has been considered as the outermost ring of Saturn consists of several rings".

Others noted similar detail in the A ring, including mirror maker James Short, who had sketched Saturn in the mid-eighteenth century and revealed that the rings apparently divided into three concentric circles. These divisions may be what later turned out to be the Encke Minima and Encke Division. It would also seem that astronomer Lambert Quetelet had identified the Encke Minima through his 25-cm refractor in December 1823. Nevertheless, it was Kater who, in a paper submitted to the Royal Astronomical Society in May 1830, was first to publish his observations.

Despite this early research, credit for the actual discovery of the Encke Minima ultimately went to German astronomer Johann Encke, who noticed a shadow near the centre of the A ring during a series of observations in April and May 1837. He also measured its exact position within the rings using a micrometer. However, by this time it was already clear that there were other peculiarities about the A ring. In

William Lassell (1799–1880), captured in 1845 in an early Daguerreotype portrait. Courtesy of the Liverpool Astronomical Society and Michael Oates of the Manchester Astronomical Society.

September 1843, English astronomers William Lassell and William Dawes reported seeing not only the Encke Minima, but also a "secondary division".

Several modern historians have argued that this secondary opening, which was officially 'discovered' by US astronomer James Keeler in 1888 and is now known as the Encke Division, may also have been previously seen by Short and Kater. Although Lassell and Dawes reported their finding, the observing conditions were not ideal. In November 1850, however, they saw it again. Lassell wrote: "I suspected a second division of the outer ring, but could not absolutely verify it. The appearance I saw, or suspected, was a line one-third of the breadth of the outer ring from its outer edge."

Subsequent observations over the next few nights led them to conclude there was indeed a narrow, high-contrast division close to the outermost edge of the A ring. Another astronomer, Phillip Coolidge, also remarked in December 1854 that he had seen "two (and at times I suspect three) divisions in ring A". It was Dawes, however, who, shortly after confirming the existence of the new division, spotted a third ring around Saturn. It was also seen at about the same time by US astronomers William Bond and Charles Tuttle of Harvard College in Cambridge, Massachusetts, and is today known either as the C or 'crepe' ring.

The C ring is a dusty band, some 17,400 km wide, lying just inside and seeming to merge with the innermost edge of the B ring. Together with the A ring, these are the three 'classical' rings that can be easily seen from Earth. It seems likely that the C ring was seen by Herschel as early as November 1793, and also by Kater in January 1828 who noted "the shading, or deeper yellow, of the inside edge of the inner ring" in his journal. However, it would seem that neither of them recognised it as a separate ring.

Until the arrival of the Voyager spacecraft more than a century later, these three were the only rings visible from Earth. A more fundamental problem in the mid-nineteenth century was that, even 200 years after Huygens had first identified the rings, scientists remained divided over their nature and dynamics. Certainly, Herschel and French astronomer Pierre Laplace thought they were solid objects, while US astronomer George Bond argued in 1850 that only a liquid structure could fully account for their long-lived stability.

Bond's theory gained some support two years later when several observers noticed that Saturn's limb was faintly visible through the dusty C ring. A solid ring, they reasoned, would be unlikely to create such an effect. Furthermore, in view of the sheer extent of the rings, they would very quickly become unstable and break up if they were solid. It was not until 1859 that Scottish physicist James Maxwell undertook a definitive mathematical analysis of them. He concluded that their observed stability was only achievable if they were composed of "an indefinite number of unconnected particles".

Maxwell's model was verified by James Keeler, who, in 1895, wrote: "Mathematical investigation shows that a solid or fluid ring could not exist under the circumstances in which the actual ring is placed." Keeler and his colleague William Campbell took spectroscopic measurements of different portions of the rings and revealed that material closest to Saturn moved more quickly than material at the periphery.

In addition to neatly vindicating Kepler's Third Law, this provided persuasive evidence that the rings were indeed composed of literally millions of separate particles. As early as 1790, Herschel had estimated the rotation period of the rings at only ten and a half hours. Although this indicates that the speed of ring particles around the Saturnian racetrack are truly phenomenal, the near-circular nature of their orbits ensures that the relative motion of one to another is small enough to make 'collisions' little more than gentle nudges.

By this time, scientific opinion about the nature and evolution of the rings had changed considerably. The realisation that they contained particles ranging from enormous boulders to specks the size of sand inevitably led many astronomers to wonder what was stopping them from coalescing into a single large moon like the others known to orbit Saturn. A few years before Maxwell's work on the rings, French mathematician Edouard Roche argued that they formed when an ill-fated moon drifted too close to Saturn and was torn apart by tidal forces.

Roche reasoned that two particles orbiting at different distances from the planet would be subject to slightly different gravitational effects. At the same time, however, their own inherent gravity would be exerted on each other. Depending on how far from Saturn they reside, the particles might either pull themselves together and possibly coalesce into a larger body or be prevented from doing so by the gravitational pull of the planet itself. The 'borderline' distance that demarcates regions where rings exist from regions where moons are at liberty to form is called the 'Roche limit'.

According to this model, rings are always found inside a planet's Roche limit. However, observations during the twentieth century uncovered peculiar phenomena in several ring systems and led some astronomers to suspect that tiny moons might be able to withstand tidal disruption and actually thrive within the Roche limit. The inherent gravitational pull of these tiny moons could then help to 'shepherd' the material of thinner rings, preventing it from drifting off into space or dropping into the planet's atmosphere, and perhaps even playing a role in shaping their delicate structure.

Today we know that all four giant planets in our Solar System – Jupiter, Saturn, Uranus and Neptune – have ring systems, although those of Saturn incomparably surpass the others in terms of beauty, intricacy and grandeur. By the close of the nineteenth century, astronomers had not only identified a plethora of detail in these rings but had also taken steps towards understanding why and how that detail occurred. As early as 1866, US mathematician Daniel Kirkwood identified the important role played by several of Saturn's inner moons in carving out the Cassini and Encke Divisions.

Kirkwood's research drew him to conclude that the motion of ring particles along the edges of these divisions must be tightly controlled by the nearby moons Mimas, Enceladus, Tethys and Dione. Each time Mimas orbits Saturn, its gravitational influence is sufficient to shove ring particles clinging to the edges of the Cassini Division around the planet twice. Likewise, the bigger Enceladus is capable of delivering even more oomph by accelerating particles three times around Saturn for every one of its own orbits. Numerous other 'resonant' relationships between moons and ring particles were identified theoretically and observationally throughout the twentieth century.

THE DISCOVERY OF URANUS

For more than three centuries, until a chance find in March 1977 and subsequent observations by spacecraft, astronomers assumed that Saturn was the only world in the Solar System to possess a ring system. Yet, although our knowledge has increased incrementally since the beginning of the Space Age, this does not mean that no other scientific advances were made in the interim. Among the most important discoveries in the eighteenth and nineteenth centuries were the surprising detection of a seventh planet (Uranus), which ultimately laid the foundations for finding the eighth (Neptune).

When William Herschel – who was busy at the time working with his sister Caroline on an all-sky catalogue of stars as faint as eighth magnitude – found Uranus on the evening of 13 March 1781, it was arguably one of the most significant milestones in modern astronomy. For thousands of years, educated people simply assumed and accepted that the Solar System had only five other planets besides

William Herschel (1738–1822), the first person credited with the discovery of a new planet.

Earth. Now, another had been added to their ranks, although Herschel could not quite believe what he had found, first presuming it to be a comet.

Four nights later, he saw the strange, far-off blob again; and although he noted that it was certainly unusual for a comet to possess such a distinct disk and completely lack a long gaseous 'tail', he nevertheless announced his observations to the Royal Society shortly afterwards. His notes from the time still survive. "In the quartile near Tauri," we can see him writing in his observation journal by the dim glow of a candle, "the lowest of two is a curious, either nebulous star or perhaps a comet".

At length, however, it became clear that the motion of the object was in no way comet-like, and Astronomer Royal Nevil Maskelyne was the first to suspect that Herschel had actually found a new planet. Six weeks after the first sighting, Maskelyne wrote to Herschel: "I am to acknowledge my obligation to you for the communication of your discovery of the present comet, or planet, I don't know which to call it. It is as likely to be a regular planet moving in an orbit nearly circular round the Sun as a comet moving in a very eccentric ellipsis."

Further work by Anders Lexell, a Swedish astronomer working at St Petersburg in Russia, and Jean-Baptiste Saron and Pierre Laplace, both in France, mathematically confirmed that the new object's orbit was more akin to that of a planet than a comet. By this time, Uranus was thought to lie at least twice as far as Saturn from the Sun, meaning that in just a single night's observing – and without really knowing it – the Hanover-born professional musician had doubled the radius of the known Solar System.

Herschel estimated the new world's equatorial diameter at around 54,700 km, or some four times larger than our own world; remarkably, this is not too far from the current measurement of 51,200 km pegged by ground-based and Voyager observations. It was a magnificent new scientific discovery and, in recognition of his efforts, Herschel was made a member of the Royal Society and awarded its prestigious Copley Medal in 1781. Moreover, his annual subscription was waived to allow him more money for research.

One important next step after the official acceptance of Herschel's object was to name it. This was a difficult decision. At first, French mathematician Joseph Lalande, who had helped to calculate the orbital motion of the new planet, diplomatically suggested that it should be named after its discoverer. However, the name that we have today came from German astronomer Johann Bode, who argued that Uranus was the mythological father of Saturn, who in turn was father to Jupiter, who in turn fathered Mars, Venus and Mercury.

A mythological name also fitted better with those of the other planets than a world called 'Herschel' might have done. Other candidates at the time included Cybele, Adrastea, Minerva and, perhaps a little ironically, Neptune. Herschel's response to whether he accepted or opposed these proposals was a long time coming, but at length he wrote to King George III and asked if he might name the planet after him: "Georgium Sidus", or George's Star.

The grand gesture assured Herschel of an annual royal pension of £200, although for various reasons the name was not received particularly well. Parts of the world where George III did not reign opposed the name. Astronomers, too, argued that Georgium Sidus was by no stretch of the imagination a 'star', and in any case the nomenclature was far too cumbersome and out of character with the names of the other planets to be acceptable. As a result, the Germans knew it as Uranus, the French (for a time) called it Herschel and the British named it "The Georgian". Bode's proposal ultimately prevailed, of course, but the planet was not officially and universally known as Uranus until 1850.

MORE PLANETS?

By this time, Herschel himself was long-since dead; coincidentally, his death in 1822 at the age of nearly 84 meant that his life ran for almost exactly the length of time Uranus takes to circle the Sun. Moreover, yet another new planet was found in the late 1840s, which turned out to be Uranus' near-twin, in size at least. Herschel's discovery had stimulated the search for other worlds beyond Saturn and attention quickly focused on an obscure concept called Bode's Law or, to give it its proper name, the Titius–Bode Rule.

This rule owes its genesis firstly to German mathematician Johann Titius who, in 1766, translated a work by Swiss naturalist Charles Bonnet called *Reflections on Nature*. As part of his attempt to relate observations of nature to God's handiwork, Bonnet referred to the Solar System and Titius inserted his own paragraph, which offered an intriguing progression in the distances of the planets from the Sun. He began with a series of numbers – 3, 6, 12, 24, 48, 96 and so on – in which successive numbers are twice their predecessor.

Titius then added four to each number and divided their sums by 10. The result was a surprisingly close approximation of the distances of the then-known planets from the Sun in astronomical units (AU). One AU measures almost 150 million km and is the mean distance from Earth to the Sun. Titius' rule worked well for all the planets, except for a gap between Mars and Jupiter, where there was no known large world. Rather than simply accept that no planet existed in this gap, Titius theorised that it might be occupied by as-yet-undiscovered moons of Mars or Jupiter.

When Bode read Titius' translation of Bonnet's work, he was intrigued by the possibility but had doubts about the gap being inhabited by unseen moons; rather, he speculated, another major planet still awaited discovery. In the meantime, Herschel discovered Uranus and Bode quickly realised that if 'his' rule of planetary distances had been carried a step further, it might have theoretically predicted the giant world's position at 19.6 AU. In fact, Uranus lies remarkably close to this distance, at about 19.2 AU, or 2.9 billion km, from the Sun.

Based upon the apparent success of the rule, and having finally credited Titius with its discovery, Bode pressed on with his own efforts to find a new planet in the Mars–Jupiter gap. Several other astronomers, including Franz von Zach and Johann Schröter in Germany, joined the effort, but it was a Sicilian priest and mathematician named Giuseppe Piazzi who saw a faint, star-like object in the constellation of Taurus on the night of 1 January 1801. It was named Ceres and its distance from the Sun – 2.77 AU – was remarkably close to the 2.8 AU predicted by the Titius–Bode Rule.

This neatly vindicated the rule, but not for long. In March 1802, German physician and amateur astronomer Wilhelm Olbers discovered another small 'planet' that he called Pallas. It lay in approximately the same patch of sky as Ceres. This complicated the Titius–Bode Rule's predictions by having two worlds in an area where there should only have been one. Furthermore, when Herschel estimated the diameters of both objects, he came up with 260 and 180 km respectively. Today, we know this to be closer to 715 and 486 km, but either way both are considerably smaller than even the smallest planet.

In a bid to 'save' the rule, Herschel suggested that both Ceres and Pallas were actually a new breed of Solar System objects – not 'planets' at all – and proposed they should instead be dubbed 'asteroids'. Nowadays, the term 'minor planet' is generally preferred by most astronomers. Although Olbers accepted Herschel's definition, he argued that maybe Ceres and Pallas were fragments of a much-bigger planet that had been ripped apart by a cometary impact. He speculated that a careful inspection of the region where their orbits came closest to intersecting might be a worthwhile area to look for more 'fragments'.

By 1807, another two asteroids had been found and named Vesta and Juno. The pitiful size of all four new worlds made it increasingly unlikely that they ever formed part of a substantial planet. Even today, with around 9,000 named asteroids known to reside in a 'belt' between the orbits of Mars and Jupiter, their combined mass is no more than 5 per cent of the mass of our Moon. Indeed, to provide a reservoir of enough fragments to build a planet as big as Earth would require an asteroid belt several thousand times more extensive than it actually is.

PROBLEMS WITH URANUS

Regardless of this apparent flaw in the Titius–Bode Rule, many astronomers in the early nineteenth century swore by it and even brilliant mathematicians like Urbain Leverrier and John Adams independently used it to predict the existence of an eighth planet at a distance 39 AU from the Sun. Although Neptune, as it was eventually named, was indeed found in the same area of the sky as Leverrier and Adams had predicted using the rule, it was actually somewhat closer to the Sun, at just over 30 AU, than it should have been; a significant 23 per cent error margin.

Taken further, the position of Pluto – the outermost known planet – is close to where Neptune *should* have been, but only halfway as far as the Titius–Bode Rule predicted a ninth planet to reside. Clearly, the eventual discovery of Neptune exposed the rule for what it was: a series of coincidences and little more than a curiosity. Yet astronomers were not just following the rule when they sought Neptune; they were concerned by the peculiar behaviour of Uranus. Although several mathematicians had successfully plotted aspects of its orbit as early as 1783, a number of problems had arisen.

First, too little was known of the mechanics of Uranus' orbit to determine its extent and how long it took to circle the Sun. Bode therefore sifted through past observations to discover if someone had accidentally spotted Uranus before Herschel, but had mistaken it for a star. He found not one, but two instances: by the first Astronomer Royal, John Flamsteed, in 1690 and more recently by Tobias Mayer in 1756. These provided observations spanning almost a century, which should have been enough to calculate Uranus' orbit. The calculations 'worked' for a few weeks, then degraded into uselessness.

Renewed attempts were made in the early nineteenth century, when French astronomer Alexis Bouvard compiled a list of 17 pre-Herschel sightings of Uranus and more than four decades of post-discovery orbital calculations. He managed to bring together over 130 years' worth – or one and a half Uranian orbits – of observations, but was frustrated in his effort to derive a path that precisely matched the planet's motion. If all calculations were used, the 'older' observations would correspond but the 'younger' ones would not, and if the most recent studies alone were taken, the older ones would not fit.

Ongoing observations of Uranus' position did little to help. It rarely appeared at the predicted point in the sky, leading German mathematician Friedrich Bessel to despair over "the mystery of Uranus". He also speculated, intriguingly, that the reason for its erratic and apparently unpredictable behaviour might be due to another planet lying further from the Sun. As astronomers began to search, Uranus continued to run too 'fast', constantly turning up 'ahead' of its expected position. Then, strangely, around 1829–1830, it dropped to a more predictable pace, and then started running 'slow' and fell behind its predicted position.

Bessel's hunch that an external force was acting on Uranus returned to the fore. Perturbation by unseen moons was possible – in fact, Herschel had found two small attendants, Titania and Oberon, in January 1787 – but considered unlikely as they would need to be much larger to account for the disturbance. Moreover, if they *were* large, they should have been detectable from Earth. Other suggestions included a comet hitting Uranus around the time Herschel discovered it, altering its orbit, but this too was discarded because the planet did not even match all the calculations made since 1781.

THE FRUSTRATING SEARCH FOR AN EIGHTH PLANET

Towards the end of the 1830s, however, most astronomers and mathematicians accepted that Uranus' orbital path was probably being affected by the gravitational

John Adams (1819–1892) at about the time Neptune was discovered.

George Airy (1801–1892).

pull of another world further from the Sun. Shortly afterwards, John Adams, a brilliant young mathematics student, read of the mystery and jotted a note to himself to solve it as soon as he completed his degree at Cambridge University. It was to be one of the most challenging problems ever faced, to such an extent that some mathematicians had already written it off as insoluble.

It would also bring frustration and eventual recognition to Adams and leave two of England's most renowned astronomical figures red-faced. The first was Astronomer Royal George Airy, who publicly stated that not enough was known of Uranus' orbital motion to predict the position of another world even further from the Sun. In response to a letter from astronomer Thomas Hussey in 1834, regarding the possibility of discovering an eighth planet using mathematics, Airy discouraged any such search and suggested waiting a few centuries until Uranus' motion was better understood!

The second person, who apparently closely paralleled Airy in his ability to retain a closed mind, was James Challis, Director of the Cambridge Observatory. He too saw little point in attempting to find an eighth planet based solely on the calculated motions of Uranus. Nor did many other mathematicians in France or Germany, at least until 1845. The solution itself, in theory, was straightforward and it was possible to calculate Uranus' orbital path by working out the gravitational effect it had on the motion of Saturn.

The relative distances of both giant planets from the Sun were well known, as were their approximate positions in the heavens. This meant that it was possible to determine the direction from which Uranus' gravitational force came. Moreover, Herschel's discovery of Titania and Oberon soon after finding the planet itself, and a subsequent determination of their orbits, had allowed a calculation of Uranus' mass to be made.

Adams solved the problem in reverse. Firstly, he subtracted the perturbations imposed on the planet by its known moons and by Jupiter and Saturn. His next step was to use the other irregularities to find out exactly the source of the perturbing force, together with how fast and along what orbital path that body was moving, its distance from Uranus and its mass. However, if the disturbance *was* coming from another planet, as many astronomers believed, then that unknown world would itself be orbiting the Sun and its distance and position with respect to Uranus would always be changing. As a result, the nature of its gravitational pull on Uranus would also be subject to constant change.

Adams, unlike Uranus, was unperturbed. He felt confident that he could calculate the position, orbit and mass of the unseen planet, making sure that it travelled along the right path to produce precisely the observed effects on Uranus. Unfortunately, Adams did not know its distance or mass, so he returned to the Titius–Bode Rule and its prediction of 38.4 AU for a planet exterior to Uranus. His next move was to assume that the new planet might not move in a perfect circle around the Sun and refine his mathematical model to take account of an elliptical path.

By the autumn of 1843, Adams was sufficiently confident with his numbers to contact Challis, asking for the best and most up-to-date observations of Uranus' position to help to identify the whereabouts of the unknown planet. Challis contacted Airy at the Royal Observatory in Greenwich, who responded with a wealth of data and invited Adams to contact him personally. Within two years, Adams had incorporated Airy's data into his calculations and determined an elliptical, rather than circular, orbit for Uranus.

Yet it was from this point and during the following years, until 1847, that triumph was snatched from his grasp and Adams' professional career nose-dived. For some reason, when Challis received Adams' predictions, he failed to undertake a telescopic search or advise the young mathematician to publish his work. Instead, he suggested that Adams send his results directly to Airy. There followed a series of unfortunate and frustrating occasions when the two men were unable to meet personally, although Adams nevertheless left a summary of his work for Airy's perusal.

The Astronomer Royal was unimpressed and some historians have speculated that this may have been linked to his inherent dislike of theoretical approaches to scientific research, or even outright snobbishness. He felt that observations were the key to understanding nature and mathematics was just a tool to explain the findings; the process, he felt, did not operate in reverse. It was at this stage that another aspect of Airy's character was exposed: he favoured individuals who were already well-established professionally and perhaps doubted or even resented young upstarts like Adams.

Airy did mention Adams' results to William Dawes – who, a few years later, would discover Saturn's dusty C ring – although the latter was unable to undertake a telescopic search himself. Nevertheless, he sent word to William Lassell, who happened to be bedridden with a sprained ankle at the time. Two promising opportunities to find the trans-Uranian planet had been lost, through no fault of Dawes or Lassell. Airy, on the other hand, seems to have completely misunderstood the point of Adams' work.

First, he made it quite clear that he thought Adams' results were 'assumptions', whereas in fact the young mathematician had only used data supplied by Airy in his calculations. He had not assumed that a trans-Uranian planet existed, nor had he guessed at its possible orbital position. Airy, however, expressed little interest and in his few letters to Adams pestered him with irrelevant questions. On the other side of the Channel, meanwhile, French, German and other European astronomers were just as perplexed by the mystery.

During the summer of 1845, while Adams was busily refining his own calculations, the respected Director of the Paris Observatory, François Arago, assigned the task of solving the Uranus mystery to Urbain Leverrier. By this time, aged 34, the latter was already well-established in scientific circles for his work on cometary orbits. The Uranus mystery would be the high-watermark of his career. Over the course of nearly a year, from November 1845 until August 1846, Leverrier presented three papers to the Paris Academy of Sciences.

The first dealt with the puzzle of the 'old' and 'modern' observations of the planet, which Leverrier felt were irreconcilable. There was, he said, no problem with the

Urbain Leverrier (1811–1877), whose calculations led Johann Galle and Heinrich d'Arrest to discover Neptune.

accuracy of the observations of Flamsteed and Mayer, but rather there was something amiss with Uranus' motion. His second paper demonstrated that these observed irregularities were almost certainly caused by an unknown planet further from the Sun. When Airy received a copy of this paper towards the end of June 1846, he noticed parallels between Leverrier's calculations and those of Adams.

The Astronomer Royal immediately wrote to Leverrier, whom he recognised as an established scientist, to express his enthusiasm and delight. He even wrote in his personal diary: "To this time I had considered that there was room for doubt of the accuracy of Mr Adams' investigations. But now I felt no doubt of the accuracy of both calculations." In spite of this admission of an error in judgement, Airy did not share his enthusiasm with Adams. Moreover, neither Airy nor Challis bothered to start a telescopic search.

At length, early in July 1846, after being pushed in the right direction by one of his old Cambridge professors, Airy finally saw the light and contacted Challis to urge a search for the new planet. The latter, for unknown reasons, dragged his heels until the end of that month, even when Adams supplied more precise data on the position of the planet and suggested that it should be large enough to show a disk. Challis appeared to lack confidence in Adams' or Leverrier's calculations, feeling, perhaps, that mathematics was still inappropriate for finding a planet.

NEPTUNE: A PLANET FOUND ON A PIECE OF PAPER

Perhaps if Challis had incorporated their predictions into his own observations, he may have found the eighth planet within a few days. In fact, incredibly, he *did* see it ... on *four* separate occasions in August 1846, exactly where Adams and Leverrier had predicted it would be, but both he and his assistant failed to recognise it. Nor was England the only country to miss it. Sears Walker, an astronomer at the US Naval Observatory in Washington, DC, read of Leverrier's work, but the facility's heavy schedule meant that a telescopic search never happened.

In the meantime, on the last day of August, Leverrier presented his third and final paper on Uranus to the Paris Academy of Sciences. He methodically laid out the orbital elements, mass and precise position of the world he thought was perturbing Herschel's planet. Apparently, to encourage astronomers to start looking, Leverrier informed them that the planet was already at its closest to Earth and visible as a disk, meaning that the long tedium of nightly star mapping was unnecessary. Still, no-one in the packed audience chamber – not even Arago – offered to start a telescopic search.

It was at this frustrating time that Leverrier came across a doctoral thesis submitted a year earlier by an assistant at the Berlin Observatory. His name was Johann Galle. When he received a reply from Leverrier on 23 September, together with predictions for locating the trans-Uranian planet, Galle could hardly contain his excitement. He pleaded with the observatory's director, Johann Encke, to permit a search. At length, the man credited with finding Saturn's Encke Division agreed to let Galle and graduate student Heinrich d'Arrest use the observatory's 23-cm telescope that night.

Johann Galle (1812–1910), the telescopic discoverer of Neptune.

The two men spent much of their time painstakingly sifting through star maps, until they came across a new chart of a region in the constellation Aquarius. While Galle kept his eye to the telescope that night and called out stars in his field of view, d'Arrest matched them with known objects on the maps. At length, during this procedure, Galle called out "Right ascension 21 hours 53 minutes 25.84 seconds; magnitude 8," and a long pause ensued before d'Arrest replied quietly, "That star is not on the map".

Astonishingly, not only had it taken them less than an hour to find the new planet, but it also turned out to be only a degree from where Leverrier predicted it should reside. Galle and d'Arrest sent for Encke – breaking up his birthday party – and continued watching the distant speck until the early hours of the next morning, when it finally set. It was not until they returned to the telescope the following evening, however, that they could establish with absolute certainty that the object had moved against its starry backdrop. It *was* a planet! It was Neptune.

Galle immediately informed Leverrier, proposing that the new discovery should be named Janus, after the two-faced Roman god of past and future. The Frenchman became an instant celebrity in his own country, but wrote modestly to Galle: "I thank you cordially for the alacrity with which you applied my instructions to your observations of 23 and 24 September. We are thereby, thanks to you, definitely in possession of a new world." He was not, however, so keen on Galle's choice of name for the planet, feeling that the German was trespassing on a 'true' discoverer's right.

Leverrier bluffed Galle by telling him firstly that the French Bureau of Longitudes had already picked the name Neptune, then reconsidered and decided to call the planet after himself. His supervisor François Arago was unhappy but agreed to put forward Leverrier's request for consideration, if Uranus could be officially renamed Herschel. This, we should recall, was the original French proposal for Uranus. Although Arago made a good case for Leverrier, the Paris Academy of Sciences flatly rejected his proposal. Uranus would remain Uranus and the new arrival would become known as Neptune.

A MAJOR SCANDAL

In England, meanwhile, Airy's reaction to the discovery is not recorded. Challis, for his part, continued to miss what was staring him in the face. When the London *Times* announced the discovery on 1 October, he confessed to Airy: "After four days of observing, the planet was in my grasp, if only I had examined or mapped the observations." In fact, he had seen it just two nights before. That autumn, everyone involved tried to make excuses and grab their own slice of the fame and recognition pie.

John Herschel, a noted astronomer himself and son of Uranus' discoverer, cited his own introductory speech at a meeting of the British Association for the Advancement of Science three weeks earlier, in which he predicted that an eighth planet would soon be found. Airy and Challis both tried to excuse their incompetence, while the latter actually wrote to Arago and told him that he had been looking for Neptune since late July and had seen it four times. He omitted to mention that he had failed to identify it for what it was.

It was at this time that Adams' name emerged into the limelight. He was mentioned in passing in a letter from Airy to Herschel, but without much in the way of fanfare. Challis, too, made no detailed mention of Adams' research; nor did he acknowledge that his own botched observations were based on the young Cambridge mathematician's calculations. In fact, when Airy wrote to Leverrier on 14 October and told him "collateral researches had been going on in England and led to precisely the same results as yours", he failed to mention Adams at all.

Up to this point, Leverrier had not even heard of Adams. He furiously asked the Astronomer Royal why neither he nor Challis had mentioned Adams before and why they had not encouraged him to publish his work. The reaction of most scientists in France was that England was outrageously trying to steal recognition from Neptune's true discoverers – as indeed was the case. The bulk of the criticism was levelled at Airy and Challis. However, as far as Arago was concerned, Adams had not published his results, and that was his own hard luck. Arago stated that Adams had no claim. Ironically, Adams himself had played no part in the furore.

Perhaps unfairly, the French scientific establishment, and eventually their press too, vented a considerable amount of venom at Adams. However, the young mathematician, for his part, gradually received a measure of acclaim thanks to his modesty and lack of disappointment or bitterness. In fact, in the November 1846 edition of the *Monthly Notices of the Royal Astronomical Society*, he wrote: "I mention these dates merely to show that my results were arrived at independently and previously to the publication of M. Leverrier, and not with the intention of interfering with his just claims to the honours of the discovery."

"There is no doubt", Adams continued, "that his researches were first published to the world, and led to the actual discovery by Dr Galle, so that the facts stated above cannot detract, in the slightest degree, from the credit due to M. Leverrier." Challis, however, was asked why, armed with Adams' calculations, he had not begun a search as early as September 1845, to which he lamely replied that he lacked confidence in the data. He also revealed that he felt the probability of a discovery was low until a broader area of the sky had been scrutinised.

In the eyes of many, the reputations of both Challis and Airy were left in tatters by the fiasco. One of Airy's colleagues, a geologist named Adam Sedgwick, told him clearly that "to say the very least, a grand occasion has been thrown away". Although Airy's self-confidence got him through the crisis (he held the Astronomer Royal job until 1881), the distribution of Royal Society awards said it all. The coveted Copley Medal went to Leverrier in 1846, but at length Adams' contribution was recognised and he received the same award two years later.

Adams was offered a knighthood and the Astronomer Royal job when Airy retired, but declined both. He did, however, supersede Challis as director of the Cambridge Observatory and became a professor of astronomy and geometry there. As for the other characters in the Neptune story, Leverrier later took over from Arago as director of the Paris Observatory, while Galle reluctantly left Berlin to become a professor of astronomy at the University of Breslau. D'Arrest's important role in the discovery, sadly, did not become widely known until 1877, two years after his sudden death at the age of 52.

PRE-DISCOVERY SIGHTINGS OF NEPTUNE?

In 1979, US astronomer Charles Kowal and NASA historian Stillman Drake revealed that Galileo Galilei had observed Neptune in the early seventeenth century, when it was quite close to Jupiter in the sky. In one of his notebooks – now preserved at the National Central Library in Florence – he sketched what he thought was a background reference star. Although he noticed that it moved slightly over a two-night period late in January 1613, he overlooked it and assumed it was merely a faint, eighth-magnitude star. The skies over Padua were cloudy during the next few nights and he could track its progress no further.

Galileo's scribbled notes show Jupiter and three of its moons connected by a solid line. A second, dashed line then runs down to the bottom left-hand corner of the page, where he labelled a star as 'a'. The dashed line then continues to the lower right of the page, where Galileo identified two asterisks as 'a' and 'b'. The Smithsonian Astrophysical Observatory's stellar catalogue has identified 'a' as the star numbered

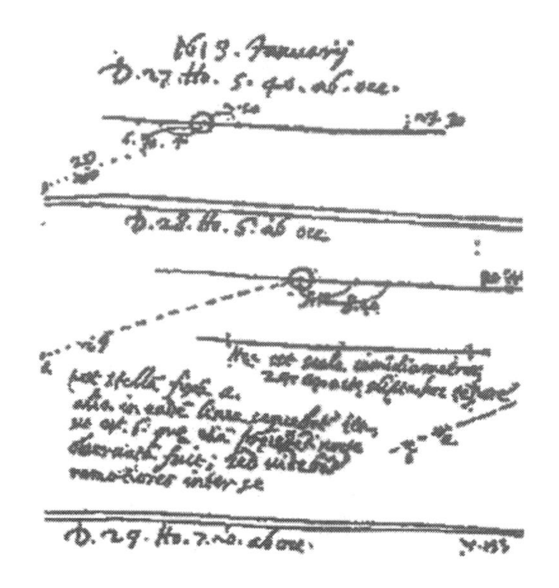

Extract from Galileo Galilei's notebook in January 1613, showing what he thought to be a faint, eighth-magnitude star (identified as 'b' on the dashed line in the bottom-right-hand corner of the page). Galileo didn't know it, but 'b' was the planet Neptune.

SAO 119234. However, although no known star lay at the same position as 'b', Neptune certainly did. If only Galileo had realised the significance of that speck of light!

Other pre-discovery sightings of Neptune include that of Joseph Lalande who, over three nights in May 1795, watched the planet, but dismissed it as a star. So too did John Herschel, who saw it in July 1830, and Scottish-born German astronomer Johann von Lamont observed it at least three times between October 1845 and early September 1846. Von Lamont's last two observations were over a four-night period and should have enabled him to recognise Neptune's orbital motion. It was not to be, and less than two weeks later, Galle and d'Arrest made their own sighting.

THE SEARCH FOR 'PLANET X' AND DISCOVERY OF PLUTO

Ironically, even the addition of Neptune to the Solar System's ranks caused problems. It simply did not follow its predicted orbit, even after experts in celestial mechanics had taken the combined perturbations of all other planets into account. This brings the story firmly into the twentieth century and US astronomer Percival Lowell's search for the famed 'Planet X'. Although his main interest was Mars – and he is particularly remembered for extensively mapping what he thought were 'canals' criss-crossing the Red Planet – Lowell also devoted considerable energy to finding an unknown world beyond Neptune.

Percival Lowell (1855–1916).

His initial photographic and telescopic searches from 1905 onwards were all fruitless. Then, towards the end of 1908, he attended a lecture by his friend William Pickering of Harvard University, who explained his use of a graphical plot of the motion of Uranus to help predict the whereabouts of a trans-Neptunian planet. Pickering dubbed the unknown world 'Planet O' and guessed that it might reside 52 AU – nearly 8 billion km – from the Sun and take 373 years to complete an orbit.

By 1909, Lowell's own research into the hypothetical object, which he called Planet X, left him sufficiently confident to predict that it was situated 47.5 AU from the Sun, took 327 years to complete a full orbit, and was less than half as massive as Neptune. Yet, despite his efforts, the ninth planet remained elusive. Lowell also knew that, since Neptune took 165 years to circle the Sun, the giant planet simply had not been observed for long enough for perturbations of its motion by another world to be apparent.

Instead, he was forced to rely on the smaller perturbations on Uranus to support his argument, although these were much reduced now that the gravitational influence had been accounted for. He began his second telescopic search for Planet X in July 1910 and the following year Pickering published his own predictions for the existence of three more trans-Neptunian objects, which he called 'P', 'Q' and 'R'. The middle object, Planet Q, said Pickering, had a mass 20,000 times greater than Earth, which would have made it 63 times more massive than Jupiter and 6 per cent solar mass. This seemed rather unlikely.

Although Pickering's prediction of the location of Planet O was very close to where Pluto was eventually found in February 1930, he lost credibility with many astronomers because of the less-than-rigorous approaches he used to calculate their 'positions'. In the meantime, Lowell spent the last few years of his life (he died of a stroke in 1916) fruitlessly searching for his ninth planet. Or, perhaps, not quite fruitless, for his second search found 515 asteroids, 700 variable stars and, in the spring of 1915, took two pictures of Pluto!

It is unlikely that Lowell even looked at the two photographic plates that showed an object somewhere between 15th and 16th magnitude – five times fainter than he had predicted for Planet X. Yet it is one of the cruel ironies of history that he never knew how close he came to finding Pluto. In fact, Lowell's brother later wrote: "That X was not found was the sharpest disappointment of his life." The young man who eventually found Pluto joined the Lowell Observatory in Flagstaff, Arizona, in January 1929. His name was Clyde Tombaugh.

Although he had no formal training in astronomy, Tombaugh enthusiastically

Clyde Tombaugh (1906–1997).

plunged into his assignment of finding Lowell's Planet X. To achieve this end, he used a device called a 'blink comparator', which alternately shone a light through one photographic plate and then another, so that any object that appeared in the same spot on both plates – such as an apparently-motionless star – would remain steady. On the other hand, an object that turned up on one plate, but not the other, or in different places on both plates, would 'blink' noticeably.

By allowing Tombaugh to switch plates backwards and forwards, the effect was akin to flicking through a picture book. It had obvious benefits for finding new planets, which would be expected to 'move' against a seemingly-fixed backdrop of stars, and his attention was drawn to a pair of plates taken six days apart in January 1930. They revealed a faint speck moving slowly across the sky. Tombaugh wrote: "On 18 February 1930, I suddenly came upon the images of Pluto! The experience was an intense thrill, because the nature of the object was apparent at first sight."

A few more nights of observing confirmed the discovery, and on 13 March – which would have been Lowell's 75th birthday – it was formally announced. Two months later, Planet X was renamed Pluto. Remarkably, the name came from an 11-year-old English schoolgirl named Venetia Burney, who happened to be learning about mythology at the time and thought it fitting that a dark and gloomy planet so far from the Sun be named after the Roman god of the underworld.

Its first two letters – PL – were also the initials of the man who came so close to finding Pluto. Even William Pickering, after initial scepticism, was pleased with the name, commenting: "That's a good name – Pickering-Lowell!"

AN INSIGNIFICANT SPECK?

Pluto turned out to be a very diminutive world and, as shall be seen later in this book, has regularly ignited debate over whether it should even be classified as a planet or something else entirely. With an equatorial diameter of only 2,300 km, it is nowhere near the size of Jupiter, Saturn, Uranus or Neptune and, indeed, many of those planets' moons are somewhat larger than Pluto. Such a small world would appear unlikely to have accounted for the discrepancies observed in the orbits of Uranus and Neptune.

However, these discrepancies vanish if the mass of Neptune, as determined by the Voyager 2 spacecraft, is used. It would appear that they arose primarily because astronomers simply did not have enough observational data on Neptune's orbital progress. In fact, it takes 165 years to circle the Sun, so has not yet completed even one full orbit since its discovery.

A 10th planet has long existed in popular imagination and, for some time, there was a realistic possibility that such a world might be found. This is now unlikely – although the Voyagers may be our best chance of locating it – and, as far as we know, our Solar System consists of nine planets, a very ordinary Sun and countless rocks and lumps of ice whose number increases each year. Yet humanity's perception of our place in the grand scheme of things has changed radically over several millennia. The mindset change enforced by Copernicanism and its implications have already become apparent.

From an Earth-centred Universe, together with a comfortingly close and personal relationship with God, to one in which the Sun was perceived as the dominant player, it was not until as recently as a couple of centuries ago that astronomers began to comprehend the true insignificance of our world in the cosmos. One of the most important contributions towards understanding Earth's place in the Universe was made by William Herschel.

After analysing the 'proper motions' of 13 stars measured by Nevil Maskelyne and Joseph Lalande, Herschel realised in 1805 that our Sun and its entourage are moving, relative to their celestial neighbours, across space in the general direction of the constellation Hercules (the Hunter) and the bright, bluish star Vega. Moreover, Herschel undertook to determine the shape and structure of the Milky Way system, in which not only our Sun, but hundreds of millions of others, reside. He made a series of exhaustive star counts, which led him to conclude that our home is a "flat disc, of very irregular contour, and with a deep cleft".

Herschel estimated the Milky Way's thickness as about a sixth of its diameter and argued that our Sun was probably located quite close to the centre. Although modern cosmologists broadly agree that it is indeed disk-shaped – or, more precisely, consists of a long, central bar with spiral 'arms' radiating outwards – it is far larger than Herschel could ever have imagined. In fact, the distance from our Solar System to the very heart of the Milky Way is thought to be around 26,000 light years.

As its name implies, this unit of distance is a measure of how far light travels in a single Earth year. Light, as far as we are presently aware, moves at a staggering speed of close to 300,000 km/s! The Milky Way is *big*! Furthermore, our Solar System resides nowhere near its centre, but somewhere along one of the long spiral arms close to its periphery. Herschel's revelation that our Sun is actually nothing more than a very ordinary star in an unfathomable sea of countless others proved almost as Earth-shattering as the work of Copernicus.

THEORIES OF FORMATION

Yet it was clear that something special had happened in our Sun's case. As early as 1745, French naturalist Georges de Buffon proposed that the planets had condensed from material left over after a comet hit the Sun. Although he was unaware that comets are actually little more than chunks of ice and rock, de Buffon nevertheless made several other insightful observations and was one of the first thinkers to suggest that Earth might actually be much older than the 6,000 years traditionally espoused and accepted by Church doctrine.

Another important effort to understand where the Solar System came from was made by Prussian philosopher Immanuel Kant, who argued in 1755 that the Sun originated in an enormous cloud of gas – which he called a 'nebula' – that collapsed under its own gravity and generated a tremendous amount of heat in the process. This heat, he reasoned, was enough to make the nebula 'glow'. Pierre Laplace pushed the idea further in 1796. He was intrigued to learn that every known planet and moon seemed to circle the Sun and rotate in the same direction and orbital plane.

Like Kant, Laplace believed that the Solar System was born from the contraction and cooling of a large, flattened and slowly-rotating nebula. However, he argued that as this contraction continued, the nebula's rotation rate – its 'angular momentum' – increased and, in so doing, threw off multiple rings of excess material around its equator. The result, when the nebula finally stabilised itself, was that material in these rings coalesced to form planets in near-circular orbits in a single plane around what later became the Sun. Similar processes then occurred on a smaller scale around the planets themselves to produce systems of moons. This suggested that every star may possess a system of planets.

Certainly, observations of the Orion Nebula and others since the 1650s strongly suggested that they were indeed clouds of glowing gas, although William Herschel's research led him to suspect that at least some nebulae were composed of vast numbers of individual stars. Later astronomers, including William Parsons – who constructed a large telescope at Birr Castle in Ireland – also noticed that the structure of several nebulae appeared to take a spiral form and thought they might represent disks of gas encircling infant stars.

However, it seemed unlikely that planets could form from rings of material cast off by the Sun. Instead, they would have hurtled away from the rotating disk and continued into space without condensing into solids. The Sun revolves much too slowly to have formed from a rapidly-rotating gaseous cloud. One puzzle was the fact that the planets possess most (98 per cent) of the Solar System's angular momentum, yet account for just 0.1 per cent of its total mass.

Additionally, if their moons formed by a similar process – as Laplace believed – it should have been the *moons*, not their parent planets, which retained most of the Solar System's angular momentum. Twentieth-century research has also shown that nearly a third of all known planetary moons rotate in opposite directions to the others. Still others, with Neptune's large moon Triton being a notable example, do not move in the same direction as their parent planet's rotation.

It was not until the twentieth century that astrophysicists began to understand how the Sun obtained its energy and how the nebula shed its angular momentum into rings of material and, ultimately, into planets and moons. Early studies focused on the possibility that the ongoing contraction of the Sun's interior provided the energy. Then, in 1938 German-born US physicist Hans Bethe realised that our parent star is actually powered by energy released by the fusion of hydrogen and helium – a discovery that later won him the Nobel Prize for Physics.

However, the problem remained as to what caused the planets to form in the first place. As early as 1901, US scientists Thomas Chamberlin and Forest Moulton argued that the Sun was once involved in a near-collision with another star. The effect of this close shave was that the gravitational attraction of the passing star tore a series of long, cigar-shaped gaseous clouds from the solar surface and ejected them into space. Clouds emanating from the side of the Sun nearest to the intruder were yanked out to distances comparable with the orbits of Jupiter, Saturn, Uranus and Neptune.

On the other hand, gaseous clouds emerging from the Sun's far side were less susceptible to the intruder's gravitational tug and were pulled less violently to distances where the terrestrial planets Mercury, Venus, Earth and Mars now reside. It was from the very heart of these multiple clouds, according to Chamberlin and Moulton, that the cores of what would later become planets formed. Meanwhile, the outer parts of the clouds expanded and cooled into a huge swarm of solid particles, and spread – thanks to the motion of the intruder – into a disk revolving around the young Sun.

Gradually, the cores drew in surrounding debris, which accreted into early versions of today's planets. Since the bulk of this debris lay in the outer regions of the Solar System, the majority of the 'growth' took place here; hence the distant planets are much larger than the inner ones. However, mathematical analyses by English scientists James Jeans and Harold Jeffreys in 1916 led them to conclude that tidal interactions between the Sun and the intruder would have caused our parent star to lose a single, cigar-shaped filament of hot gas, rather than the multiple clouds envisaged by Chamberlin and Moulton.

Jeans and Jeffreys argued that this filament would then condense directly into planets, in effect avoiding the need to draw in and accrete debris onto initial cores. They envisaged that the fat central section of their 'cigar' would condense into the giant planets, while its tapering ends would provide building blocks for smaller terrestrial worlds. This hypothesis later led Jeans to suggest that if planetary systems came about purely because of chance encounters between grazing stars, there must be relatively few worlds outside our Solar System capable of supporting life; a gloomy prospect indeed.

Although the work of Jeans and Jeffreys was adopted by some scientists, others, including US astronomer Henry Russell, doubted that a close encounter with another star could have left the Sun – which is, after all, a *thousand times* more massive than all the planets combined – with such a small percentage of the Solar System's angular momentum. The Sun would barely have been affected. Russell and, in 1939, his student Lyman Spitzer were convinced that after extraction from the Sun, far from cooling and condensing into planets, the gaseous cigar would actually have been so hot that it would have rapidly dissipated and become even more tenuous.

Numerous attempts were made to work around the problem. In 1964, English physicist Michael Woolfson argued that the intruder was a less substantial – and cooler – 'proto-star' and that most of the gaseous material that eventually condensed into planets came from this source. Another suggestion, from English astrophysicist Fred Hoyle, proposed that the Sun was once the junior member of a binary star system. Its companion, said Hoyle, had evolved much more quickly, and exploded into a supernova, leaving a cloud of gas and dust that eventually condensed to form a planetary system. As always, however, the issue remained how the planets had ended up with the system's angular momentum.

THE MODERN NEBULAR HYPOTHESIS

However, by this stage, other astronomers had already returned to variations of Kant and Laplace's nebular hypotheses. In 1944, German physicist Carl von Weizsacker accepted that the Solar System probably began with a large, rotating cloud of gas and dust, which gradually contracted and flattened out into a disk. Yet this was where the similarities ended. Von Weizsacker suggested that, instead of spinning out vast rings of excess material, the central part of the cloud collapsed to form our Sun, which created turbulence in its outermost reaches.

This turbulence fragmented the cloud into separate vortices, which later became planets and moons. Von Weizsacker's next step was to explain how the system's angular momentum could be delivered from the infant Sun to the outer parts of the nebula, and he proposed that cylindrical 'energy waves' shot out from our parent star and accomplished this task. Later, Swedish cosmologist Hannes Alfvén suggested the cloud's intrinsic magnetic field helped to transfer angular momentum to the disk and slowed the rotation rate of the central mass.

Alfvén's work neatly removed one of the key flaws in Laplace's original theory. As the temperature of the contracting cloud gradually increased and its atoms became ionised to form an electrically-conducting 'plasma', the magnetic field was 'locked in' and intensified as the process of contraction continued. As rings of excess material were cast off, the plasma carried the magnetic field into the outermost reaches of the disk. In effect, the Sun was magnetically linked to the disk from which the planets and their entourage of moons would later condense.

In 1951, Dutch-born US astronomer Gerard Kuiper revamped von Weizsacker's theory. He thought a "chance eddy" of swirling gas provided the impetus for a chain of events that led the cloud to condense into solid, planet-sized objects. Although problems remained, the basic idea has persisted until the present day. Most scientists now accept the Solar System is about 4.5 billion years old and formed from a cloud of gas and dust, although what event(s) set the process in motion remains a subject of debate.

During the formation process, which may have taken as little as 100,000 years, the cloud's temperature increased sharply and superheated the core. At length, the incredible temperatures and pressures in the core were sufficient to form a proto-star from the bulk of the gas, while the remainder flowed around it in a disk. Over time, this disk radiated away its own heat and cooled down. It was within the disk that gas and dust coagulated to form primordial versions of the planets we know today. At length, they grew big enough to pull in more and more material.

The fact that some planets ended up larger than others probably had much to do with the density and composition of the nebula at specific distances from the central proto-star. Evolutionary theorists have argued that in the inner nebula, the primordial planets varied in size from large asteroids to roughly the present size of our Moon, while those further out could be as much as 15 times larger than Earth is now.

Due to the thermal gradient across the disk, heavier refractory elements condensed towards the inner nebula and the bulk of the lighter, more volatile ones ended up further from the core. This is thought to have led to two breeds of planets. The innermost ones were predominantly rocky, or 'terrestrial', in nature: Mercury, Venus, Earth and Mars. Far from the young Sun, the outermost planets ended up much larger, growing over a few hundred million years to become several times bigger than Earth, but also less dense because their chemical ingredients were mostly hydrogen and helium.

There are two possible ways in which the nuclei that became Jupiter, Saturn, Uranus and Neptune gathered these ingredients. One is that large solid cores accreted first and when these became massive enough, their gravity began to attract and 'hold' onto nebular gas. As they accumulated more and more gas, the cores grew heavier and their attraction for more gas – particularly light hydrogen and helium – increased. Another possibility is that instability within the nebula itself led to the separation and pulling-together of gaseous, self-gravitating proto-planets that were massive enough to resist being dispersed by later tidal forces.

This catastrophic process of collision, accretion and the gradual increase in size and mass of these early proto-planets started from around 4.5 billion years ago. The formation of the terrestrial planets was probably almost complete within 10–20 million years, although there remained a significant number of Mars-sized 'planetesimals' roving through the inner Solar System during this time. On the other hand, the giant planets took much longer to evolve and required several hundred million years to assemble their deep gaseous atmospheres and assume their present sizes.

Eventually, the proto-star ignited – the Sun, at last, was born – and generated a very powerful 'solar wind', which swept away all the remaining gas in the nebula. In fact, observations of young T Tauri variable stars – the first of which was found in 1852 – have led many astronomers to conclude that these objects may be 'infant Suns' caught in these final stages of formation. Tantalisingly, recent Hubble Space Telescope images have shown that several T Tauri stars possess accretion disks around their equators, which may hold the seeds of future planetary systems.

Today, we live on a stable world in orbit around a stable, regular star just on the brink of middle age and about halfway through its 10-billion-year period on the 'main sequence', after which it will evolve into a red giant and, in doing so, consume the inner planets.

Humans, and any other creature with the curiosity to look upwards and marvel, have watched it and the other planets for thousands of years and wondered about their nature. Our knowledge has increased incrementally over the centuries. Then, in the summer of 1977, something happened that would teach us more about the giant planets and the outer Solar System in a few short years than we had learned in the whole of previous recorded history.

Two unmanned robot emissaries called Voyager 1 and Voyager 2 set off to explore Jupiter, Saturn, Uranus and Neptune up close and personal, and in unprecedented detail. All four planets would at last be unveiled, not as distant specks in telescope eyepieces, but as *worlds* in their own right. The results would be stunning and unexpected, the pictures breathtaking, the adventure exhilarating. In that heady summer, the Voyagers left Earth on what would become the greatest journey of exploration ever undertaken in human history.

2

A chance of three lifetimes

A REMARKABLE ACHIEVEMENT

Sailing through an uncharted ocean of emptiness, more than 10 billion km from the world that sent it, the Voyager 2 spacecraft continues to transmit a stream of data back to expectant scientists on Earth. Approximately 20 billion km away, roughly in the opposite direction and also heading for the very edge of our Solar System, its sister ship Voyager 1 continues its own lonely journey.

It is a journey that these robotic explorers have kept up for more than 25 years, relentlessly pushing our understanding of our celestial backyard to new heights. Already, they have picked up intense radio emissions thought to be coming from beyond the Solar System; but more on those later. In a few years' time, all being well, they should locate and cross over the outermost boundary of the Sun's vast empire – the as-yet-elusive 'heliopause' – and become the first artefacts fashioned by human hands to venture into the Milky Way Galaxy.

Yet exploration of the Milky Way was not their original mission. By the time they reach the mysterious region beyond the Solar System, called the 'interstellar medium', both Voyagers will – in terms of electrical power at least – be close to death. Indeed, if they are lucky enough to return useful information from this unique location, they will do so on borrowed time. When they were launched in 1977, they were meant to survive just four years. A quarter of a century later, managers confidently expect another 15–20 years' service before power levels run too low to sustain any scientific instruments.

The Voyagers' advertised mission was to fly past the giant planets Jupiter and Saturn and perform detailed, close-range observations of these strange worlds and their retinues of attendant moons. They were also given other tasks more familiar to explorers of old: measuring shapes, sizes, colours, textures, chemical make-up and magnetic fields. Had this epic venture ended after leaving Jupiter and Saturn, they would easily have supplied enough new material to rewrite the astronomy textbooks. But for both Voyagers, their mission would involve more than this; and for one of them, its mission would feature two more planetary encounters.

REACHING THE OUTER PLANETS

The Voyager project (the twin spacecraft only received this name a few months before launch) underwent considerable development after NASA approved it in mid-1972. Originally, the space agency proposed an ambitious Grand Tour of the five outer planets, with two launches each to Jupiter, Saturn and Pluto in the 1976–1977 'window' and two to Jupiter, Uranus and Neptune around 1979. Unfortunately for the tour, which was estimated to cost $750 million, its direct competitor for NASA dollars at the time was the development of the problem-prone, but politically protected Space Shuttle.

Of course, most of the engineers, scientists and trajectory specialists assigned to the tour could not have foreseen that the Shuttle would rob them of such an important venture. In fact, this particular space science mission is one of the most important ever conceived, for reasons associated as much with the orbital motions of the outer planets as with exploration. As early as the eighteenth century, astronomers and mathematicians began to develop an awareness of the perturbing effects of planets on cometary orbits and the powerful ability of Jupiter in particular to modify the trajectory of passing objects.

It was not until the dawn of the twentieth century, however, that several pioneers of space vehicle orbital theory started to examine the usefulness of multiple-planet rendezvous as a means of staging ambitious exploratory missions deep into the Solar System. Unfortunately, it also became clear that relying entirely on chemical propulsion would severely restrict what these missions could achieve. The German civil engineer Walter Hohmann realised this and, in 1925, theoretically demonstrated that placing a spacecraft into an ellipse tangential to the orbits of both Earth and a 'destination' planet would expend the least amount of energy.

This 'transfer' has provided the basis not only for numerous deep-space and planetary missions since the mid-1960s, but also for allowing trajectory specialists to transfer payloads from low orbits to higher (geosynchronous) locations around Earth using a minimum amount of propellant.

Hohmann's seminal work was pivotal in its importance because it identified not only the most formidable obstacle in the path of future Solar System exploration – the requirement for huge amounts of energy to get anywhere – but also the scope for using a planet's gravitational field to alter the trajectory of a passing spacecraft.

During the three decades between Hohmann's work and the advent of the Space Age, several other developments were made in understanding how the concept that we know today as 'gravity assist' could be practically applied to deep-space missions. One key personality in this effort was German physicist Krafft Ehricke, whose 1957 paper, *Instrumented Comets – Astronautics of Solar and Planetary Probes*, studied the way in which cometary orbits are exquisitely modified by Jupiter as well as exploring in depth the main characteristics of gravity assist.

Ehricke identified the usefulness of hyperbolic encounters between planets and spacecraft as a means of decreasing or increasing the orbital size of the latter. He showed that the shape ('eccentricity') and orientation ('inclination') of a spacecraft's trajectory and its velocity vector on leaving the planet could be similarly adjusted.

"In general," he wrote, "perturbative gravitational fields can have the effect of increasing or decreasing the orbital energy and changing the eccentricity of the vehicle's orbit, depending on the conditions of approach to the field. If these conditions are correctly selected, the perturbative force can be utilised for astronautical benefits."

A considerable amount of controversy has arisen over who 'discovered' the gravity assist concept, which has even led to the filing of several lawsuits based on the perceived loss of Nobel Prizes and theft of intellectual property! Although it is not the purpose of this book to consider this issue in depth, many scientists have given credit to Ehricke and English physicist Derek Lawden – who explicitly described the amount of velocity change afforded by gravity assist as early as 1954 – for authorship of the concept. It is clear, however, that its origins can be traced back even further.

By the 1960s, therefore, it was known reasonably well that to rendezvous with another celestial body it was necessary to alter a spacecraft's velocity by placing it into an elliptical orbit that intersected Earth's own velocity vector at the time of launch. Such a spacecraft would typically be launched in the same direction as Earth's own orbital path to reach a planet further from the Sun and in the opposite direction to get to Mercury or Venus.

Although the Hohmann transfer minimised propellant consumption, it could take years to reach the closest giant planets and an entire human generation to get to Neptune. Ordinary chemical rockets simply did not have enough impulse to aim directly for worlds beyond the asteroid belt and, if they had, the journey would have been much too long to be practicable. It seemed unlikely that a spacecraft could travel three decades to Neptune and still be sufficiently acute to return useful data when it got there.

Early plans for planetary missions, therefore, concentrated on reaching the 'local' terrestrial worlds of Mercury, Venus and Mars. This imposed limits on how far humanity could venture from the Sun; at least, that is, until two PhD researchers uncovered a rare geometric oddity. In 1961, mathematician Michael Minovich was hired by the mission design branch of NASA's Jet Propulsion Laboratory (JPL) in Pasadena, California, and developed a detailed analysis of a gravity-assisted mission trajectory to Venus and Mercury. This work ultimately led to the Mariner 10 spacecraft, launched in 1973, which became the first mission to visit two worlds using gravity assist.

Minovich realised that planetary gravitational fields could profitably be used to change the velocity of an incoming spacecraft relative to the Sun. If the spacecraft caught up with a planet from 'behind', it gained energy and was flung outward at an increased speed, rather than returning to the inner Solar System. On the other hand, if it passed in 'front' of the planet, it lost energy and fell closer to the Sun. By 1962, Minovich was using this knowledge to design exploratory missions as far afield as Jupiter.

Since Jupiter is by far the most massive planet in the Solar System, it struck Minovich that using its gravitational field as a 'slingshot' could help to reduce the flight time to more distant worlds. He realised that carefully planning the trajectories

Gary Flandro.

of spacecraft launched in either the 1962–1966 or 1976–1980 timeframes could theoretically achieve multiple rendezvous with planets beyond the asteroid belt. Then, in July 1965, Gary Flandro, an aeronautics postgraduate also working at JPL, modified Minovich's calculations to make a crucial breakthrough.

He found that a spacecraft launched between 1976 and 1978 could utilise Jupiter's gravitational field to reach not only Saturn, but also Uranus and Neptune. In effect, all four giant planets would be 'spread out' along the ecliptic plane and the spacecraft's course would employ a succession of gravity-assisted arcs to travel from world to world. Flandro called his venture a 'Grand Tour', drawing the name from one used by Italian mathematician Gaetano Crocco, who had proposed a gravity-assisted mission to Venus and Mars in 1956.

THE GRAND TOUR AND PROJECT VOYAGER

In what he described as "a rare moment of great exhilaration", Flandro sketched out his own Grand Tour. A spacecraft sent aloft in 1977, he suggested, would reach Jupiter two years later. It would then be accelerated by about 11 km/s by the planet's enormous gravitational field and its course deflected by 97 degrees towards Saturn, which it would encounter in 1980. After departing Saturn, it could proceed to Uranus in 1984 and Neptune in 1986. However, Flandro noticed that if the 1976–1978 launch opportunity was missed, it would not appear again for an astonishing 176 years.

In fact, the last time the quartet had been aligned so favourably had been in 1801, during Thomas Jefferson's presidency! Flandro's tour offered a fantastic chance to see four unknown worlds in a single mission: truly a chance of not one, but three, lifetimes. He received encouragement from his supervisor Joe Cutting and from Homer Joe Stewart, one of his professors at the California Institute of Technology (Caltech) in Pasadena, and within days JPL issued a press release which officially kicked-off the design phase.

The enthusiasm of many in the scientific community was audible, to such an extent that two studies were funded by the National Academy of Sciences to evaluate it. Both were broadly supportive, indicating "widespread eagerness" to undertake such a venture and remarking that "an extensive study of the outer Solar System is one of the major objectives in space science". Moreover, the Neptune journey time, which ran to 30 years without gravitational assistance, could be greatly reduced.

Unfortunately, a number of problems were identified and concerns raised. One of the most serious was a requirement for the spacecraft to travel dangerously close to Saturn in order to pick up sufficient gravitational impetus to reach Uranus. This passage, it was realised, might expose the spacecraft to impacts from ring particles. In 1966, James Long of JPL's advanced projects office conceived a more conservative mission design, which employed two identical spacecraft on courses that were kept well clear of Saturn's rings.

The first Grand Tour mission, which Long dubbed 'GT-1', would be sent aloft in August 1977 to reach Jupiter 17 months later and Saturn in August 1980. This latter encounter would then be used to direct the spacecraft onto a five-year path to distant Pluto, with which it would rendezvous in December 1985. However, rather than significantly increasing GT-1's outward velocity, Saturn's main role would be to 'tilt' its trajectory almost 25 degrees out of the ecliptic to reach Pluto, which, in the winter of 1985–1986 would be over 1.25 billion km above the ecliptic and actually closer to the Sun than Neptune.

Then, in November 1979, GT-2 would head into space and employ a Jovian rendezvous two years later to direct it towards Uranus (1985) and Neptune (1988). Long argued that 1977–1979 were the best years for launching the tour because they offered the greatest reduction in flight times to the outer planets without having to employ rockets beyond existing capabilities.

Meanwhile, working on the blissful assumption that the tour was safe and much too scientifically valuable to be cancelled, by June 1969 JPL had laid preliminary plans for four spacecraft, loosely based on the design of Mariner, a highly-successful series of NASA planetary missions staged between 1962 and 1975. Although three Mariners failed, the others journeyed to Venus, Mars and – in the case of Mariner 10 – even barren Mercury, returning reams of highly-prized data. In fact, Mariner 10 remains the only spacecraft to have visited Mercury, until the MESSENGER orbiter makes its first flyby of the tiny world in July 2007.

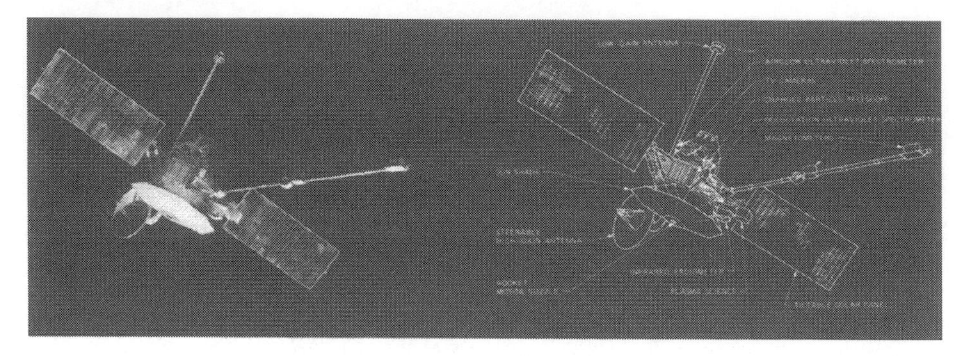

The Mariner 10 spacecraft.

Each Mariner was subtly different, so we shall briefly look at Mariner 10 to build a picture of the Grand Tour's heritage. This last Mariner consisted of an octagonal 'bus', 46 cm deep and 1.4 m wide, weighing about 430 kg; each of its eight faces contained all the electronics and equipment needed to run the spacecraft. Two large solar panels, each 2.7 m long and almost a metre wide, were attached to the top and supported over 5 m^2 of solar-cell area.

Mariner 10 also carried a liquid-fuelled rocket engine, two sets of reaction-control jets to stabilise it through three axes, a star tracker, Sun sensors and a 1.4-m-diameter high-gain antenna capable of transmitting at both X- and S-band frequencies. Obviously, the Mariners were only designed to explore the terrestrial planets, and when the time came to plan souped-up versions to go further afield, several tricky obstacles reared their heads. One of these was power. In its 1969 summary, JPL revealed that each of the four Grand Tour spacecraft would weigh 640 kg and be known as Thermoelectric Outer Planet Spacecraft (TOPS).

PLUTONIUM POWER SOURCE

Solar panels would be inadequate in the outer Solar System. Even in the asteroid belt, before the TOPS mission had really started, the energy available in sunlight would be too small to be an effective power source. It is worth noting for clarity at this juncture that the amount of energy available in sunlight falling on a spacecraft's solar panels at Jupiter is just 4 per cent as much as would be available in Earth orbit.

By the time they reached Saturn and beyond, it was predicted that the apparent size of the Sun in the sky and the amount of energy deposited on solar panels would have decreased still further. As a result, each spacecraft would utilise several Radioisotope Thermoelectric Generators (RTGs), from which the TOPS name partly derived. Today, these devices have been used successfully for more than two dozen space missions spanning more than three decades, and consist of small plutonium-fed units bolted end-to-end on extendable 'booms'.

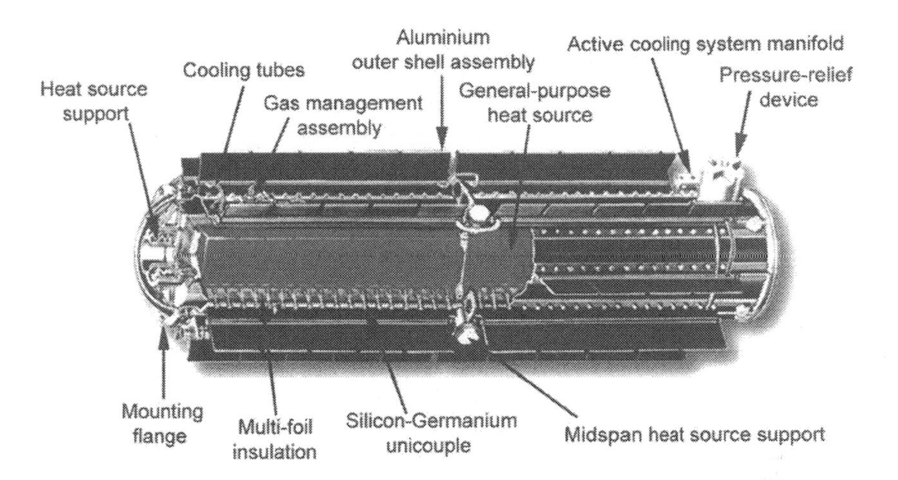

Cutaway diagram of an RTG power unit. Devices similar in design flew on board both Voyager spacecraft.

Each RTG contains a small quantity of fissionable plutonium dioxide, whose gradual radioactive decay produces heat. This is then converted into electricity using a silicon–germanium bimetallic thermocouple to run a myriad of instruments and subsystems. Despite public controversy which arises when RTG-powered missions are sent aloft, the devices are exhaustively tested pre-flight and have proven to be exceptionally safe. Versions of these devices were used to run experiments packages left on the Moon by some of the Apollo landing crews, as well as the Pioneer, Viking, Voyager, Galileo, Ulysses and Cassini deep-space missions.

Their safety is attributable in part to their lack of moving parts. Three accidents have occurred in 40 years, only one of which released radioactive pollutant. This incident involved the Transit 5-BN-3 navigation satellite, crippled by a launch vehicle failure in April 1964; its RTG was destroyed during re-entry and dissipated plutonium into the upper atmosphere. As part of the investigation into the accident, design changes were implemented to enable future RTGs to survive re-entry and be recovered intact by ground personnel.

When the Nimbus B-1 weather satellite was lost shortly after launch in May 1968, it dutifully released no plutonium. Nor did the experiments package on board the Apollo 13 lunar module, which broke up as it re-entered Earth's atmosphere in April 1970. Not only have safety procedures for operating RTGs changed over time, but so have the capabilities of the devices themselves. The earliest ones produced just 2.7 watts of electrical power, whereas modern versions – like those currently running the Cassini spacecraft, due to reach Saturn in July 2004 – are capable of generating up to 290 watts each.

Early RTGs carried their plutonium in the form of a metal, but those that flew on the Voyagers in 1977 consisted of small powder 'pellets'. Each Voyager RTG produced about 157 watts of electrical power at launch. A key benefit of using these devices is sheer longevity. Despite having been launched a quarter of a century ago, the

Voyagers' power supplies are not expected to be exhausted until 2020. Yet slowly, like an eggtimer, their power output dwindles by an average of 7 watts per year, due primarily to the decaying plutonium and degradation of the bimetallic themocouples.

Each Voyager carries three RTGs, but the ensemble is only about 6 per cent efficient. Between them, they released 7 kilowatts of heat at launch, which was converted into about 470 watts of electricity, although by early 1997 that had fallen to less than 335 watts. Typically, their on-board scientific instruments can run on just 400 watts, but several devices – such as the cameras, which no longer have any targets to inspect – have now been turned off to ensure that demand does not exceed supply.

Had TOPS gone ahead, each spacecraft would have been equipped with four RTGs, with a total weight of 117 kg. These were expected to produce 439 watts of electrical power at the end of the planned mission and would have been positioned at least 1.5 m from the main spacecraft bus on extendable booms. Although this practice of keeping the RTGs at a safe distance from TOPS' sensitive scientific

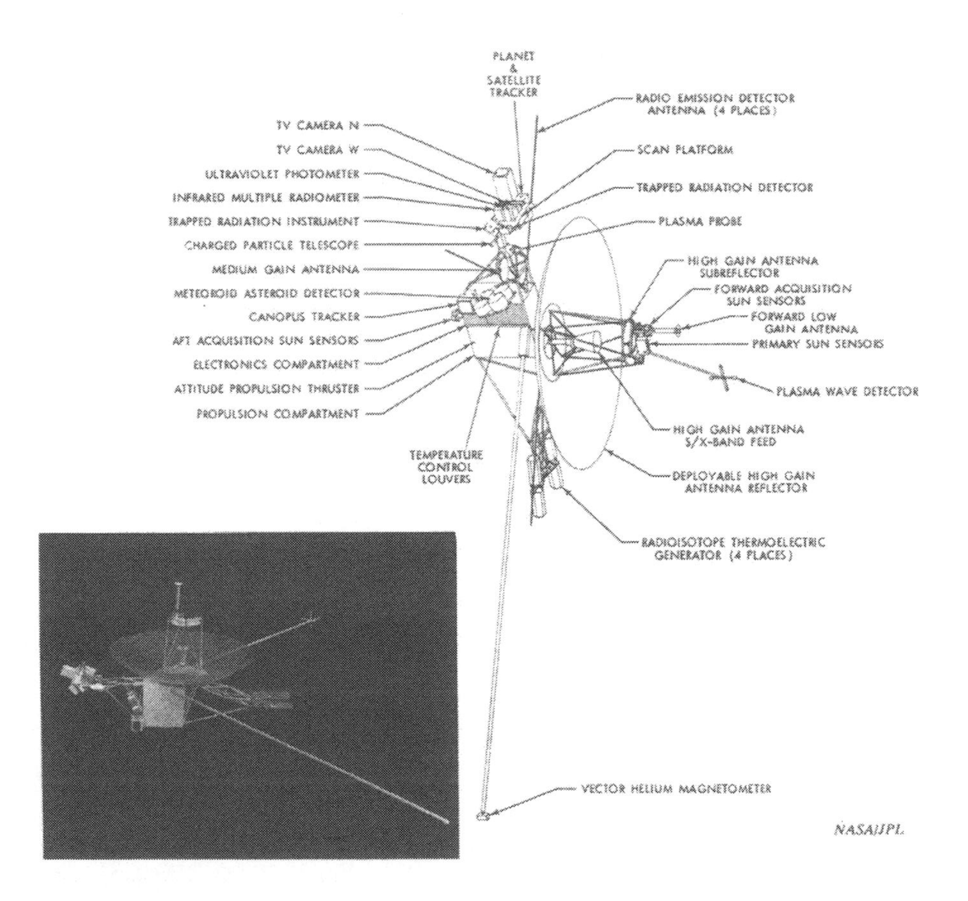

Line drawing and model of TOPS spacecraft.

instruments minimised the risk of radiation damage, additional protection was afforded by extra shielding and 'hardening' of the electronic components.

Other features included a 4.3-m-diameter high-gain communications antenna, which would have been carried into orbit in a stowed configuration and deployed soon after TOPS left the final stage of its launch vehicle. Studies identified the need for at least 48 antenna 'ribs' to ensure that it deployed properly and maintained the correct shape throughout its long journey. Temperature variations and micrometeoroid impacts were cited as factors that could potentially foul the antenna. The TOPS antenna would have been capable of returning 400 pictures from Neptune over an 11-day period.

AN AMBITIOUS – AND EXPENSIVE – VENTURE

One part of the TOPS operating philosophy that was eventually implemented with the Voyagers was duplicating each mission. This was actually a common strategy at the time and several other planetary missions were also launched in pairs. Of the original four TOPS, two would have been launched in 1977 to undertake the Jupiter—Saturn–Pluto mission and two in 1979 to accomplish the Jupiter–Uranus–Neptune venture. Likewise, the two Voyagers were designed and built identically. This duplication was partly to maximise the overall scientific harvest, but chiefly to provide backup in case one was lost at launch or failed *en route*.

However, gradually spiralling project costs would ultimately sound the death-knell for the Grand Tour. An initial estimate, completed in December 1968, calculated that the research and development phase of the mission alone would cost $17.5 million, but this included numerous frills such as the fabrication of a 'feasibility model' precursor spacecraft to demonstrate its advanced technology components. As predicted costs and budget requirements increased towards the end of 1969, a less ambitious plan was chosen which deleted the feasibility model. Yet the complexity of the mission itself – which included advanced computers and state-of-the-art instruments – was gradually taking its toll.

It was known from the outset that a mission as ambitious as the Grand Tour would require spacecraft to be operational for up to a decade in very inhospitable conditions. With this in mind, the reliability of TOPS was considered of paramount importance. As they journeyed further from Earth, it was expected that communication times would increase to such an extent that the spacecraft should be capable of operating unsupervised for long periods of time. This required a computer several steps ahead of anything previously used on an unmanned spacecraft; one capable of identifying and correcting failures within itself.

The TOPS computer came to be known as Self Testing and Repair (STAR) and was originally devised by Alverez Avizienes and his colleagues at JPL in the mid-1960s. By the end of the decade, they had developed and built a prototype that was capable of diagnosing errors within itself and effecting cures. The downside, however, was that the unit was huge: it completely filled three 1.8-m equipment racks, which meant that it would have to be miniaturised to fit into the confines of a TOPS-sized spacecraft bus.

STAR was designed so that, in the event of problems within the computer, messages generated by a failing component would become deliberately 'corrupted'. This corruption would take the form of illegal words not recognised by the system and these would be picked up by a Test and Repair Processor (TARP) that would identify and isolate the failed unit and switch to a backup. In order to ensure that it did not itself become corrupted, TARP included three monitors that would 'vote' on each decision to be taken by the system.

If one monitor disagreed with its two peers, it would be tested and – if it continued to dissent – STAR would assume that it had failed, remove it from the system and replace it with a backup unit. The end result was that, in addition to checking on the performance of spacecraft systems, TARP would be able to perform periodic health checks on itself.

At about the same time Avizienes and his team were working on the development of STAR, the Martin Marietta Corporation was carrying out detailed studies into the feasibility of dropping instrumented probes into Jupiter's atmosphere. TOPS was one spacecraft bus under consideration to deliver such a payload to the giant planet. Had this plan gone ahead, it was expected that the probe would weigh only about 8.6 kg, but would have been enshrouded in a protective aeroshell to withstand the tremendous temperatures and pressures as it penetrated the outer Jovian atmosphere.

It would probably not have been too dissimilar to the probe dropped into Jupiter's atmosphere by NASA's Galileo spacecraft in December 1995. TOPS would have jettisoned the probe about six weeks and 26 million km before its closest approach to the giant planet. The probe would have hit the atmosphere at about 50 km/s and taken a couple of hours to plunge to a depth of 400 km below the cloud tops. At this point, it would be crushed by a pressure 300 times that at sea level on Earth.

During its brief foray into the Jovian atmosphere, the probe would carry out measurements using a mass spectrometer, accelerometers, a photometer, temperature and pressure sensors and perhaps even an optical flash detector to pinpoint lightning. However, the proposal turned up problems. Due to the different approach trajectory required by TOPS to achieve its other planetary rendezvous, the Jovian probe could only survive to an atmospheric depth equivalent to 10 times sea-level pressure on Earth. A two-stage probe was suggested instead, but this would increase the weight on board TOPS, which in turn would jeopardise the gravity assist needed to reach Saturn.

CANCELLATION AND AFTERMATH

Gradually, as the mission proposals became more ambitious – one alternative even advocated orbiters and atmospheric-entry probes for both Jupiter *and* Saturn – the price tag climbed rapidly. Congress predictably balked at the steep cost, which peaked at close to $900 million, and pressed NASA to come up with a cheaper option. The Grand Tour was officially cancelled on 24 January 1972. NASA, for its part, had already begun to lay the groundwork for developing the Shuttle and expected to have little money to spare in the years leading up to the Grand Tour's 1977 launch window.

More attractive politically and economically, the Shuttle offered 50,000 new jobs and a much-needed boost to the fortunes of the United States' aerospace industry. The tour, in comparison, was seen by some as a frivolous waste of taxpayers' dollars. Perhaps, as had happened during the Moon race in the 1960s, the outcome would have been very different if the Soviets had been in a position to stage their own tour. Unfortunately, little serious consideration seems to have been given to missions to the outer planets from behind the Iron Curtain, either during this period or later.

This did not, however, deter Soviet space scientist Timur Eneyev from claiming in the *New York Times* in May 1971 that his team was studying the feasibility of delivering unmanned, instrument-laden balloons into the atmospheres of the giant planets. Although these plans never reached fruition, the Soviet angle does at least invite speculation on what might have been achieved. Certainly, from the late 1960s, the 12th Department of TsNIIMash – the leading research institute in the Soviet rocket industry – proposed several ambitious ventures, including missions to Jupiter and Mercury.

More recently, in 1986, Vladimir Perminov, a designer at the Lavochkin bureau in Moscow, proposed a 2,000-kg mission codenamed 'Tsiolkovsky', which would head firstly for Jupiter to get a gravitational shove towards the Sun for an extended programme of solar observations. During its brief time near Jupiter, it would drop a 500-kg probe into its atmosphere. Had Tsiolkovsky flown, Perminov's plan called for it to be launched by a medium-lift Proton booster in 1995–1996, about the same time that Galileo dropped its own instrumented probe into the murky Jovian clouds.

Presumably, the fall of communism and the Soviet Union in the early 1990s, together with the harsh economic downturn that followed, put paid to this first – and last – serious attempt to send a spacecraft bearing the Hammer and Sickle to one of the giant planets. Likewise, TOPS also breathed its last on a draughtsman's board towards the end of January 1972.

Regardless of the disappointing loss of the tour, JPL was partially appeased by a smaller funding package to develop and launch a two-spacecraft mission to Jupiter and Saturn in mid-1977. In an effort to keep its costs from spiralling, NASA imposed a $250 million spending cap and opted to adapt the Mariners that had worked so well at Mars and Venus. To underline this heritage, the new project, officially unveiled on 1 July 1972, was dubbed 'Mariner Jupiter–Saturn' or 'MJS'. In some quarters, the names Mariner 11 and Mariner 12 were even used for a short time.

Uranus and Neptune, however, were conspicuously absent from the new mission's résumé. Neither had yet been visited by a spacecraft and both worlds were little more than fuzzy specks to even the most advanced Earth-based telescopes. Yet, although many scientists recognised that it should still be possible to reach Uranus and

Neptune by carefully planning the trajectory of at least one MJS spacecraft, such a plan was officially off the agenda. It was rationalised that dropping the option to visit these far-off worlds would reduce the length of the mission, development costs and stresses and strains on the MJS vehicles.

In reality, scrapping a grand, four-spacecraft flotilla to visit five worlds and short-changing the scientific community with a two-spacecraft effort to visit just two worlds seemed sheer lunacy. America was being offered, virtually on a plate, a once-in-three-lifetimes chance to have four distant, unknown planets in alignment for one spectacular space shot. The intention to send two spacecraft on what was pretty much the old Grand Tour trajectory with no intention of considering Uranus or Neptune made it all the more unpalatable.

Nevertheless, MJS proceeded and by mid-July 1972, 77 proposals for scientific instruments to fly on the two spacecraft were received in response to NASA's Announcement of Opportunity. This list was eventually winnowed down to 28 finalists, nine of whom were charged to develop actual instruments and the others to participate in observations using the on-board imaging and radio science gear. As a result, when MJS Project Manager Harris "Bud" Schurmeier and Project Scientist Ed Stone held their first Science Steering Group meeting at JPL in December 1972, they were joined by 11 Principal Investigators – one supervising each instrument proposal.

Artist's concept of Pioneer 10 during its December 1973 encounter with Jupiter.

The decision as to what form these instruments would take was aided by the experience of another NASA spacecraft – Pioneer 10 – which, in December 1973, became the first human-made machine to reach and journey beyond Jupiter. It was in for a shock. As it sped past, it recorded 10 'hits' from minute particles on its micrometeoroid detectors and suffered significant damage, including fried circuits and darkened optics, from the savage radiation belts. In light of the Pioneer discovery, one of the instruments planned for MJS to search for such particles was deleted.

The Pioneer experience gave NASA a chance to 'harden' the MJS spacecraft's instruments and electronics to better withstand the increased dosage expected near Jupiter. Around 1975, the Science Steering Group approved the inclusion of a Plasma Wave Subsystem (PWS) to characterise the radiation environment around both Jupiter and Saturn. By this time, MJS was changing beyond the wildest dreams of Mariner or even Pioneer: clearly its mission would be much longer and more arduous. As a result, on 4 March 1977 the project was renamed 'Voyager'.

TRAJECTORIES

As the spacecraft took shape, there remained one nagging problem. Regardless of their improved capabilities, both Voyagers were still tied to only Jupiter and Saturn. The two outermost giants, Uranus and Neptune, were politically out of bounds. However, presumably hoping that the political winds might change once Voyager had shown what it could do, both spacecraft were designed to support operations beyond Saturn and the trajectory of one was engineered such that it would offer the option of a free ticket to Uranus. The carrot was too tempting to ignore, and it was to prove a wise move.

During this time, trajectory specialists sifted through over 10,000 potential roadmaps through the Solar System for two that would maximise the harvest from Jupiter and Saturn. Of particular importance were paths that would fly close to the large moons Io and Titan and traverse the entire radial extent of Saturn's rings. The original Grand Tour flight-path *was* still possible with the Voyagers, but only by sacrificing close-up views of Io and Titan. If Voyager 1 flew close to Io, its velocity would be increased so much by Jupiter's gravity that it would reach Saturn too early to be slung towards Uranus.

On the other hand, if Voyager 1 flew close to Titan, it would leave Saturn in the wrong direction to reach Uranus. However, if Voyager 1 relinquished its chance to fly to Uranus and undertook both the Io *and* Titan encounters, Voyager 2 could be placed onto a suitable trajectory to visit the two outermost giants. Of course, if Voyager 1 failed at Saturn, the trajectory of its sister ship would have to be altered in mid-flight to inspect Titan. But if Voyager 1 succeeded, its partner could broaden the coverage of the Saturnian system and go on to complete the Grand Tour.

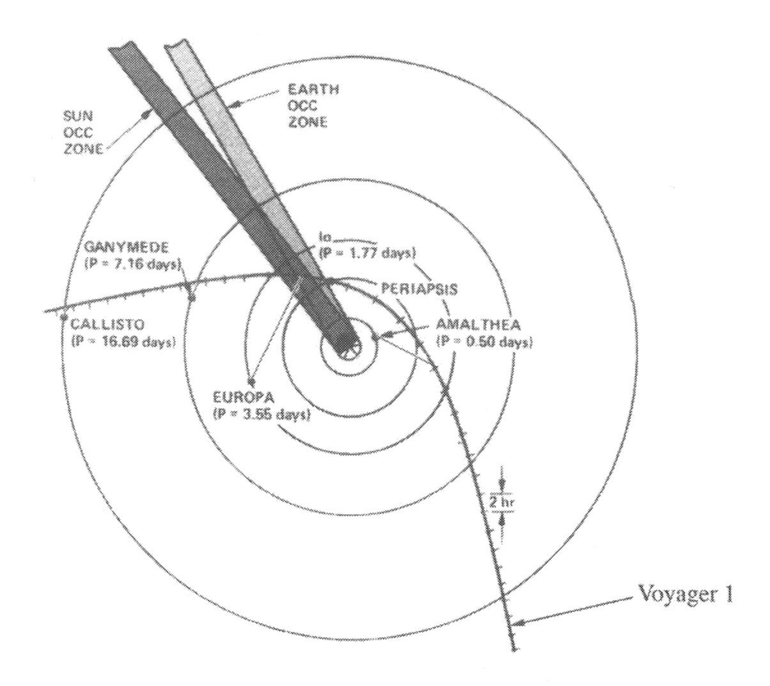

Voyager 1's trajectory through the Jovian system.

THE VOYAGER SPACECRAFT

By the spring of 1977, JPL engineers were putting the finishing touches on the two spacecraft. Both were identical, each decked out with 10 scientific instruments to capture as much of the essence of the alien worlds as possible in the short time available. They would, after all, be travelling past the giant planets without stopping, and would have at most only a few weeks to conduct the bulk of their critical observations.

In appearance, both Voyagers are the epitome of the cumbersome: boxy, 10-sided buses with 3.7-m dish-like high-gain antennas, three spindly booms and a pair of whip antennas, bristling with scientific sensors that stretch like gnarled branches into the void. Each spacecraft weighs 722 kg. The main bus, which carries engineering and control subsystems, is 47 cm high and 1.8 m from flat to flat. The centreline of the bus, on which the high-gain antenna is mounted, is called the 'Z-axis' and always points towards Earth. The Voyager rolls about the Z-axis by firing tiny thrusters affixed to the bus.

The antenna – part of the Radio Science Subsystem (RSS) – receives commands from Earth and transmits scientific and engineering data back home. It was also to be used during each planetary encounter as a research instrument in its own right and employs both the X-band (8.4 GHz) and S-band (2.3 GHz) frequencies to carry telemetry data. When the Voyagers drew near to the giants, the RSS was utilised to

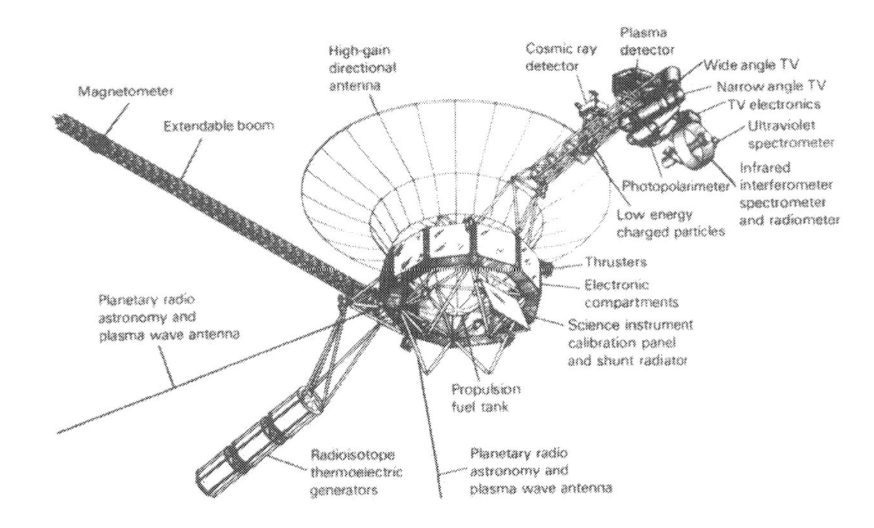

Diagram of the Voyager spacecraft, showing the positions of each of its scientific instruments, subsystems and RTG power plants.

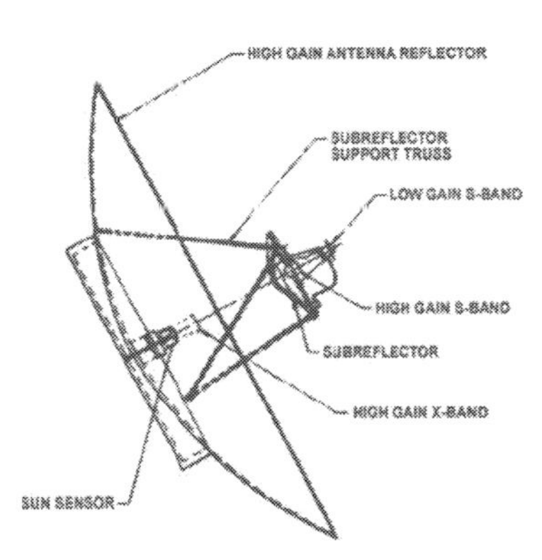

Structure of Voyager's high-gain antenna.

study atmospheric and ionospheric composition, provide insights into gravitational fields, masses and densities and determine the existence and extent of rings.

These data were acquired by means of a technique known as 'radio occultation', in which the changes effected on the spacecraft's radio signal as it was passed through each planet's upper atmosphere helped to construct chemical, pressure and temperature profiles. Furthermore, by carefully tracking this signal during close encounters with the planets and their moons, it was possible to detect gravitational fields, determine masses and measure physical densities. Such information would prove invaluable for understanding the internal structure and geological evolution of the giant planets' moons.

Primarily, however, the high-gain antenna was designed to transmit information back to Earth. The S-band channel was reserved for delivering engineering data at the relatively slow rate of 40 bps and was used until Voyager 2's Neptune encounter. Both scientific and engineering data are currently returned on the X-band channel at a speed of 7.2 kbps. The usable rate of the X-band channel has diminished significantly with increasing distance from Earth: data rates of 115.2 kbps were possible during the Jupiter encounters, but dropped to 44.8 kbps at Saturn, then 21.6 kbps at Uranus and 14.4 kbps at Neptune. In doing so, the Voyagers have become the first deep-space missions to employ X-band as their primary telemetry link frequency. Both spacecraft also have smaller low-gain antennas in reserve.

COMMUNICATIONS

The 70-m DSN antenna at Goldstone in California's Mojave Desert.

The Voyagers' antennas are controlled and their data managed through the Deep Space Network (DSN), a global series of tracking stations run by JPL on NASA's behalf. Three 70-m antenna complexes are situated at 120-degree longitude intervals across the world, at Canberra in Australia, Madrid in Spain and Goldstone in the

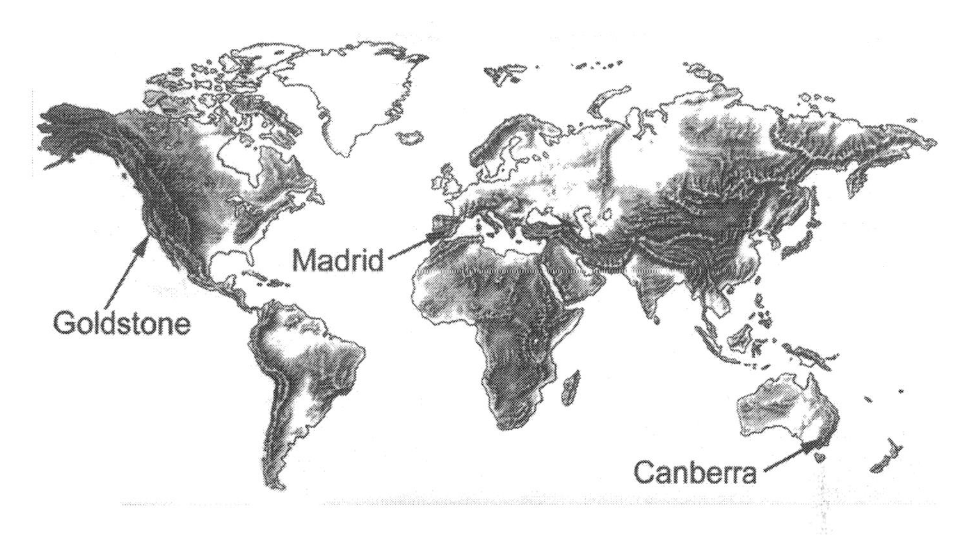

Map of the positions of the Goldstone, Madrid and Canberra DSN stations. Spaced at 120-degree intervals across the globe, the DSN allows near-continuous communication with distant spacecraft as Earth rotates.

United States. Not only does this make the DSN the biggest and most sensitive scientific telecommunications facility in the world, but it also allows mission controllers to monitor the Voyagers near-continuously as Earth rotates.

After Voyager 2's Neptune encounter, NASA aimed to acquire at least 16 hours' worth of data per day from each spacecraft, although this is no longer possible due to intense competition for DSN resources by other deep-space missions. All three DSN antennas are steerable, high-gain parabolic reflector dishes. This enables them to receive telemetry and scientific data from both deep-space and a few Earth-orbiting missions, transmit commands to them, track their positions and velocities and perform research tasks in support of radio science and other investigations.

All three DSN complexes are situated in rugged, bowl-shaped regions of semi-mountainous terrain; the sheer remoteness of these locations helps to shield them from unwanted terrestrial radio interference. The Australian station lies about 40 km southwest of Canberra, close to the Tidbinbilla Nature Reserve in New South Wales. Its Spanish counterpart is located 60 km west of Madrid at Robledo de Chavela, while Goldstone is part of the US Army's Fort Irwin Military Reservation, 72 km northeast of Barstow in California's inhospitable Mojave Desert.

Each complex has at least four deep-space stations, each equipped with ultra-sensitive receiving systems and antennas. These are a 34-m high-efficiency antenna, a 34-m beam waveguide antenna, a comparatively small 26-m antenna and a large 70-m antenna. Some DSN complexes have more antennas than others: Goldstone, for example, has no fewer than three high-efficiency dishes, while a grand total of five beam waveguide assemblies – three at Goldstone and one each at Canberra and Madrid – were added to the network in the late 1990s to handle the profusion of small spacecraft operating in the inner Solar System.

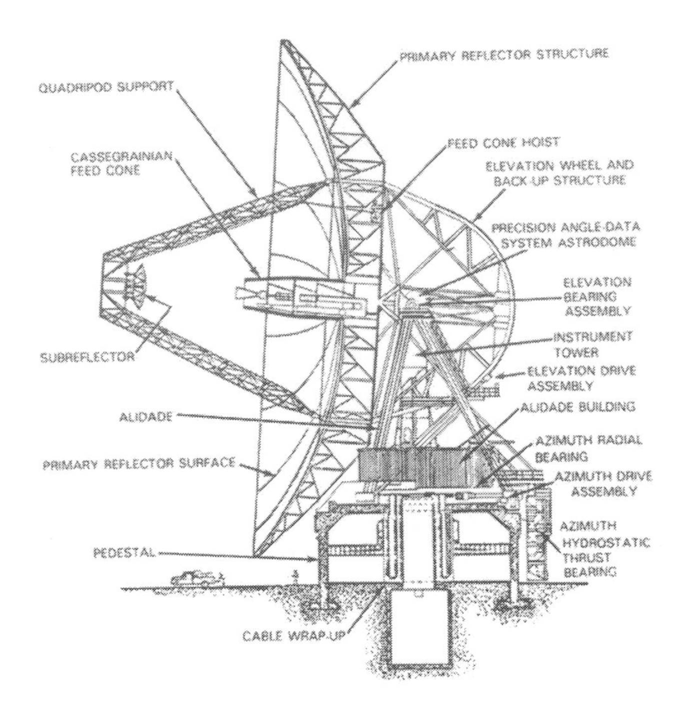

Line drawing of a DSN antenna station.

The entire network is constantly undergoing renovation and overhaul (in fact, another antenna is currently under construction in Madrid) and significant advances have been made in recent years to link the stations together electronically. The demands placed on the DSN by the Galileo mission to Jupiter, for example, have obliged NASA to link together – or 'array' – several of the antennas at Goldstone and Canberra.

Each DSN station has its own signal-processing centre to remotely point and control the antennas, gather telemetry, transmit commands and initially process incoming data. When the data has been processed at the antenna complexes, it is transmitted to JPL for further work and eventual distribution to scientists over a modern ground-based communications network.

The roots of the DSN go back to January 1958, when JPL developed portable tracking stations in Nigeria, Singapore and California to receive telemetry and plot the orbit of the United States' first artificial satellite, Explorer 1. Following the establishment of NASA later that year, the concept of a global telecommunications network was explored in greater depth. This relieved mission planners from having to devise their own means of acquiring and managing data. It also left the DSN free to research and develop some of the most advanced low-noise receivers, tracking, telemetry and digital signal-processing systems in the world.

NASA began to plan the building of antennas larger than 26 m across in 1960, expecting to send missions deep into the Solar System towards the end of that decade.

Its plans at the time included grand ventures like lunar satellites equipped with high-resolution multi-spectral mapping cameras, planetary orbiters and manned missions to the Moon. Following the evaluation of several proposals to develop more sophisticated deep-space missions, NASA awarded the Rohr Corporation a $12 million contract in June 1963 to build a 64-m complex at Goldstone. Two years later, it successfully received signals from the Mars-bound Mariner 4 spacecraft.

A second 64-m antenna was subsequently completed at Canberra in July 1972 to provide continuous coverage for deep-space missions requiring the support of large-aperture antennas. By this time, advances in antenna-building technology meant that this enormous antenna could share the same valley as another, older station called DSS-42 without any radio interference between them. In December 1972, the new Canberra antenna acted in a backup role to the 64-m Parkes radio telescope, also situated in New South Wales, to help track the Apollo 17 lunar landing mission.

In readiness for the Voyagers, from 1976 to 1979 NASA increased the size of its smaller tracking antennas at Canberra, Madrid and Goldstone from 26 m to 34 m in diameter. Each antenna would have high efficiency at X-band (8.4 GHz), which was the primary data channel to be used by the Voyagers. The reasons for this particular conversion were two-fold. Obviously, larger receivers were essential for JPL to track two spacecraft travelling billions of kilometres from Earth. Moreover, the new 34-m size meant that engineers could modify the existing structure, rather than dismantle the whole complex and rebuild it from scratch.

Further advancements in DSN capabilities – including the 'arraying' of antennas to receive ever-fainter radio signals from the distant Voyagers – were completed throughout the 1980s in preparation for the Uranus and Neptune encounters. Those modifications will be discussed in greater depth later. One notable benefit of these improvements, aside from the Voyagers, was that they extended the range of ground

A view of the 34-m and 70-m antennas at Tidbinbilla, near Canberra in New South Wales.

communications with the Pioneer 10 spacecraft from 4.6 to 5.6 billion km from the Sun. NASA finally lost contact with this spacecraft towards the end of February 2003, marking the end of a truly illustrious mission.

CAMERAS

It has been this impressive worldwide network of antennas that has, for more than a quarter of a century, helped to collect reams of data and countless thousands of pictures that we have from each of the four giant planets and their multitude of moons. On the Voyagers, the 10 instruments that were used for these observations are mounted either on a movable 'scan platform' – precariously perched at the end of a 2.4-m science boom – or at the even more dizzying extremes of a 12-m magnetometer boom or on two 10-m whip antennas.

All of these insect-like appendages stick out at crazy angles from the central bus; the magnetometer like a long lattice structure, the whip antennas like a pair of radio antennas forming a V-shape. (The third boom holds the RTGs and their plutonium pellets at a safe distance from the scientific payload.) Many of the Voyagers' instruments have served the scientific community exceptionally well, returning by 1989 alone, an amount of data equivalent to that contained in 6,000 editions of the *Encyclopedia Britannica.*

The Voyager mission has cost approximately $865 million, which would at first glance appear significantly higher than the 'too expensive' $750 million Grand Tour. However, it must be borne in mind that the Voyager price tag is spread over a quarter of a century and, as we shall see, the two spacecraft accomplished far more than their 'basic' mission and actually completed most of the original Grand Tour objectives. The three quarters of a billion dollars that might have financed the tour, on the other hand, were spread across only a decade's worth of operations.

Although the data produced by each Voyager instrument has provided more than enough in the way of scientific discovery to pay for the mission's price tag, it was undoubtedly their on-board cameras – part of the Imaging Science Subsystem (ISS) – sold the venture to the public. Without them, we would have none of the spectacular pictures of Jupiter and Saturn and would have missed humanity's first-ever close-up glimpse of Uranus and Neptune. Each Voyager carries a 38.2-kg imaging system, which contains low-resolution, wide-angle (200 mm focal length) and high-resolution, narrow-angle (1,500 mm) cameras. Both are mounted on the scan platform boom.

Each camera's output forms an 'array' of 800 lines of video by 800 pixels per line; or 640,000 pixels per picture. Pixel brightness levels were represented by an eight-digit binary number and exposure times varied from planet to planet, increasing as the twins journeyed further from the Sun. By the time Voyager 2 reached Neptune in 1989, it took between 0.005 and a full 61 seconds to take a picture. The images then took between a few seconds and 8 minutes just to be 'read out' on the ground, depending on whether they were recorded or transmitted in real time.

The cameras were expected to return pictures of unprecedented clarity from each of the giant planets. Indeed, they were advertised before and during the mission as being capable of reading a newspaper headline from a distance of a kilometre. The system is basically an upgraded version of the slow-scan 'vidicon' devices first used on Mariner and consists of television-type cameras, each with eight filters in a commandable wheel mounted in front of each unit. Unlike the other instruments, their operation was not automatic; instead, it was run by a table of parameters pre-loaded in the Voyagers' computer memory.

Broadly, the tasks of the cameras were to observe and characterise the circulation of the giant planets' turbulent atmospheres, with a particular focus on monitoring the growth and dissipation of clouds, storms and mysterious spot-like phenomena like those on Jupiter. They were also designed to help determine the chemical constituents of each planet's atmosphere, measure wind speeds, map the distribution of ring particles, obtain detailed multi-spectral pictures of as many moons as possible and generally support the other scientific instruments. Obviously, as the Voyagers drew closer to each celestial target, these objectives could be refined to encompass detailed geological and geophysical surveys.

REMOTE-SENSING EQUIPMENT

Also attached to the scan platform are two other instruments that have proven invaluable in furthering scientists' understanding of the giant planets' atmospheres. These are the Infrared Radiometer, Interferometer and Spectrometer (IRIS) – a mouthful by anyone's standards – and the Ultraviolet Spectrometer (UVS). The former, as its name implies, comprises an interferometer sensitive to the mid-infrared portion of the electromagnetic spectrum, coupled with a radiometer and spectrometer for measurements in the visible and near-infrared range.

The IRIS, which has a 50-cm-diameter Cassegrain telescope and dual-channel Michelson interferometer, protected by a dish-like sunshade, therefore comprises three instruments, all rolled into one. Firstly, it is a sensitive thermometer, capable of working out the distribution of heat energy a celestial body emits to determine global temperatures. Secondly, its spectrometer portion was designed to measure whether certain elements or compounds were present in an atmosphere or on a surface. Finally, its radiometer mapped the total amount of sunlight reflected by a celestial body at ultraviolet, visible and infrared wavelengths.

Primarily, the IRIS helped to determine the vertical thermal structure of each giant planet's atmosphere, in order to improve theoretical models of how they circulate and where their internal heat comes from. This proved particularly important at Neptune, which emits much more internal heat than it theoretically should, in view of its immense distance from the Sun. Additionally, the IRIS has aided efforts to check whether the relative abundances of hydrogen and helium in the giants' atmospheres are consistent with ideas about their ratios in the primordial solar nebula. This has far-reaching implications for our understanding of how the Solar System evolved.

Although at the time of their construction, Uranus and Neptune were officially off the Voyager flight plan, many scientists realised that if the mission *was* extended to visit these far-off worlds, the capabilities of the IRIS might be inadequate to return useful data. Although the IRIS would probably be able to hold its own when mapping Uranus' thermal radiation, it was expected to have poor sensitivity at near-infrared wavelengths, due to the lower intensities of reflected sunlight there. Although NASA agreed to fund a Modified IRIS (MIRIS) for Uranus, problems with the new instrument meant that it never flew.

It has already been shown that, due to the gradual radioactive decay of plutonium in the Voyagers' RTGs and the subsequent reduction in their electrical power output, it is becoming increasingly necessary to turn off some of the scientific instruments to keep the spacecraft running as long as possible. Over the next few years, sheer necessity will force reluctant managers to either shut instruments down permanently or turn them 'off' and 'on' as part of a power-sharing regime to squeeze as much from them as possible before they can be operated no longer.

The UVS, which has been used as an invaluable tool for studying distant stars and galaxies in addition to its duties at each of the giant planets, has already been given the death sentence on Voyager 2. It was 'retired' – though not without drama – towards the end of 1998. This instrument was the first to be decommissioned because it relied on the scan platform's heater to keep its ultraviolet sensor warm enough to operate. In an effort to gather as much data from the instruments as possible, managers decided to turn off the least essential subsystems first.

One of these, unfortunately for the UVS, was the scan platform heater. Within weeks of switching it off, the UVS became too cold to run and was itself turned off. However, during the effort to deactivate Voyager 2's scan platform early in November 1998, a minor scare kept engineers and managers alike on the edges of their seats. Shortly after the commands were sent to the spacecraft to switch off the scan platform, contact was abruptly lost and for almost three days ground controllers heard nothing.

At length, they guessed that during the shutdown procedure, an X-band 'exciter' – part of Voyager 2's communications system – might also have been inadvertently turned off. Correctional commands were hurriedly transmitted to the spacecraft, telling it to turn the exciter back on, and after 66 tense hours relief was evident on faces in the Voyager control room at JPL as contact was re-established.

It was an unglamorous end to a glorious career. Both UVS instruments more than proved their worth, particularly when scientists wanted precise measurements of chemicals in the giant planets' thick, soupy atmospheres. The UVS is mounted next to the IRIS on the scan platform. By using the technique of 'atomic absorption' – a process of elimination whereby specific colours of sunlight are absorbed by particular elements or molecules – the instruments were able to detect the presence, or conspicuous absence, of various key chemicals in their atmospheres.

The UVS was also able to work out abundance ratios of one chemical to another, which greatly helped scientists to refine their ideas of planetary-formation processes, and measure their distribution with altitude. This, in turn, also enabled them to better understand some of the dynamic phenomena they were to see in the giants' atmospheres. Other UVS tasks included spotting aurorae and cloud-top lightning bolts and trying to unveil the mysterious forces responsible for heating the planets' dense atmospheres.

Adjoining the IRIS and the UVS is the 2.5-kg Photopolarimeter Subsystem (PPS), a device designed to explore the physical properties of particles in both the giants' atmospheres and their rings. Originally, it was only expected to perform ring-plane investigations at Saturn, but after scientists found Uranus' narrow rings during a stellar occultation in March 1977, searches carried out at Jupiter and Neptune also yielded rings. To study such minute particles, the PPS measured the intensity and polarisation of scattered sunlight across eight spectral wavelengths from 235 to 750 nm.

In addition to its duties at the giants themselves, the PPS also proved useful when mapping the texture and chemical composition of their many moons. At the Jovian moon Io, it was instrumental in helping to determine the properties of a sodium cloud first identified by US physicist Bob Brown in 1974. It was also used to search for optical evidence of lightning and high-energy aurorae in the giants' upper atmospheres. Unfortunately, Voyager 1's PPS failed shortly before its arrival at Jupiter and although the instrument on board its sister ship was operating it was nevertheless significantly damaged. Nevertheless, it still provided useful data on particulate matter in the atmosphere of the Saturnian moon Titan.

The cause of the PPS problems appeared to derive from three commandable filter wheels mounted inside each instrument. The first of these wheels contains a series of coloured filters, the second carries drilled holes of various sizes to control the angular view of the PPS, and the third has polarising filters in several orientations. It would appear that excessive use of the wheels prior to the Jovian encounter, combined with radiation damage as it ventured into the Jovian magnetosphere, caused Voyager 1's PPS to fail.

The instruments on the scan platform were boresighted so that they all inspected the same target. All these instruments on the scan platform, and others to be described later, could be pointed with an accuracy of better than a tenth of a degree, although software modifications effected after Voyager 2's Saturn encounter allowed even greater precision when observing Uranus and Neptune. This was essential in order to properly centre their high-resolution, narrow-angle images. To avoid smearing the television pictures, both Voyagers' angular rates were 15 times slower than a clock's hour hand. They also carried 'target plates' of known colour and brightness – a bit like Munsell charts used by archaeologists and geologists – for calibration.

THE PROBLEM OF DISTANCE

Despite their blocky appearances, the Voyagers have sailed gracefully through the heavens, further, faster and smoother than any other ships in history, achieved staggering propellant efficiency and returned ground-breaking data about our planetary siblings. This is not to suggest that faults have never reared their heads;

nor that they will not do so in the future. Not only were the spacecraft built identically to provide redundancy, but each carried backups for all of their most critical components, from radios to star trackers and computers to Sun sensors.

Backup radio receivers were, and still are, particularly important, standing by in case their primary counterpart fails or encounters trouble. Obviously, a failed receiver might be rendered unable to react to commands sent from Earth, so both Voyagers' computers were programmed to restart a Command Loss Timer each time the radio sensed a command receipt. This timer could be set to any duration, but initially was set for a week. If that time elapsed and the timer ran down without receiving a single command, the computers would assume that the primary radio receiver had failed and switch automatically to the backup.

The need for the Voyagers to run on their own wherever possible, with little interaction from ground controllers, was essential, mainly due to the ever-increasing time delay that went hand-in-hand with the ever-widening distance from home. Travelling across space at the speed of light (almost 300,000 km/s), in 1979, just two years after launch, a radio signal from Earth took 35 minutes to reach the spacecraft at Jupiter – a mere 630 million km away – and a similar length of time for a reply to be picked up by the DSN receiving stations.

However, when Voyager 2 flew past Neptune a decade later, that time lag had more than quadrupled. In response to such delays, which will inevitably grow longer, each Voyager carries three computers: one to move the instrument-laden scan platforms (now shut down), another to supervise the scientific instruments and a third to store and deliver commands to the others at appropriate times.

The two Attitude and Articulation Control Subsystems (AACS) have for more than a quarter of a century handled the Voyagers' attitude. These computers have kept the high-gain antenna pointing towards Earth by using Sun and star sensors for celestial reference and firing the small attitude-control jets when required. Reprogramming has been required several times over the years to improve the stability of the spacecraft and maintain the instruments' pointing accuracy. The AACS was also responsible for manoeuvring the scan platform.

The two Flight Data Subsystem (FDS) computers have kept track of instrument timing and control functions, as well as formatting all engineering and scientific data before transmitting it to Earth through the high-gain antenna. Although portions of the FDS are no longer usable, some clever use of what is left has helped to minimise the impact of any memory loss. The Computer Command Subsystem (CCS) is responsible for supervising the AACS and FDS and delivering second-by-second timed command sequences to them for instrument observations and calibrations.

Another important role of the CCS is providing protection against the faults or failures mentioned earlier. Both Voyagers have seven 'top-level' fault-protection routines stored in their CCS memories, and these can handle a whole range of potential problems. The spacecraft can place themselves into self-preserving 'safe modes' within seconds, pointing their instruments away from the blinding Sun and their high-gain antennas towards Earth until the problems have been solved. This is of absolutely critical importance when one considers that round-trip communication times in September 2001 extended to almost 23 hours for Voyager 1 and 18 hours for its twin.

A SHAKY START

Problems encountered at these long distances could quite easily develop beyond the point of recovery before ground-based engineers even know about them. The loss of communications following the Voyager 2 scan platform scare in November 1998 is a notable example. However, in general, the precautionary thinking paid dividends both before and during Voyager 2's launch at 10:29 am Eastern Standard Time (EST) on 20 August 1977. The spacecraft suffered a double computer failure whilst sitting on its seaside launch pad at Cape Canaveral, and endured yet another as its Titan 3E-Centaur launch vehicle thundered into the clear Florida sky.

This problem appears to have been due to Voyager 2 suffering 'vertigo' as it rolled and pitched in its ascent trajectory. It also separated from the final stage of the Titan too quickly, which led its computers to assume that the primary attitude-control sensors had failed and it switched to the backup set. Luckily, its attached Centaur upper stage remained in control, automatically correcting the error before releasing the spacecraft. At length, things settled down. Seventy-one minutes after launch, Voyager 2 fired its solid-propellant rocket motor for 45 seconds to heave it out of Earth orbit and onto a path for Jupiter.

Still, none of the controllers in the tense JPL control room could breathe easy, for more trouble was afoot. Less than two minutes after the trans-Jupiter burn was over, Voyager 2 sensed and frantically tried to correct another orientation error. It had been pre-programmed that, if it encountered problems, it was to 'safe' itself and ensure that its high-gain antenna faced Earth. For another hour and a half, it struggled to stabilise itself. The procedures NASA had put in place to protect the spacecraft billions of kilometres from home had been inadvertently tripped only hours after launch!

It seemed that Voyager 2's sophisticated computer brain was just a little *too* sophisticated for its own good, at least on this heart-stopping occasion. Other unwanted surprises that day included disturbing telemetry that suggested the scan platform had not deployed properly. Luckily, it was a false alarm, caused by a troublesome sensor. Voyager 1, on the other hand, was not quite so lucky when it roared into space on an identical launch vehicle at 8:56 am EST on 5 September.

Although launched 16 days later – and numerically in the wrong order – its shorter, faster trajectory allowed it to overtake its mate, be first to reach Jupiter and then, upon reaching Saturn, fly close to Titan. The decision to launch Voyager 2 first was entirely related to keeping open the option to continue to Uranus and Neptune. It was realised that by sending Voyager 2 aloft 'early', it might just be possible for it to catch the Uranus launch window for the Grand Tour, which Gary Flandro had predicted would close towards the end of August 1977.

In that way, at least NASA was in with a fighting chance of getting to Uranus. Voyager 1 could then be launched a fortnight later to Jupiter and Saturn only. Its

One of the Voyagers undergoes final preparations in the workshop.

faster trajectory meant that, by December 1977, it had overtaken its twin. It would reach Jupiter four months and Saturn nine months, respectively, ahead of Voyager 2; hence it was dubbed 'Voyager 1' because it was first to reach its planetary target rather than being first in the launch pecking order.

Two weeks after Voyager 1's launch, from a distance of almost 12 million km, it took a remarkable 'departure shot' of Earth and the Moon (*see colour section*). For the first time in human history, our home planet and its only natural satellite – both crescents at the time – were captured in a single photographic frame. It was during this early part of the mission that Voyager 1's scan platform jammed. Although it was successfully freed, mission controllers handled it with kid gloves for some months afterwards.

While Voyager 1's controllers fought to keep their spacecraft on the straight and narrow, the team working on its sister ship were not faring much better. One of Voyager 2's hydrazine manoeuvring thrusters, for some reason, ended up pointing in the wrong direction with the result that it kept spraying volatile propellant at the spacecraft itself! Although this did not cause physical damage, it had the same effect as shooting a high-pressure jet washer and caused the ship to drift off-course. This required more hydrazine to be wasted to correct the error. Eventually, this problem, too, was solved.

Voyager 2's scientific instruments were also not immune to trouble. Although the IRIS had successfully jettisoned its protective launch cover, by early December 1977 its infrared sensitivity had significantly degraded. The cause was traced to crystallisation of some bonding material which held the mirrors inside the spectrometer portion; this caused the mirrors to become warped and misaligned and, consequently, the overall sensitivity of the instrument dropped. Fortunately, the IRIS was equipped with a flash-off heater that successfully reversed the evaporation and returned the instrument to its pre-flight sensitivity.

These niggling glitches were soon eclipsed by something much more threatening. In April 1978, nearly eight months after leaving Earth forever, a series of minor problems led Voyager 2 to lose control of its main radio receiver, which effectively made it deaf in one ear. The first problem developed when the spacecraft's seven-day timer ran down without successfully receiving a command from the ground. Voyager 2 responded appropriately to its pre-programmed routine by switching to its backup receiver. Correctional commands were transmitted from Earth to return to the primary receiver, but the spacecraft at first failed to respond.

At this juncture, another glitch materialised. Each receiver is fitted with circuitry to automatically detect the frequency of the incoming radio signal and tune the receiver to accept it. One component of this circuitry, called a 'tracking loop capacitor', had apparently failed, presumably due to an electrical short in Voyager 2, with the result that the spacecraft's backup receiver would only respond to commands transmitted at precisely the right frequency. Its 'good' ear was literally tone deaf!

Calculating this 'right' frequency was no easy task, because engineers had to account for (1) the motion of the transmitting ground station due to Earth's rotation, (2) the motion of our planet around the Sun and (3) Voyager 2's own

Voyager 2 leaves Earth forever on 20 August 1977, on board a Titan 3E-Centaur launch vehicle.

velocity relative to our parent star. Moreover, the spacecraft's receiver frequency actually 'slipped' with changes in ambient deep-space temperatures: a difference of only 1°C shifted it by as much as 96 Hz. At length, a breakthrough was made and contact re-established. However, during the backup-to-primary switching process, an electrical short blew both fuses in the primary receiver, effectively killing it.

This forced managers to confront the very real possibility of an impaired or even failed mission. "Thrusters fired at inappropriate times, data modes shifted, instrument filter and analyser wheels became stuck and the various computer control systems occasionally overrode ground commands," observed imaging team member David Morrison at the time.

"Apparently, the spacecraft hardware was working properly, but the computers on board displayed certain traits that seemed almost humanly perverse – and perhaps a little psychotic. In general, these reactions were the result of programming too much sensitivity in the spacecraft systems, resulting in panic over-reaction by the on-board computers to minor fluctuations in the environment. Ultimately, part of the programming had to be rewritten on Earth and then transmitted to calm them down."

SAILING IN THE SOLAR WIND

During the course of all these troubles, the bulk of both Voyagers' scientific equipment was powered up and began observing the apparently empty void of interplanetary space all around them. But, far from being empty, this strange environment actually turned out to be a hive of invisible activity. This activity is dominated by the action of the gusty 'solar wind' – a relentless torrent of charged particles streaming from the Sun at between 1.6 and 3.6 million km/h – which forms a huge magnetic bubble, called the 'heliosphere', around our parent star and its entourage.

In an effort to understand this activity and better comprehend the workings of the solar powerhouse, each Voyager carries five 'fields and particles' instruments to study various types of subatomic particles – minuscule bits of atom – throughout the Solar System. These five were called the Low-Energy Charged Particle Detector (LECP), the Cosmic Ray Subsystem (CRS), two pairs of Magnetometers (MAG), a Plasma Wave Subsystem (PWS) and a Plasma Investigation System (PLS).

Of these, the LECP, CRS and PLS were designed to look for charged particles (ions and electrons) across a broad range of energies. The 9.9 kg PLS, equipped with two Faraday cup detectors, looked at the lowest-energy particles in plasma, as well as helping to pinpoint their speeds and directions of origin. In doing so, the PLS aided physicists' understanding of the properties and radial evolution of the solar wind and its interaction with each of the planets and their moons. This became particularly important when the Voyagers reached moons like Io, which is deeply embedded in Jupiter's magnetosphere.

Ultimately, it is hoped that some day the PLS will be able to determine the precise whereabouts of the 'termination shock' – a theoretical region of space where the influence of the solar wind begins to slow down to as little as a quarter of its previous speed and become considerably more dense. When this region is encountered, scientists will brace themselves to meet the 'heliopause' – the outermost edge of the Sun's empire – just a few years later and for the first time sample plasma from beyond our Solar System.

Sensitive to higher energies than the PLS is the LECP, a 7.5-kg instrument fitted with a charged-particle telescope and low-energy magnetospheric particle analyser. Its sensitivity ranges from 10 keV to 11 MeV for electrons and 15 keV up to at least 150 MeV for protons and 'heavy' ions. Although its work overlaps with that of the CRS, it is the LECP that has the broadest energy-detection range of the three. Very crudely, the instruments might be envisaged as chunks of wood, in which charged particles play the part of 'bullets'. The faster the bullets travel, the deeper they penetrate the 'wood' and thus their penetration depth records their kinetic energy.

Furthermore, the number of bullet holes accumulated over time gives an approximate figure for charged particle densities in various parts of the inner and outer Solar System, including those in both the solar wind and around the planets themselves. The environments surrounding the giant planets, as far as we know, are controlled primarily by magnetic fields generated deep within their interiors. These fields carve out gigantic cavities – known as 'planetary magnetospheres' – within the solar wind. This cavity extends sunward about 12 planetary radii in the case of Earth and 50-100 planetary radii in the case of giant Jupiter.

As well as carving out a large area of space in the solar direction, magnetospheres also wind backwards even further 'behind' the planet, like a corkscrew. This is called the 'magnetotail'. Magnetospheres are known to be prodigious accelerators of charged particles; in the case of Earth, the most well-known consequences are polar aurorae and the Van Allen radiation belts, which are known to present significant hazards to orbiting spacecraft and astronauts.

The third charged particle detector on board each Voyager is the CRS, which is fitted with both high- and low-energy telescopes. These looked for only very energetic particles in space plasma, such as those in the intense radiation fields around Jupiter. Sensitive to energy ranges from 3 to 110 MeV, it has the highest sensitivity of the three instruments. The CRS makes no attempt to 'slow' or capture the particles; they simply pass through its detectors, but nevertheless leave signs that they were there.

As well as measuring the energies, directions and speeds of charged particles, the three instruments have tried to pinpoint their sources. Were they from the realms of the gas giants themselves, physicists wondered, or from monstrous flares erupting millions of kilometres into space from the Sun's hellish surface, or perhaps from even more alien climes beyond the Solar System?

The answers turned out to include all three. Other Voyager instruments kept watch, and continue to watch ever more intently today, for tiny changes in the solar wind itself as both spacecraft ventured deeper into the unknown. Scientists expect the wind speed to slow down dramatically as the Voyagers draw closer to the heliopause, that marks the boundary of the heliosphere. This substantial change – maybe a reduction in speed of up to 75 per cent – should easily be detectable using their sensitive plasma instruments and magnetometers.

The magnetometers on board both Voyagers are mounted on spindly, fibreglass booms, which were deployed from 60-cm-long canisters shortly after launch. After they had telescoped and rotated their way out of their canisters to a length of nearly 13 m, their orientations were controlled to a precision of better than 2 degrees. Their main tasks have been (1) to measure changes in the Sun's magnetic field over increasing distance and time, (2) to determine if each of the outer planets has a field and how their moons and rings interact with them and (3) to investigate how they interact with the solar wind.

The final fields and particles instrument is the PWS, a 1.4-kg device that shares the two 10-m whip antennas jutting at right angles into a V-shape from the spacecraft's main bus with another instrument called Planetary Radio Astronomy (PRA). It covers a frequency range from 10 Hz to 50 kHz and was designed to measure the density of electrons in the vicinity of each of the outer planets, as part of efforts to better understand the complex workings of their magnetospheres. At long range, the PRA investigated radio emissions from each planet in its search for a magnetic field, and during the period of closest approach it listened for atmospheric lightning bolts.

CLOSING IN ON JUPITER

The findings of the fields and particles instruments have proved groundbreaking in their own right, but as 1979 dawned the excitement within NASA and the scientific community had reached fever pitch as both spacecraft neared Jupiter. Named after the chief god of the ancient Roman pantheon, it is easy to see from the stunning Voyager imagery that the nomenclature is entirely appropriate. With an equatorial diameter of almost 143,000 km and capable of gobbling 1,300 Earths, Jupiter is by far the largest and most massive planet in the Sun's family.

Jupiter is also one of the most majestic, and even today remains one of the strangest worlds in the solar system, looking more like an enormous balloon of gas that might dissipate at any moment than a living, breathing, violent planet. Some have called it a failed star, arguing that despite having a substantial atmosphere in which hydrogen and helium predominate, its mass and the temperatures and pressures at its core are woefully insufficient to trigger nuclear fusion. In order to undergo the metamorphosis to turn itself into a star, Jupiter would need to be at least 100 times more massive than it is now.

Although the Pioneer 10 and 11 spacecraft had visited it earlier in the 1970s, the mighty Jupiter guarded its secrets jealously and yielded them grudgingly over many years. Despite the tremendous advances made by the Pioneers, and also by the Voyagers, our knowledge of this strange world has taken significant strides forward in recent years, thanks to two other remarkable spacecraft: the Hubble Space Telescope (HST), launched in 1990, and Galileo, which eased itself into orbit around Jupiter at the end of 1995 and began a detailed survey that was to last almost eight Earth years, or two-thirds of a Jovian 'year'.

For the Voyagers, on the other hand, the encounter was relatively short and sharp

An important tool in the exploration of Jupiter since the Voyagers, NASA's Hubble Space Telescope is here seen in orbit after being serviced by the STS-109 Shuttle crew in March 2002. This servicing mission was also the last wholly-successful flight of the Space Shuttle Columbia.

NASA's Galileo spacecraft, which explored Jupiter and its entourage in unprecedented detail from 1995 until 2003, is here shown on its Inertial Upper Stage (IUS) after deployment from the Space Shuttle Atlantis during the STS-34 mission in October 1989.

and more than one writer has compared it to a symphony of four movements. Early in January 1979, with two months still to go before closest approach, the Observatory phase of the Voyager 1 encounter got underway. Throughout this month-long period, scientists monitored the planet from afar and took a near-continuous stream of pictures. At one stage, the spacecraft returned a picture every 96 seconds for four days! Each picture was of unprecedented clarity compared to what had been transmitted before.

As the spacecraft drew nearer, the pictures became correspondingly more stunning, revealing the mesmerising action of Jupiter's swirling atmosphere. At the same time, Voyager 1's other instruments were also working continuously. By 13

February, a mere three weeks before closest approach and only 20 million km from Jupiter, the UVS was scanning the giant planet eight times a day, the IRIS was measuring heat emissions from its upper atmosphere and the PRA/PWS combination was searching for plasma and radio bursts that might yield clues about the magnetosphere.

Late February 1979 saw the Far Encounter phase, by which time Jupiter was so close that it overwhelmed Voyager 1's field-of-view. It was here that some of the hardest decisions of the mission had to be made. With only days to gather the most valuable data of their careers, competition became fierce as scientists sought more time to focus 'their' instrument on a specific peculiarity or feature. Although 'sequencing' – the precise order in which the instruments were to operate – had been planned months before, the sheer volume of new discoveries made it difficult to stick to timelines.

"When you're far away, it's no big problem," said Lonne Lane, who worked on the Voyager Science Integration Team at JPL until just before Voyager 1's Jupiter encounter. "However, when you get in close – within, say, about five days – you end up with intense competition. That is, each instrument has an optimal time to make a certain kind of measurement because of the geometry, or because of the light level, or the flux of radiation coming out of the planet, or a certain satellite happens to be in a certain position as seen from the spacecraft."

One example of this competition came when Voyager 1 passed between Jupiter and its large moon Io. One team of scientists was eager to investigate Pioneer 10 and 11 reports that there were charged particles around Io, which would have been an unexpected finding. However, another team was simultaneously absorbed in the dynamics of Jupiter's atmosphere and wanted to take colour pictures. Unfortunately, the instruments needed to do these tasks – the UVS and one of the cameras – were mounted on the scan platform, which could only face one direction at a time.

Lane had the unenviable job of weighing the merits of each observation and adjudicating conflicts; the effort, he said, required "a Solomon and a half". When neither side would budge, the final decision went to Project Scientist Ed Stone. Although the conflicts never came to fisticuffs, they were to become a familiar aspect of the Saturn, Uranus and Neptune encounters. Sadly, it was the inevitable drawback of a flyby mission: the option to hang around and explore a planet simply was not available to either Voyager. No sooner had they arrived, than they were gone.

After the first two movements of the Jupiter symphony, the climax was the Near Encounter, which kicked-off on 1 March 1979, four days before Voyager 1's closest approach. By this time, JPL was crawling with journalists and daily press briefings were held to discuss the latest pictures and incoming data. The floodgates were opening as scientists braced themselves for a new and unprecedented torrent of information from the realm of Jove. It was a torrent that would bring some of the most astounding discoveries of the entire mission and leave everyone gasping for breath.

3

Into the realm of Jove

AN EXPLOSIVE DISCOVERY

On 9 March 1979, a few days after Voyager 1 swept past Jupiter, one of the most significant discoveries of the entire mission began to unfold before the eyes of an astonished young astronomer. By this time, most of the journalists had left JPL and 26-year-old navigation engineer Linda Morabito was checking the spacecraft's line of departure from Jupiter by measuring the positions of stars off the limb of the giant planet's moon Io. At this stage, Voyager 1's next planetary target – the ringworld Saturn – was a mere 20 months and 650 million km away.

As she sifted through a series of stunning pictures of the Jovian family – which had been deliberately overexposed to reveal faint reference stars that would be used for navigation purposes – she came across a parting shot of Io. "This one image, which had been taken the day before, had been put up like all the others on the monitors at JPL for everyone to see," Morabito recalls. "But I suddenly noticed an anomaly to the left of Io, just off the rim of that world. It was extremely large with respect to the overall size of Io and crescent-shaped."

Today, Morabito is an education programmes manager for The Planetary Society, but the 'anomaly' she found remains the highpoint of the Jupiter encounter. After initial speculation that the faint bluish crescent might be a blemish on Voyager 1's lens, or even a new moon peeking out from behind Io's limb, it soon became clear that it was actually a volcanic plume rising 280 km from the surface! The source was traced to a hotspot subsequently named Pele, after the Hawaiian goddess of volcanoes, and scrutiny of other Voyager 1 pictures quickly uncovered another eight eruptions elsewhere on Io.

Morabito's chance find not only marked the first direct observation of active volcanism on a world besides our own, but recent research by HST and Galileo has since confirmed it as the most volcanic place known. Although it has fewer volcanoes than Earth – about 120, compared to our 600 or more – it is the heat they emit, rather than sheer numbers, that makes Io so active. "Io is only about one-third as big as Earth, but puts out twice the energy," says volcanologist Rosaly Lopes-Gautier of JPL. "One of Io's volcanoes, Loki, is more powerful than all of Earth's volcanoes combined."

Such is the ubiquity of this volcanism that when the International Astronomical Union (IAU) came to choose a 'theme' for place-names, the decision was easy: allusions to fire deities and references to volcanism or thunder dot our maps of this weird world. Although one volcano, Prometheus, has belched violently and with unbroken continuity for over two decades, others have shut down and restarted and still more have burst into life without warning. When Voyager 2 flew past Io in July 1979, one of the nine eruptions (Pele) seen by its sister ship was quiet, yet the others remained in full swing.

Strictly speaking, the Ionian eruptions are not 'volcanoes' like those on Earth, but bear a closer resemblance to Hawaiian fire fountains or the Old Faithful geyser at Yellowstone National Park in Wyoming. One important difference, however, is that Io's lava is much hotter and its enormous volcanic craters – known as 'calderas' – are far larger than any of their Earth-based counterparts. Indeed, globally, Io spews a hundred times as much lava per year as does Earth.

The Amirani lava flow on Io, photographed by Galileo in July 1999. Extending some 300 km, this is the longest-known active lava flow anywhere in the Solar System.

VOLCANOES OR GEYSERS?

When Voyager scientists first gazed awestruck at the Ionian plumes, many were convinced they were seeing violent eruptions. However, terrestrial volcanic eruptions tend to be violent, but brief, whereas those on Io were remarkably stable and long-lived. Some of the eruptions seen by Voyager 1 in March 1979 remained active when its twin arrived four months later, and others were still running when Galileo turned its attention to them in June 1996. Most scientists now think, in some cases at least, that they are seeing vast geysers rather than scaled-up versions of terrestrial eruption plumes.

Although Io's weak gravity – about one-sixth of Earth's, roughly comparable with that of our Moon – and thin atmosphere certainly help gas and dust to ascend to fantastic heights above the surface, theoretical estimates have shown that even the most violent, 'Plinian' style eruptions would be insufficient to hurl material to altitudes of up to 500 km, which was seen by Galileo on one occasion in August 2001. Terrestrial eruption plumes rarely achieve velocities above 100 m/s, yet the Ionian plumes were evidently rising 5 or 10 times faster.

However, in view of the small-scale nature of terrestrial geysers, how could they be behind the explosive, high-velocity phenomena on Io? The fact that they *are* of such minor significance on Earth is due to our gravity and substantial atmosphere, but the lack of both factors on Io changes the picture considerably. For example, the

A wintertime shot of the Old Faithful geyser in Yellowstone National Park. Several geologists have interpreted some of Io's 'volcanoes' as larger, more powerful versions of terrestrial geysers.

Old Faithful steam plume rises no higher than 52 m on Earth, but would ascend an astounding 35 km if relocated to Io! Unlike Old Faithful, however, Io has no subsurface water supply and the working fluid for its geysers is probably sulphur or sulphur dioxide.

In 1982, geophysicist Sue Kieffer, of Ontario-based Kieffer Science Consulting Inc., showed that sulphur-driven geysers can produce sufficient ejection velocities for a typical Ionian plume. Her model involved liquid sulphur coming into contact with hot rocks at fairly shallow depths beneath the surface. Upon forcing open fractures in Io's crust, the superheated fluid rises rapidly, boiling and gaining energy as it does so. When it finally breaks the surface, Kieffer argued, it begins to 'snow out', shooting a high-speed column of cold gas and frost into the near-vacuum of Io's tenuous atmosphere.

Kieffer's theory also seemed to fit some of the confusing results from Galileo when it first arrived in orbit around Jupiter in December 1995: the spacecraft saw surprisingly few eruptions, in comparison to the nine photographed by the Voyagers. Although Prometheus was still in action and a new plume had appeared over a feature known as Ra Patera, several hundred kilometres south-west of Loki, the first-found Ionian volcano, Pele, was eerily silent.

In her model for sulphur-fed geysers, Kieffer suggests that such features might be supplied by enormous subsurface 'reservoirs'. This could also help to explain why Galileo initially saw so few volcanic features: they *were*, in fact, there, but just hard to see. According to Kieffer, these 'stealth' plumes might be created if the sulphur dioxide reservoir needed to run them formed at very high temperatures. In such cases, erupting gas would expand so quickly that it would not produce snow, but rather a high-velocity jet of cold gas that would be difficult to see in visible light.

Observations by HST in July 1996 lent credence to this idea by clearly spotting an erupting plume from Pele, albeit only in silhouette against the dramatic backdrop of Jupiter's tumultuous clouds. Galileo, meanwhile, was watching Io at the same time, but saw nothing in scattered sunlight. Other pictures taken while Io was being eclipsed by Jupiter revealed glowing aurorae – energised by charged particles from the giant planet's magnetosphere – over regions that displayed no obvious eruptions in visible light.

In addition to the nine erupting features seen by the Voyagers, upwards of 200 calderas – each more than 20 km across – have been observed in the past quarter of a century. By contrast, Earth has only 15 calderas in this size range, and has nothing to compare to the largest of Io's calderas, which are several hundred kilometres across. Altogether, more than 80 new volcanic hotspots were identified by Galileo alone, whose eight-year mission in orbit around Jupiter is due to end in late 2003 with a kamikaze plunge into the giant planet's atmosphere. The sheer volume of discoveries made by Galileo is far higher than the few dozen new features anticipated before its arrival at Jupiter. Moreover, long before Galileo's mission ended, scientists had gained new insights into the nature of Io's plumes.

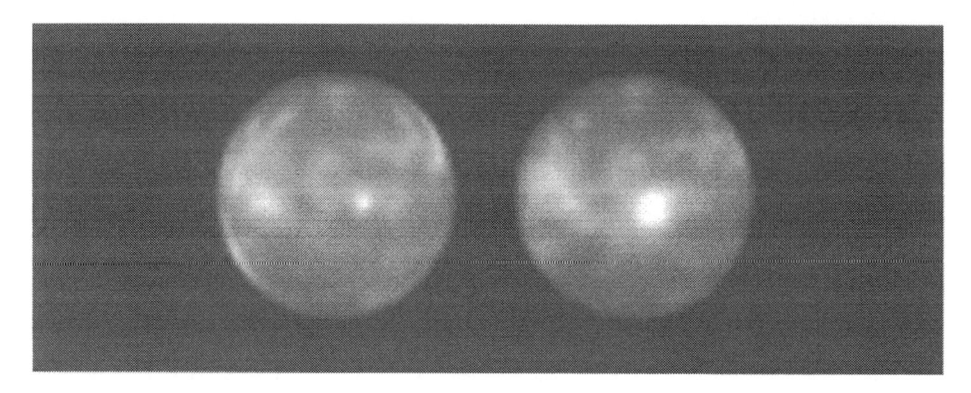

HST images of Io from March 1994 (*left*) and July 1995 (*right*), showing the appearance of a large yellowish-white spot in the middle of the moon's disk.

AN ACTIVE SURFACE

Not only has the discovery of new volcanic regions come thick and fast since 1979, but the behaviour of Io's surface as a whole has changed markedly. Ra Patera, named after the ancient Egyptian sun god, is a 'shield volcano' right in the middle of Io's Jupiter-facing disk and was first seen by Voyager 1. Shield volcanoes are very large structures which, on Earth, can measure many tens of kilometres in diameter with relatively gentle slopes. They are usually produced by successive outpourings of fluid lava and often comprise more than one vent, as well as fissures along their sides.

Ra Patera is Io's largest shield volcano, with a diameter of 450 km, although its summit rises a mere kilometre above the surrounding terrain. It appeared to be inactive in Voyager pictures, although the twins did identify yellow flows indicative of sulphurous lava, and images from HST's newly fitted Wide Field Planetary Camera (WFPC)-2 in March 1994 confirmed that it had undergone little change since then.

Then, follow-up pictures from July 1995 picked out a vast, 320-km-wide yellowish-white spot (about the size of New Jersey) in the region. It had clearly been deposited by Ra Patera during the 16-month interval, but HST's resolution was insufficient to study it in detail. Observations during Galileo's first orbit of Jupiter, however, revealed a 100-km-tall plume from the region, just barely visible from behind Io's limb. A comparison of Voyager and Galileo images revealed that the new eruption had resurfaced an area of nearly 40,000 km^2.

The size of Ra Patera has allowed volcanologists to make inferences about the nature of its lava. Its ability to flow readily down very gentle gradients points to extremely low viscosity and close parallels have been drawn with the low shields and overlapping flows of the Snake River volcanic plain in Idaho. Many scientists think the 1994–1995 event is one of a new breed of 'transient' features and, in the words of John Spencer of Lowell Observatory at Flagstaff, Arizona, is "probably composed of frozen gas, ejected from Ra Patera by a large volcanic eruption or fresh lava flows".

SILICATE OR SULPHUR FLOWS?

Although the discovery of volcanism was surprising, it was a moment of triumph for three scientists – Stan Peale of the University of California at Santa Barbara and colleagues Pat Cassen and Ray Reynolds from NASA's Ames Research Center at Moffett Field, California – who published a paper in the journal *Science* shortly before Voyager 1 reached Jupiter, predicting volcanism! They argued that the gravitational pull of Jupiter and two of its moons might be enough to generate enormous tidal forces, which would heat Io's interior.

To this day, their work represents one of the shortest intervals on record between a theoretical prediction and its observational confirmation. Almost as soon as the first volcanic features were spotted, geophysicists began to debate their nature and composition, and one of the first questions was whether they were sulphur-driven or, like those on Earth, 'silicate' (molten rocks). Crude density measurements as Voyager 1 sped past Io suggested that the moon was made from rock and iron.

It was initially speculated that the gravitational stresses placed on Io by Jupiter had turned the moon's silicate lithosphere into the floor of a global ocean of molten sulphur several kilometres deep. The 'visible' surface, it was proposed, is actually a veneer of frozen sulphur a few hundred metres thick that caps this ocean. Volcanism was seen as a localised phenomenon as a means of relieving periodic build-ups of sulphur dioxide gas within this ocean. Plumes erupted explosively and the vents essentially sealed themselves when the gas pressure had been relieved.

There was a clear precedent for this in the case of our Moon. During its accretion process, the infall of vast 'planetesimals' kept its outer few hundred kilometres so hot that it formed a 'magma ocean'. As a result of thermal differentiation, lightweight aluminium-based silicates rose to the surface and crystallised to form a crust of anorthosite. When this had thickened sufficiently to resist further impacts, the final stages of planetesimal bombardment formed the large basins, some of which were later flooded by upwellings of dark lava. Although the circumstances were obviously different in Io's case, the result, it was argued, was strikingly similar.

As far as scientists can tell, Io is desiccated. Any water and ice that it may originally have possessed has long since been lost into space. Water molecules would have been dissociated by solar ultraviolet radiation, hydrogen swept away and oxygen partly recovered by the crust and undergone chemical reactions. Earlier in its development, while it still possessed some of these hydrated minerals, Io's volcanic plumes were probably driven by water steam. When this supply was exhausted, the moon would have been deprived of a means of 'cooling' itself. Its crustal temperature would have risen until the next most readily vented gas was boiled off. In this way, over millions of years, Io has lost its water, nitrogen, carbon dioxide and neon.

If it were not for the gravitational effects of Jupiter on Io, it would almost certainly have retained these volatiles and may well have resembled the ice-encrusted moon Europa. At present, Io is in the process of boiling off its sulphur dioxide which, as it condenses and falls back, 'paints' a frosty halo on the surface around the volcanic vent. Other volatiles, including gaseous sulphur, also condense out to produce very chemically-diverse 'blankets' of material.

At 'room' temperature in a terrestrial environment, sulphur takes the form of a pale yellow solid, but at 400 K it melts and turns orange. Incremental temperature changes alter its colour and viscosity significantly. It turns pink at 435 K, red at 465 K and becomes a black substance with the consistency of tar at 500 K. Gradually its viscosity falls, turning into a runny fluid at 650 K and becoming least viscous just before it boils. This last stage occurs at 715 K in a terrestrial environment, but varies elsewhere depending on the pressures under which it is placed.

Galileo has revealed parts of Io that closely match sulphur-driven plumes, but has also identified regions that are much hotter, suggesting silicate volcanism. In December 1995, gravity-field data hinted that Io may possess a giant iron sulphide core up to 900 km across, together with a rocky mantle. These results led to several persuasive theories in favour of silicate volcanism. For a start, Io's crust needs to be sufficiently strong to support the weight of the 5- to 10-km-high mountains and deep, steep-walled calderas seen by both Voyagers. Solid sulphur, particularly when heated, would not be rigid enough. On the other hand, it was possible that vents opened by silicate volcanism may have subsequently been exploited by sulphurous volcanism.

Some scientists have argued that temperatures inferred from Voyager IRIS data for 'hot' regions are too low for molten silicate rocks – which form at 1,270 K or more – but offer a better match for molten sulphur. However, the 'footprint' of the IRIS sensor was very broad and it could only measure the *average* temperature of a particular area. If this contained a 'hotspot', its exact temperature was not accurately represented. The result was that, while the average temperature might rule out silicate volcanism and lend support to the pro-sulphur camp, the reality might be very different.

Many scientists who favour sulphur volcanism point out that this element is abundant in both our Solar System and the Universe. Although Earth has large reserves stored in its core as iron sulphide, it actually plays only a minor role in terrestrial volcanism and generally manifests itself as gases or deposits close to volcanic vents. In Io's case, large concentrations of sulphur dioxide 'frost' were detected spectroscopically on its surface and sulphur gas was identified in the thin atmosphere by both Voyagers' IRIS instruments. This gas could then easily be separated into oxygen and elemental sulphur by ultraviolet sunlight.

The pro-sulphur camp had already gathered substantial support from ground-based observations of Io's surface in the early 1970s, which revealed it to be much more reflective than would be deemed 'normal' for silicate rocks. Furthermore, its bizarre yellowish-brown hue is indicative of sulphur, which absorbs sunlight at the blue end of the spectrum. Subtle colour variations of caldera floors and along lava flows could, say the pro-sulphurites, be attributable to different kinds of sulphur – called 'allotropes' – that crystallise at different temperatures.

Emakong Patera, a few hundred kilometres east of Prometheus, is surrounded by yellow-white lava flows. Most flows, both on Io and other planets, are dark, which has prompted speculation that Emakong's caldera is producing moderately hot sulphurous lava rather than very hot silicates. There are, however, dark, sinuous 'channels' in the light flows, which would appear to be black sulphur that evolves to yellow as it cools.

Io's surface colour is extremely varied, indicative of subtle sulphur chemistry. Journalist Mike Hanlon, among others, has likened it to a "giant pizza covered with melted cheese and islands of tomato and ripe olives". Even hardened planetary scientists can barely comprehend what they are seeing, more than 20 years after the moon's volcanism was first discovered. "We think the bright colours are due to sulphur in various forms, but that the very dark colours are due to active lavas," says Lopes-Gautier. "Every place we see high temperatures, if we look at the surface we see black materials. That would be the olives on the pizza. The reds are deposits from the plumes of volcanoes. With time, the reds become yellow because of changing to a different form of sulphur. We're still quite puzzled by what some of the very small green areas are. We joke and call them golf courses. They may be areas rich in sulphur, but contaminated by another material. Another possibility is they are very olivine-rich lava." One region, Tupan Patera, named after an ancient Brazilian thunder god, displays all four colours cited by Lopes-Gautier, perhaps due to interactions between silicate and sulphurous compounds.

Ground-based observers have helped to resolve the Io conundrum, even turning the debate in favour of the pro-silicate camp. Much of the moon's surface is quite cold and the only regions that emit energy at 'warm', mid-infrared wavelengths are the volcanic hotspots. Clearly, although these cover only a small fraction of the surface, they are responsible for producing most of its radiation at these wavelengths. In 1986, a team at the Mauna Kea observatory in Hawaii saw a dramatic increase in Io's infrared brightness compared to its 'normal' volcanic output. They concluded that the increase was due to a brief eruption with a temperature over 1,170 K. Temperatures in this range are much too hot to be molten sulphur, implying that silicate lava is involved in at least some of Io's volcanism. Further observations by ground-based astronomers and Galileo have measured hotspots with temperatures so high (more than 1,800 K) that, in some regions, Io has the hottest surface in the Solar System, save that of the Sun. Infrared thermal-emission data acquired by Galileo throughout the late 1990s identified numerous locations that are hotter than is reasonable for sulphur-fed volcanoes.

Although these results suggest that silicate lava is actually a key player in most, if not all, Ionian eruptions, there remains persuasive evidence that cooler, sulphur-driven activity exists and is responsible for the striking surface colours. Galileo images from 1996–1997 revealed astonishing red deposits around several active vents, which seemed to fade back to the prevailing yellow-brown surface colour over time. This is precisely what geologists would expect to see if high-temperature sulphur allotropes were erupting, cooling and reverting back to 'normal' sulphur.

"Most of the volcanic plume deposits on Io show up as white, yellow or red, due to sulphur compounds," says Alfred McEwen of the University of Arizona at Tucson. Yet HST and Galileo pictures of a 120-km-high eruption from Pillan Patera in June 1997 revealed a pronounced grey patch, possibly caused by the interaction of its plume overlying the distinctly reddish deposits from neighbouring Pele. McEwen thinks this could point to a composition rich in silicates and future exploration might identify new types of silicates that are uncommon on Earth.

In fact, Pillan Patera is quite similar to another feature known as Babbar Patera,

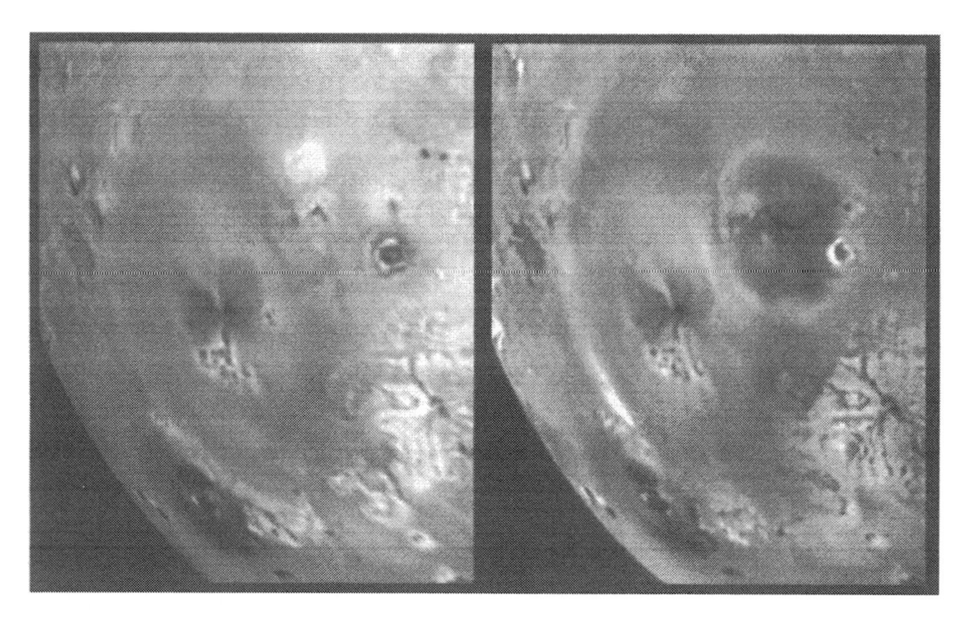

Galileo's pictures of Io from April (*left*) and September 1997, showing the deposit (*upper right*) from the Pillan Patera eruption, superimposed on the 1,000-km-diameter ring around Pele (*centre*) and the older, dark patch around Babbar Patera on the opposite side.

which had disfigured the far side of halo-shaped Pele with dark material at some time in the past. A pair of hotspots observed by Galileo while Io was in eclipse corresponded to the vent and tip of the Pillan flow, revealing temperatures as high as 2,000 K. This marked the first conclusive evidence of silicate volcanism on Io. As the hottest-known lavas on Earth are around 1,500 K, those of Io are hotter than anything that has erupted on our world for several billion years.

'Ultramafic' flows – lavas composed of magnesium-rich igneous rocks such as dunite, peridotite or pyroxenite – are known to have reached these kind of temperatures early in Earth's history. Today, of course, our planet's lithosphere (its crust and upper mantle) is dominated by plate tectonics and the melting point of the hydrated lava produced when an oceanic plate is subducted by a continental margin is about 300 K cooler. "It's very surprising to see lava flows as hot as the ancient flows on Earth," said Bill Smythe, a member of the Galileo Near Infrared Mapping Spectrometer (NIMS) team.

"After Voyager, we thought that *all* the lava flows were sulphurous, but sulphur vaporises at about 700 Kelvin. These hotter lavas have to be basaltic. Now the question is – are *any* of the lavas sulphurous? Galileo has detected areas on Io with temperatures between 300 Kelvin and 600 Kelvin. That's about right for sulphur, but these could also be places where tiny volcanic vents at around 1,800 Kelvin are surrounded by cold ground." Perhaps, as its most profound contribution, Io will tell us a great deal about our own world's infancy.

SULPHUR DIOXIDE PLUMES

During one of Galileo's final flybys of Io, in August 2001, its NIMS instrument confirmed sulphur dioxide deposits near the source of a tall plume hanging over a previously inactive volcanic feature. A key objective of this encounter had been to take another look at the site of a recent spectacular eruption in the northern hemisphere called Tvashtar Catena; although the region itself was now quiet, a new volcano 600 km away erupted suddenly, just as Galileo flew overhead. Its plume, which soared a record-breaking 500 km high, allowed the spacecraft's plasma instrument to analyse its sulphur dioxide 'snowflakes'.

"Galileo smelled the volcano's strong breath and survived," exulted Louis Frank of the University of Iowa. A landing vehicle, of course, would be the most direct and desirable means of ascertaining the precise composition of Io's volcanism, although such a venture is unlikely in the near future. Yet, although Galileo is nearing the end of its operational life, astronomers and geologists can still rely on HST – which has pinpointed volcanic clouds rich in sodium – to provide new insights into how Io ticks. One common factor, however, is the enormous power and velocity of its eruptions.

HST and Mauna Kea images have revealed solid material blasted to altitudes in excess of 160 km, which trajectory specialists think would require ejection velocities from the volcanic source somewhere in the range of 560 m/s. Indeed, in late 1996 Galileo watched sulphur dioxide gas rise to a height of 50 km above Prometheus at a speed of nearly 500 m/s. Interestingly, although both Voyagers had seen Prometheus in action 17 years previously, Galileo's pictures revealed that the location of its active site had moved!

The Voyagers saw a plume rising from near Prometheus' caldera – a large, D-shaped volcanic crater measuring 28 km long by 14 km wide, near the north-eastern edge of the feature – but Galileo found it rising instead from the western end of a new lava flow, 75 km away. "It appears that Prometheus has characteristics remarkably similar to those of Kilauea in Hawaii," said Laszlo Keszthelyi of the University of Arizona at Tucson, "with flows that travel through lava tubes and produce plumes when they interact with cooler materials."

Like Prometheus, Kilauea's lava tubes transport lava from the vent to the front of the flow, but they are only 10 km long. The lava then disgorges into the Pacific Ocean and generates clouds of steam. It has been suggested that, in the case of Prometheus, the deep reservoir lies beneath the caldera, but broached the surface 15 km south of the rim, which corresponds to the eastern hotspot. The lava then travels west through lava tubes to the front of the flow, where it disgorges onto the surface and produces the plume by vaporising sulphur dioxide frost.

This suggested that, although Kieffer's model of sulphur-driven geysers was generally accepted before Galileo arrived at Jupiter, other mechanisms responsible for generating plumes are also present on Io. Whereas the Pele plume seemed to be composed of gases rising from the primary vent, that of Prometheus would actually appear to be caused by the sublimation of surficial sulphur dioxide as silicate lava flows across the surface.

MOUNTAIN-BUILDING

Such observations, obviously, point to a very dynamic interior. Yet very few of Io's volcanic features resemble the crater-topped, 'stratovolcanic' peaks found on Earth. Instead, most of its erupting craters are in quite flat regions, not near mountains; but, on the other hand, nearly half of its mountains are situated close to volcanic vents! This has led a number of scientists, including Elizabeth Turtle of the University of Arizona at Tucson, to speculate that volcanism may be the driving force behind the formation of mountains on Io, even though they are not themselves volcanic edifices. A remarkable feature of the mountains on Io is that, unlike Earth, they all stand in isolation and do not form 'ranges' or even boast foothills.

"It appears that the process that drives mountain-building – perhaps the tilting of blocks of crust – also makes it easier for magma to get to the surface," says Turtle. Certainly, Galileo's images of a mountain called Tohil Mons show that material collapsing down its slopes apparently did not pile up in the volcanic crater below. This implies that the crater had been in a molten state much more recently than any landslide could have occurred. Infrared spectra from Galileo revealed the crater itself was hot, meaning it was either active or had been so recently.

"On Earth, we have large-scale lateral transport of the crust by plate tectonics," says McEwen. "Io appears to have a very different tectonic style, dominated by vertical motions. Lava rises from the deep interior and spreads out over the surface. Older lavas are continuously buried and compressed until they must break, with thrust faults raising the tall mountains. These faults also open new pathways to the surface for lava to follow, so we see complex relations between mountains and volcanoes, like at Tohil."

Paul Schenk of the Lunar and Planetary Institute in Houston, Texas, and Mark Bulmer of the Center for Earth and Planetary Science at the National Air and Space Museum in Washington, DC, conducted a topographical analysis of the mountain Euboea Mons to assess a possible landslide.

Euboea Mons is situated in Io's southern hemisphere and has a prominent curving ridge crest peaking at about 10.5 km high. The crest serves to divide the steep southern flank from the much smoother and shallower northern slope leading to a thick and heavily ridged deposit possessing a scalloped outer margin. It appeared that the entire north face had slumped and the northern deposit represented an 'apron' of rock and debris from an avalanche. If it was an ancient mountain, then its steep southern flank suggested it was structurally strong, but the collapsed north flank pointed to a structural failure.

In their 1998 paper in *Science*, Schenk and Bulmer argued that Euboea Mons rests on a shallow, overthrusting fault dipping to the north. They suggested that the massif itself was rigid, that the structural failure was triggered by the act of uplift and the debris is actually volcanic material which had been deposited on the block. The smoothness of the northern flank implies that it is a formerly horizontal surface. The six-degree slope presumably matches that of the fault, but the southern flank – a 10-km-high exposure – provides an excellent 'window' into Io's crustal structure.

Io is so volcanically active that it may be resurfaced at a rate between several millimetres and a centimetre per year and Schenk and Bulmer suggested that as layer after layer of 'new' material is constantly erupted, older, buried material is forced to subside. As it does so, it undergoes lateral compression, which may induce the thrust-faulting that seems to have produced Euboea Mons. Their opinion was that such massifs are relatively young and, despite extensive evidence of slumping, their underlying structures are rigid silicate blocks; it is the much weaker sulphurous lavas that have slumped.

Other mountains surveyed by Galileo revealed topography ranging from angular peaks to smoother plateaux surrounded by gently-sloping debris aprons; the former appear to be younger than the latter. The ridges parallel to the margins imply progressive slumping under the force of Io's weak gravity. Although the mountains seem to be at various stages of deterioration, they seem to be collapsing under their weight, which again suggests they are fairly recent creations. The mystery of Io's mountains, first posed by the Voyagers, was therefore resolved by Galileo.

LAVA LAKES

Dramatic shot from Galileo, showing a fire fountain eruption in Tvashtar Catena in February 2000.

The activity seen on Io takes many forms. Some eruptions bear some hallmarks of terrestrial volcanoes. In November 1999, Galileo watched an eruption in Tvashtar Catena hurl molten magma like a fiery curtain more than 1.5 km into its thin atmosphere. When Galileo returned in February 2000, this eruption had ceased and a new one, even larger, was underway. This soon petered out but in late December

2000 – when both Galileo and the Saturn-bound Cassini spacecraft flew close to Io – it stirred again, ejecting sulphurous material 385 km high. When this material finally 'snowed' back onto the surface, it left an enormous, blood-red ring, 1,400 km across which was remarkably similar to that around Pele.

By the time Galileo next passed Io, in August 2001, Tvashtar was quiet again, but infrared data from the spacecraft showed it to be covered with a new cluster of hotspots, and its temperature remained very high. The behaviour of this dynamic region over nearly two years is similar to Earth-based volcanism in many ways, in that activity starts vigorously, then declines slowly. Still other sites, such as Loki – Io's most 'powerful' volcano, named after the prankster god of Norse lore – seem to brighten, then fade, over periods of months, possibly in periodic cycles. This pattern is unknown to our world.

Some scientists think the Loki region, and possibly also Pele, harbour either active lava lakes or regularly flooded volcanic calderas. During Galileo's last wholly-successful flypast of Io in October 2001 (an encounter three months later was hampered by Jupiter's savage radiation belts, which knocked the spacecraft into safe mode and returned only limited data) concentrations of high-temperature lava, likened to a 'glowing shoreline', were seen near Loki. A similar feature had also been seen at Pele in 1999. These might represent hotter lava beneath the surface peeking through where a cooler lava crust breaks up as it hits the crater wall.

Indeed, huge black lakes, possibly of molten rock and sulphur, with temperatures as high as 470 K, have been found dotted all over Io's surface. One in particular, Loki Patera – just south-west of Loki's main fissure – might be as much as 200 km across; larger, in fact, than the entire above-ocean portion of the island of Hawaii.

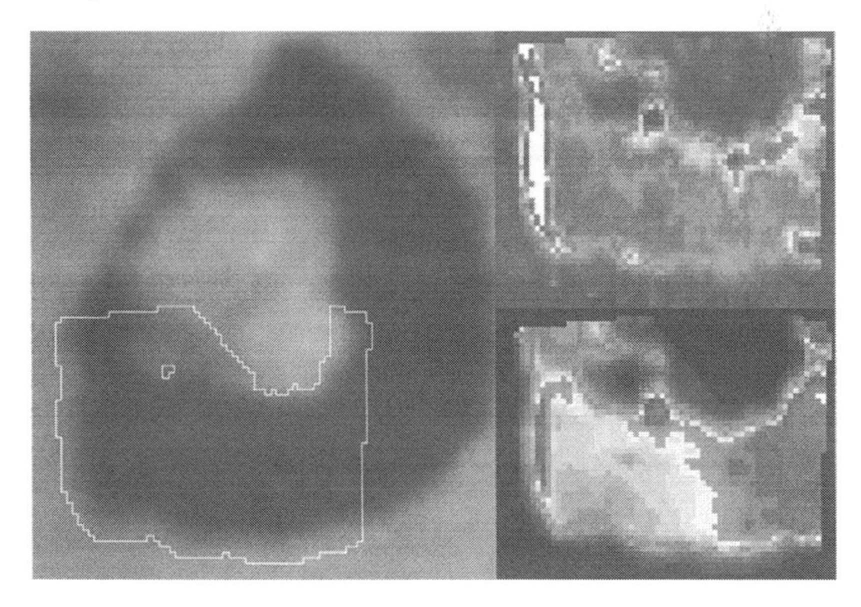

A Galileo infrared map of Loki, taken in October 2001, showing a 'glowing shoreline' created by concentrations of high-temperature lava.

Other dark, spot-like features, which could conceivably also possess lava lakes, include Pillan Patera, whose erratic behaviour and dramatic colour changes was closely monitored by Galileo, HST and ground-based teams over five months in the spring and summer of 1997.

A YOUNG SURFACE

Clearly, Io is fundamentally different from any other known world in the Solar System. Its volcanism, which churns out hundreds of cubic kilometres of lava every year, has yielded a surface that is hard to 'date' by traditional means because it is completely devoid of impact craters! This came as a shock to Voyager 1 scientists, who expected it to be covered with them. Io orbits only 350,000 km above Jupiter's cloud-tops, which is about the same distance as between Earth and the Moon. However, Jupiter is 318 times more massive than Earth and a more prodigious accelerator of impact material.

In general, surfaces with a higher density of craters are considered 'older' than those with very few scars. Io, however, is subject to such relentless, ongoing volcanic activity that lava flows constantly replenish its surface, erasing craters almost as soon as they form. So convinced were geologists that Io *must* have them that the early, low-resolution Voyager 1 images of roughly circular splotches on the surface were interpreted as impact craters. However, as the spacecraft drew closer, it became clear that they were actually volcanic features.

This distinct lack of impact craters therefore implies that Io's present surface is extremely young by Solar System standards. But *how* young exactly, and how can scientists determine the rate at which its surface is being replenished? Galileo Project Scientist Torrence Johnson of JPL and Laurence Soderblom of the US Geological Survey have estimated that if craters are created on Io at the same rate as on our Moon, a global surface layer at least a millimetre thick must be produced each year to account for the craterless Voyager images. Faster impact rates would require surface replenishment of a centimetre or more per year.

Others have measured temperatures of active regions to work out how much lava would be needed to bring a certain amount of heat to Io's surface. Their results equate to roughly the same resurfacing rate as that proposed by Johnson and Soderblom. In any event, Io produces a hundred times as much lava per year as Earth. Furthermore, its total heat flux shows that even its oldest surface areas were produced so recently that they are still cooling off! As a result, taking into account differences in lava-flow composition, colder areas should be 'older' than hotter ones.

TENUOUS ATMOSPHERE

With such vast quantities of gas and dust being ejected by Io's volcanoes each year, a substantial atmosphere might be a logical byproduct. Yet the veil surrounding this strange world is remarkably thin, with a pressure a millionth of that at sea level on

Earth. The earliest attempt to detect this atmosphere was during a stellar occultation in 1971. As astronomers measured the quivering brightness of the subject star as it slipped behind Io's limb, they noticed that its light was cut off almost instantaneously. If Io had a significant atmosphere, they reasoned, the starlight should have dimmed gradually before vanishing.

Io's first Earthly visitor, Pioneer 10, offered more precise data when it flew past in December 1973. Changes to its radio signal as it hurtled behind Io revealed high numbers of electrons 50–100 km above the surface; a tell-tale sign of an ionosphere and, by inference, the atmosphere necessary to supply it. In December 1995, Galileo flew through this ionosphere and revealed it to be more substantial: extending 900 km above the surface! Louis Frank believes the difference between what Pioneer 10 saw and Galileo's results indicate that Io's atmosphere may actually shrink and grow, depending on the severity of volcanic activity.

What could be the composition of this atmosphere? IRIS measurements taken by Voyager 1 pinpointed sulphur dioxide gas over Loki Patera and ground-based observations also detected sulphur dioxide frost covering most of the surface. Together with more recent HST and Galileo surveys, these results point to sulphur dioxide being an important constituent of the Ionian atmosphere. However, Voyager data indicated a very 'patchy' atmosphere, whose density increases noticeably above active volcanic centres. This notion was reinforced by ultraviolet observations from Galileo in May 1997, which showed the glow of dense atmospheric gases over several Ionian hotspots.

Many of these observations strongly support Caltech atmospheric specialist Andy Ingersoll's proposal that both the density and composition of Io's gaseous envelope are linked to volcanic eruptions and the sublimation of surficial 'frost'.

Subsequent Galileo observations also hinted at the presence of oxygen in Io's atmosphere – almost certainly derived from dissociated sulphur dioxide – and ground-based studies by physicist Bob Brown in 1974 pointed to the existence of a sodium cloud around the moon. More recently, in 2000, chlorine was also discovered, and in January 2003, scientists Emmanuel Lellouch of the Observatoire de Paris and Nicholas Snyder of the University of Colorado at Boulder argued in *Nature* that Brown's cloud was most likely a result of sodium chloride – salt – sprayed up by volcanic eruptions.

A TORMENTED WORLD

Io's volcanism comes – as Peale, Cassen and Reynolds suggested in their prophetic *Science* paper just before the Voyager 1 encounter – from tidal forces induced by Jupiter and its large moons Ganymede and Europa. This is primarily because all three moons are locked in 'resonant' orbits, such that Io circles Jupiter twice for each orbit of Europa, which in turn circles twice for each orbit of Ganymede. These resonances make Io's orbit slightly elliptical and, although the eccentricity is small, it has a significant effect. All four Galilean moons orbit Jupiter synchronously, with one hemisphere always facing the parent planet. On Io, however, a 'tidal bulge' rises and relaxes in response to changing gravitational fields. When Io is at its closest point to Jupiter, the bulge rises up to 100 m above the surface.

In the meantime, like a rubber ball, Io is constantly squeezed and contorted by Jupiter's enormous gravity. This stress manifests itself as heat, which melts the rock. The effect is the tormented place that we see today.

According to Rosaly Lopes-Gautier, if Io did not occupy its peculiar orbit – so close to Jupiter, yet also partaking in this celestial dance with Europa and Ganymede – it probably would not have active volcanism. "It would have cooled off long ago," she says. Instead, the friction caused by this never-ending tug-of-war is enough to heat and melt rocks inside Io, producing volcanoes, lava flows and huge geysers. In fact, infrared observations in the early 1980s suggested its surface may emit as much as a hundred *trillion* watts of thermal energy.

Although Io's ambient surface temperature is only 125 K, volcanic regions can be up to 10 times hotter and these are primarily why thermal emissions are so high. At first, some scientists felt these emissions represented a kind of balance between heat released from Io's interior and heat generated by tidal stresses imposed by Jupiter. However, recent theoretical estimates have thrown such 'equilibrium' models into question, to such an extent that some dynamicists now think Io's interior might actually heat up and cool sporadically, or even cyclically.

PLASMA 'TORUS'

When the Voyagers flew past Jupiter, they found that its immense magnetic field has profound effects on Io. As the field rotates with the giant planet in just under 10 hours, it sweeps past Io, tearing away 1,000 kg of material every second from its surface. This then forms a doughnut-shaped 'torus' of charged particles, which glows in the ultraviolet and was easily detected by the UVS instrument on both Voyagers. It was even suggested that plasma oscillations inside the torus might be responsible for the huge amount of radio noise from Jupiter, which makes it the 'noisiest' planet in the Solar System.

Ground-based observations and results from the Pioneer 10 flyby in December 1973 first identified this torus, which extends all the way around Io's orbit. Astronomers noticed that Io tended to 'brighten' slightly when it emerged from Jupiter's shadow and this was taken to mean that a gas was settling on its surface during the chilly darkness, then sublimating to gas upon the return of sunlight. Telescopic spectroscopy first identified the presence of a neutral sodium 'halo' around Io, a significant amount of which is piled in front of the moon while the rest trails behind.

It was first thought that Io's surface was probably salty in nature and that charged particles within Jupiter's magnetosphere were 'sputtering' neutral atoms from the crust. The real process by which the torus is maintained was only realised when Voyager pictures turned up the first clear evidence of volcanism. Both spacecraft identified ultraviolet emissions from ionised oxygen and both singly- and

False-colour image of Io's sodium cloud, taken by Galileo in December 1997.

doubly-ionised sodium in the torus. The plasma density of the torus was evidently related to Io's volcanism, because although Voyager 1 measured a plasma temperature of 100,000 K, this had decreased to 60,000 K by the time its mate arrived.

In fact, Io literally acts as a huge electrical generator as it ploughs through its parent planet's magnetic field, producing a current of a trillion watts or more, which flows along the field lines in a so-called 'flux tube' to Jupiter's ionosphere. This current actually proved to be five times more powerful than had been predicted before either Voyager departed Earth ... and led to one of the secondary objectives of the Jovian flypast being missed. Originally, scientists wanted to fly one Voyager right through the flux tube, but the stronger-than-expected current had twisted it more than 7,000 km from its predicted location.

The Io plasma torus provides nothing less than a 'reservoir' of charged particles to inflate Jupiter's already-huge magnetosphere. In fact, it is more than 1,200 times larger than Earth's magnetosphere and, if it could be seen in the night sky, would appear significantly larger than a full Moon. Since the mid-1950s, Jupiter had been known as a source of intense radio noise – a tell-tale sign of a substantial magnetosphere – but it was only when Pioneers 10 and 11 flew past in the early 1970s that electron densities several times greater than those in our magnetosphere were first recorded.

JUPITER'S HARSH MAGNETOSPHERE

Data from both Pioneers also provided an invaluable advance warning for Voyager mission planners. During their brief encounters, the two spacecraft absorbed a thousand times the human-lethal dose of high-energy radiation. Although they nevertheless survived the perilous passage, several of their transistor circuits failed. This experience gave the Voyager team some breathing space to 'harden' the electronic components on their own machines to better withstand such damage.

Not only was Jupiter's magnetosphere potentially lethal to its Earthly visitors, it was also vast: the Pioneers revealed it to extend at least 8 million km sunward and many scientists suspected the existence of a huge, bullet-shaped 'magnetotail' trailing many millions of kilometres behind the planet. When they arrived in the Jovian neighbourhood, both Voyagers confirmed that such a tail does indeed exist, but astounded the scientific community by revealing that it extends at least as far out as Saturn's orbit: more than 740 million km!

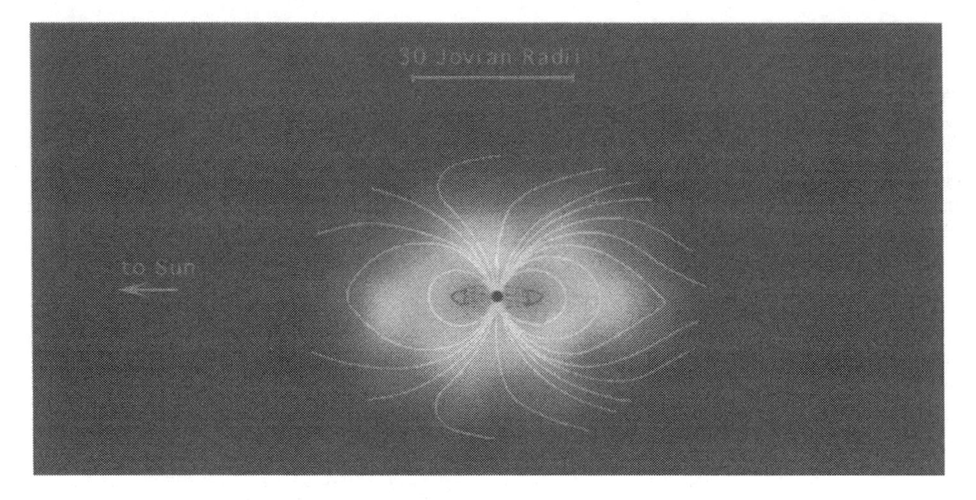

Diagram of the Jovian magnetosphere, revealed by Cassini's Ion and Neutral Camera during its flyby in December 2000.

Credit for the magnetotail discovery should go to all five fields and particles instruments, but one in particular here. The LECP has often been likened to a piece of wood, with charged particles playing the role of bullets that embed themselves into it. Their speed and direction can then be deduced by measuring the depth and orientation of the 'holes' they leave in the LECP's detector surfaces. Furthermore, the number of holes accumulated over time has been used to reveal particle densities in various parts of the Solar System and near all four gas giants.

Clearly, from its dimensions, the term 'magnetosphere' is something of a misnomer, in Jupiter's case at least, for it is far from spherical. Its enormous size in proportion to our own world's magnetosphere is due in part to that planet's strong magnetic field, but also to the fact that the solar wind pressure is considerably weaker by the time it reaches the Jovian system, 778 million km from the Sun. In fact, Jupiter's magnetosphere is home to a rich tapestry of unusual features, some of which have been seen nowhere else in the Solar System.

Firstly, vast amounts of trapped energetic plasma continuously travel back and forth along the field lines, inflating it like a gigantic, air-filled balloon, particularly around its equator. Secondly, the field actually 'co-rotates' with Jupiter's interior – once every 9 hours and 55 minutes – and all the plasma interacting with it is also forced to encircle the planet with this period. The consequences of such a rapid rotation rate are that plasma is pushed 'outwards', creating a massive 'sheet' around the planet's magnetic equator, which is inclined 9.5 degrees to the planet's equator, which is in turn inclined 3 degrees to the ecliptic (the plane in which the spacecraft flew).

Pioneer 10 and, more than five years later, the Voyagers measured energetic particle levels that rose and fell substantially as this tilted plasma sheet flopped first north, then south, over the spacecraft twice per Jovian rotation period. Centrifugal forces allow the co-rotating plasma to diffuse gradually outward against the field, filling the magnetosphere over a period of a few weeks. At this stage, by a process that remains imperfectly understood, the plasma is somehow accelerated and heated to inflate the middle and outer magnetosphere, before travelling inward as it is 'recycled' back to the inner magnetosphere.

When they eventually return to the outermost region of the plasma torus, these particles are redirected along the magnetic field and into Jupiter's atmosphere. This process deposits something between 10 and 100 trillion watts and, in so doing, almost certainly affects the temperature and dynamic behaviour of the giant planet's polar regions. How, exactly, the particles are accelerated and heated is still unknown, but could be connected to Jupiter's rapid rotation and the relationship between the magnetic field and the trapped plasma, as well as Io's influence.

Whereas the solar wind compresses the magnetosphere in the sunward direction, it drags it into a distended tail down the planet's magnetic shadow, far from the

Io appears to drift against a backdrop of Jupiter's clouds in this picture taken by the Saturn-bound Cassini spacecraft in December 2000. (see colour section)

planet. The magnetic field lines do not run from pole to pole, but distort and extend down the tail. This part of the magnetosphere does not co-rotate and therefore the trapped radiation flows downstream either as a steady wind or more sporadically, perhaps like the magnetic storms found in our own magnetosphere. According to measurements taken by Cassini, which flew within 10 million km of Jupiter on 30 December 2000, a large number of high-energy particles seem to 'bleed' from one side of the magnetosphere. Cassini's findings have helped to explain how electrons escape from Jupiter, some of which have even ended up in Earth's neighbourhood.

KING OF THE PLANETS

Not only is Jupiter's magnetosphere enormous and far-reaching, but so is the planet itself. In fact, with an equatorial diameter of almost 143,000 km, it is by far the biggest world in our Solar System. Its volume could swallow 1,320 bodies the size of Earth and it is more than twice as massive as the other eight planets combined. Yet, according to some predictions, it would encounter difficulties were it to grow any larger. If more material was added, says planetary scientist Bill Hubbard of the University of Arizona at Tucson, it would be so compressed by its own gravity that its overall size would increase only slightly.

A star like our Sun, of course, can swell to a greater diameter thanks to its nuclear heat source, and until the early twentieth century some scientists thought that Jupiter was just a failed star, lacking the mass and internal heat needed for full stellar ignition. Although it is three times as massive as the next largest planet, Saturn, it is actually only 23,500 km bigger in terms of equatorial diameter. This is due to tremendous compression in its interior, where pressures climb to more than 10 million bars, and consequently if more material was added its interior would start to 'collapse' and radial expansion would slow down.

If this line of thought is carried further, Jupiter could nevertheless conceivably

Dan Darda's concept of the brown dwarf Gliese 229B, together with a hypothetical moon, in orbit around its parent red dwarf star.

expand to a greater diameter and become somewhat more massive. Some scientists have argued that a planet of Jovian composition – principally hydrogen and helium – could attain a diameter of up to 160,000 km and a mass four or five times that of Jupiter, before internal pressures cause its interior to collapse and subsequent expansion to slow down. Yet, although no such 'super-Jupiters', or 'brown dwarfs', exist in our Solar System, they could occur elsewhere in the Universe.

Indeed, two extrasolar super-Jupiters have already been found and their masses determined. One is Gliese 229B and, like Jupiter, never became hot enough at its core or massive enough to initiate hydrogen fusion and become a 'true' star. It is roughly 40 times more massive than Jupiter, and yet is remarkably similar to it in terms of size. More recently, in December 2002, George Benedict and Barbara McArthur of the University of Texas at Austin reported HST observations that pegged the mass of a second super-Jupiter, Gliese 876B, at between 1.9 and 2.4 Jovian masses.

A STRANGE INTERIOR

Generally, for reasons associated with thermonuclear activity in the interior of these giant worlds as they grow more massive, most scientists conveniently use 13 Jupiter masses as a dividing line between 'large Jovian planets' and 'small brown dwarfs'. Exactly how similar or different Jupiter's interior is compared to the extrasolar worlds mentioned above is unknown, although Gliese 229B seems to possess large amounts of 'liquid metallic hydrogen' in its atmosphere. This bizarre form of hydrogen also exists inside Jupiter – in fact, it probably accounts for the bulk of the planet – but many of its properties are still poorly understood.

It has been known for many years that Jupiter's density of only 1.33 g/cm^3 required it to be predominantly composed of hydrogen. In 1860, George Bond determined that it radiates twice as much energy to space as it receives from sunlight. Jupiter must, he theorised, be in the process of contracting and transforming gravitational potential into heat. This led some scientists to conclude that the Jovian interior probably consisted of very hot gas and in the late nineteenth century several arguments were put forward that the planet may represent a 'failed star'.

In 1870, Richard Proctor suggested that "Jupiter is still a glowing mass, fluid probably throughout, still bubbling and seething with the intensity of the primeval fires, sending up continuous enormous masses of cloud, to be gathered into bands under the influence of the swift rotation of the giant planet." Half a century later, Harold Jeffreys argued instead for a cooler Jupiter. He posited that the planet comprised a rocky core overlaid firstly by a mantle of ice and solid carbon dioxide and then by a very deep, tenuous gaseous envelope.

Subsequent developments in spectroscopy led to the identification of ammonia and methane in the Jovian atmosphere during the 1930s, which prompted Rupert Wildt to propose a rocky core surrounded by a thick layer of water-ice and an outer 'ocean' of condensed gas. The possibility that a substance called 'liquid metallic hydrogen' may be a key player in Jupiter's atmosphere can be traced back to 1951, when it was suggested as a possible source of the planet's magnetic field. The field itself was hypothetical at the time, but was confirmed a few years later when strong radio emissions from the planet were detected.

Ground-based laboratory experiments have successfully produced tiny samples of liquid metallic hydrogen at Lawrence Livermore National Laboratory in California, using reverberating shockwaves to 'squeeze' the gas into such a state that it behaves more like a metal. The only *in-situ* data that scientists have for Jupiter's interior to date is from Galileo, which dropped a small instrumented probe into the atmosphere in 1995. The probe, however, survived only to a depth of 150 km beneath the cloud tops – a region dominated by 'ordinary' molecular hydrogen – before being crushed by steadily increasing pressures.

Nevertheless, theoretical physicists have probed Jupiter's interior and suspect that it probably has a rocky core overlaid by the thick liquid metallic hydrogen envelope and finally an outer layer of molecular hydrogen and helium. The temperatures needed to produce and sustain liquid metallic hydrogen are less than that at the Sun's surface, but the required internal pressures are in excess of 4 million bars. It is probably something akin to a molten metal and is an electrical conductor; as a result, it is almost certainly responsible for Jupiter's magnetic field.

There are probably no 'sharp' divisions between hydrogen layers within Jupiter. According to atmospheric scientist Andy Ingersoll of Caltech, the giants' atmospheres "grade downward until the distinction between gas and liquid becomes meaningless". Molecular hydrogen was first detected in Jupiter's atmosphere in the 1960s, by which time methane and ammonia had also been identified. In fact, the planet is composed of 87 per cent hydrogen and 11 per cent helium by number, with smaller amounts of methane, water, ammonia and rock.

The instrument both Voyagers used to make these measurements was the IRIS, which determined not only chemical abundances, but also atmospheric and surface temperatures for each of the giant planets and their many moons. From the IRIS data, it was possible for the first time to plot, layer-cake-style, the unique thermal layout of each world's atmosphere. The instrument also allowed scientists to calculate the balance of energy radiated by the planet to that absorbed from sunlight.

A COLOURFUL ATMOSPHERE

Jupiter reveals itself to even a casual observer as a world like no other, with a façade bizarrely reminiscent of a hellish scene straight from the imagination of Dante himself. Colourful latitudinal bands and zones of vivid reds, yellows and browns, dramatically different in their thicknesses, swaddle the planet and enormous, ever-changing storms and eddies intermingle to make this one of the most active places in the Solar System. Due to its short day – less than 10 hours – Jupiter's atmosphere rotates 'differentially', at different rates at different latitudes, giving rise to its 'banded' look.

One particularly curious feature of Jupiter's atmosphere is that its clouds, storms and eddies usually stay organised into these bands: although they move rapidly around the planet, they generally remain 'fixed' to their particular latitude. At the same time, however, they do tend to drift with respect to one another within their 'band'. This trend is primarily due to Jupiter's lack of a solid surface. On Earth, such phenomena cannot persist because they would be disrupted by large transient storms at temperate latitudes and by long pressure waves anchored to our continents. Without a solid surface, Jovian storms can endure for years or even centuries.

In fact, these psychedelic patterns are so persistent (more than 90 years of 'modern' telescopic observations have revealed few changes) that astronomers have given names to individual bands, calling the dark ones 'belts' and the bright ones 'zones'. In effect, they are the Jovian equivalent of 'jetstreams'. Both the Voyagers and Galileo revealed small eddies, which appear suddenly along the boundaries between eastward- and westward-moving streams; these only last a few days before they are torn apart by counter-flowing winds. During their encounters, the Voyagers tracked thousands of 100-km-wide cloud features.

THE GREAT RED SPOT

Of all the storms and clouds in the atmosphere, none is more prominent or long-lasting than the famous Great Red Spot, a colossal hurricane twice the width of Earth, which has unleashed its fury in the southern hemisphere since at least the seventeenth century. Voyager time-lapse movies of the spot revealed it to rotate counter-clockwise, although the clouds within it move at various speeds. Those clouds nearest to its outermost edge were found to complete a full lap of the spot in four to six days, while material closer to the centre moved more slowly and erratically.

The spot's discovery is usually attributed to either Giovanni Cassini – the celebrated Franco-Italian astronomer, after whom the spacecraft currently heading for Saturn is named – or English scientist Robert Hooke, more than 300 years ago. In 1664, Hooke wrote in the *Philosophical Transactions of the Royal Society* about his discovery of "... a spot in the largest of the three observed belts of Jupiter, and that, observing it from time to time, within two hours after, the said spot had moved east to west about half of the diameter ... its diameter is one-tenth of Jupiter".

Whether Hooke's discovery and the spot of today are one and the same is unclear, but certainly Cassini made similar observations in 1665 and the object was seen intermittently until around 1713. After this last sighting, there was a lull for more than a century, until German astronomer Heinrich Schwabe saw it in 1831; since then, although it has faded from view occasionally and changed dramatically in both size and colour – increasing to 40,000 km (almost twice is current length) at one stage and displaying an astonishing 'brick-red' hue at another – it has nevertheless remained Jupiter's constant, faithful companion.

Today's spot is distinctly oval, roughly 26,000 km along its east–west axis and 13,000 km along its north–south axis, and the counter-clockwise motion of its winds – some of which reach 400 km/h at its outermost edges – suggest that, far from simply being a Jupiter-sized version of a low-pressure Caribbean hurricane, it is in fact a high-pressure region whose cloud-tops are significantly colder and higher than surrounding areas. In fact, the central column of the spot is so far above Jupiter's ammonia cloud-tops that it is most likely the coldest part of the atmosphere.

Infrared mosaic of the Great Red Spot, taken by Galileo in June 1996. The clouds nearest the centre of the spot lie up to 30 km higher than those around its edge. The small white patch to the north-west of the spot is actually a vast thunderhead.

During the course of its mission, Galileo identified numerous storms along the periphery of the spot. The top of one enormous thunderhead was found to rise almost 25 km above most of the surrounding wispy ammonia clouds at the 0.7-bar level. The base of this storm was about 50 km deeper, at a pressure of almost 4 bars, where the only available condensate is water.

Whereas low-pressure systems operate in the same way on Earth and on Jupiter, their high-pressure counterparts are somewhat different. A terrestrial high-pressure system forms when dry air, which has emerged from the top of a cyclonic system, moves clear, cools and descends, thus increasing the surface pressure. Due to the 'dry' nature of the descending air, it cannot produce clouds. As it radiates outwards near the surface, it gives rise to anti-cyclonic circulation. On Jupiter, however, high-pressure regions serve as convergence centres which consume cyclonic systems and this inflow is responsible for the increase in pressure. Since pressure is inversely proportional to altitude, the centre of the convergence is obliged to rise to relieve the pressure.

The result is an 'upwelling column' and, as the air emerges from the top of this column, it spreads out and descends. Consequently, instead of being a descending airflow, high-pressure systems on Jupiter actually tower far above the cloud-tops. Infrared temperature mapping by Galileo supports the view that the Great Red Spot is indeed a high-pressure region. Its distinctive reddish hue remains a mystery, although the presence of several chemicals, including phosphorus, have been proposed over the years. Perhaps the greatest riddle until recently was how such spots endure for so long.

As shall be seen in later chapters, fairly long-lived spots have been identified by the Voyagers and HST on Saturn and Neptune, albeit not in the same league as their Jovian counterpart. One key obstacle with maintaining a rapidly rotating oval storm like the Great Red Spot is how they remain stable over long periods. If unstable, they would be expected to break into waves and eddies and disperse in a few days; if stable, on the other hand, they would need an energy source within the atmosphere to keep themselves going or, inevitably, they would run out of steam.

Two main models have been proposed to explain where the Great Red Spot and other planetary spots obtain their energy. The 'hurricane' model suggests that they are gigantic convective 'cells' that extract energy from below, while the 'shear instability' idea argues that they draw it in from the sides.

Other scientists have countered that they simply gather energy by gobbling up smaller spots and eddies circulating in the atmosphere. This last possibility has received support from Galileo observations in recent years, although computer simulations and experiments that injected red dye into rapidly spinning tanks to mimic Jovian-type conditions have shown that all three models, in fact, can work and produce long-lived spots. The longevity of the Great Red Spot derives from the sheer depth of Jupiter's atmosphere. In the absence of a solid surface to dissipate a storm's energy – as happens when an Earth-based storm makes landfall and is deprived of warm moisture from the ocean – the Great Red Spot has settled into a semi-stable state.

WHITE OVALS

Perhaps our best clue to understanding the nature and origin of the Great Red Spot comes from strange white ovals, which were first observed at their formation in 1939. At that time, three dark features were seen in one of Jupiter's southern zones, stretching all around the planet. After recording the longitudes of their leading and trailing edges, observers labelled them 'AB', 'CD' and 'EF'. Gradually, their shape changed until, by 1948, they had begun to evolve into distinct white ovals. Over the next three decades, until the Voyager encounters, they diminished in size to less than 11,000 km long.

However, because the ovals were anti-correlated with the dark features, the white ovals were subsequently redesignated 'FA', 'BC' and 'DE'. Although they are much smaller than the Great Red Spot, they are nevertheless anti-cyclonic and rise far above Jupiter's ammonia cloud-tops. They have remained within the latitude range 31–35 degrees and drifted in an easterly direction. They were also observed to 'jostle' one another, which caused something of a stir in February 1998 when Galileo saw the after-effects of the BC and DE ovals first squashing a low-pressure storm between them and then consuming it as they merged into a single system, known as 'BE'.

"The merged white oval is the strongest storm in our Solar System, with the exception of the Great Red Spot," said Glenn Orton of JPL. "This may be the first time humans have ever observed such a large interaction between two storm systems." Indeed, Galileo data from July 1998 indicated that the cyclonic system squashed between the two ovals had apparently 'lost power' and allowed them to continue towards each other. Two years later, in March 2000, the two remaining storms – BE and FA – collided with each other over a three-week period to form a single, larger one.

Unfortunately, the 1998 event could not be seen from Earth, although Galileo witnessed its after-effects, but the 2000 collision was observed by HST. It is quite possible that the sole remaining white oval may disappear entirely at some point in the near future and so too, eventually, might the Great Red Spot itself.

TUMULTUOUS WEATHER

In addition to this tumultuous region, there are many other cloud structures in the atmosphere that showed Jupiter to be a world of chaotic weather. In less than five years between the Pioneer and Voyager encounters, the planet's appearance became noticeably more turbulent and 'dramatic'. In the broadest white zone, for example, Voyager images revealed violent motions that were not seen in the low-resolution scans taken by their predecessors.

Neither Pioneer was equipped with 'cameras' in the same sense as those on board the Voyagers. Instead, they carried an imaging photopolarimeter which scanned Jupiter's disk repeatedly as the spacecraft rotated. Images were subsequently assembled on Earth by stacking these 'strips' side by side. However, this turned out to be a very time-consuming process and the spacecraft were only able to build up a couple of dozen images during the final days before closest approach to Jupiter. Nevertheless, the Pioneers revealed the Jovian atmosphere in unprecedented detail that would not be surpassed until the Voyagers arrived.

Yet, beneath the apparent chaos of Jupiter's incredibly active atmosphere, there seems to be order. Both Voyagers revealed that, despite their differing sizes, most features tend to move at fairly uniform speeds. This suggests that the motion behind them is 'mass motion' rather than 'wave motion'. In other words, it seems probable that the movement of material, rather than energy, is at work deep inside Jupiter. Furthermore, the Voyagers found that the rapid brightening of atmospheric features was typically followed by a 'spreading' of cloud material. Scientists speculated that this is probably a result of disturbances that trigger convective activity.

Another surprise was the discovery that the zonal (jetstream) wind pattern extends as far as 60 degrees north and south latitudes – and perhaps as high as 75 degrees – which was much closer to the poles than atmospheric scientists had predicted before the Voyagers' arrival. It was previously suspected that convective 'upwelling' and 'downwelling' would dominate the regions nearer to the poles, particularly above 45 degrees latitude. The winds typically follow an east-to-west direction, roughly the same as those in the more temperate regions where the colourful belts and zones are visible.

Prior to the arrival of Galileo, the leading theory was that, like Earth's own weather system, atmospheric circulation on Jupiter was driven by energy in sunlight

Artist's concept of the Galileo probe, hanging underneath its main parachute, as it plunges into the Jovian clouds. The cone-shaped heat shield has just been jettisoned.

and the latent heat of condensation, which led many scientists to suspect that it would be confined to a shallow zone in the upper atmosphere. Galileo's atmospheric-entry probe, however, established that winds extended to a depth of at least 150 km and that the deep atmosphere is highly convective. A rising column of heated moist air cools with expansion and clouds of rain condense out.

On Earth, this condensate is water, but on Jupiter the rising air also condenses out ammonium hydrosulphide as a higher, 'wispier' cloud and a layer of ammonia at the top. By the time the rising air emerges from the ammonia cloud layer, all the volatiles will have already condensed. Remote-sensing observations by the main Galileo spacecraft confirmed the air above these updrafts to be very cold, very clear and very dry. As this air flow 'turns over', north or south, it is deflected by the Coriolis effect. This serves to sustain the alternating latitudinal atmospheric circulation.

As the air starts to settle, it is compressed and heated. The descending cloud-free air appears visually dark, but because it is arid it is transparent at a wavelength of 5 microns; downdrafts therefore appear bright in the near-infrared and can offer a useful 'window' into the deeper, hotter regions. Galileo's infrared observations revealed that downdrafts at the probe's entry site 'soaked up' volatiles when they reached a so-called 'hot mixing zone' several hundred kilometres down, and were recycled into an updraft. However, as Ingersoll speculates, the strength of winds in the deep atmosphere "may be evidence that Jupiter has high-speed wind currents extending many thousands of kilometres deep into its hot, dense atmosphere".

In penetrating the region beneath the level to which sunlight can reach, Galileo's entry probe sampled what may be a fairly uniform 'interior atmosphere' in which one continuous circulation pattern extends from the visible surface to a depth of about 16,000 km. This convection is now thought to be powered by heat 'leaking' from the interior. By the time it halted, underwent mixing and was recycled, such a downdraft would have penetrated far beneath the level observable at 5 microns. A planetary convection cycle on this scale may yield alternating zones of rising air super-saturated with water and other volatiles and descending, arid air.

Perhaps, as Toby Owen of the University of Hawaii has suggested, Jupiter's internal heat comes out only in certain regions, where ascending currents bring up hot material from the planet's interior. The enormous thunderheads seen by Galileo, as well as the Voyagers' photographs of clouds that erupted every 10 days or so, provide clear evidence of the rising energy. Such storms are analogous to terrestrial thunderstorms, with the tall, bright portion comparable to familiar 'anvil' clouds on Earth, although very different in terms of size. Terrestrial anvils rise 18 km and are 200 km across, whereas those on Jupiter were 75 km tall and 1,000 km across the base. If such storms could somehow be transplanted into our atmosphere, they would jut out into space!

THE 'JOVIAN LIGHTS'

The existence of Jovian auroral activity was first inferred from the particles and fields data collected by Pioneers 10 and 11. It was subsequently observed in depth by the International Ultraviolet Explorer (IUE) – a joint US/European astronomy satellite operated between 1978 and 1996 – as well as the Voyagers, HST and ground-based infrared telescopes. The Voyagers' UVS and LECP instruments helped to identify ions in Jupiter's enormous magnetic field and the physical processes responsible for causing them. Additional, optical measurements came from the PPS instrument.

The aurorae themselves are quite similar (albeit much more energetic and far brighter) to the Northern and Southern Lights, which grace our skies when charged particles of solar origin excite the thin upper atmosphere. Voyager 1 detected an aurora which extended to 30,000 km and its PWS instrument noted 'whistling' radio emissions akin to those from terrestrial displays. However, the Voyager results implied that these 'Jovian Lights' are triggered primarily by the torus material from Io, which spirals down magnetic field lines into the planet's ionosphere.

More recently, Galileo's energetic particle detector has measured intense, two-directional 'beams' of electrons – nicknamed an 'electron highway' – that are aligned with Jupiter's magnetic field close to Io. They are similar to those that impinge on Earth's thin gaseous veil to generate auroral displays and, according to Donald Williams of the Johns Hopkins Applied Physics Laboratory in Baltimore, Maryland, may pour up to a billion watts into the Jovian atmosphere.

This has led to a number of unusual and striking observations made since the Voyager encounters. In a series of images taken between October 1996 and September 1997, HST witnessed aurorae several times brighter than even their most powerful terrestrial counterparts. Subsequent images from the Earth-orbiting

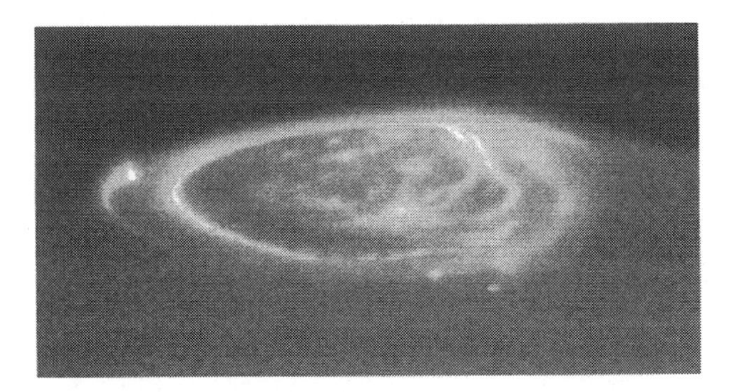

Like a glowing lasso wrapped around Jupiter's north pole, this impressive auroral display was recorded by the Hubble Space Telescope in December 2000.

telescope, released in December 2000, revealed an eerie blue 'curtain' of fluorescent gas wrapped like a lasso around Jupiter's north pole. Moreover, these auroral displays extended several hundred kilometres into space and were visible projecting beyond the giant planet's limb.

Each of Io's auroral 'footprints' has a comma-like 'tail' because the charged particles continue to excite the atmosphere for some time after the moon has 'passed overhead'. In fact, of course, it is actually Jupiter which turns beneath the slowly orbiting moon. These features were first identified by HST in 1994.

THE JOVIAN RINGS

Latitudinal banding, peculiar spots and powerful aurorae are just a few of the phenomena noted on more than one of the gas giants; another strange trend is that all four mysterious worlds possess ring systems. Each one, however, is unique and the thin ring of Jupiter was discovered serendipitously by Voyager 1 in March 1979. Scientists first began to suspect the existence of rings in December 1974, when Pioneer 11 plunged into the Jovian magnetosphere and measured a drop in the numbers of high-energy particles when it was closest to the planet; something was absorbing the radiation flowing along the field lines.

It was theorised that this finding might indicate that Pioneer had travelled either underneath the path of a hitherto-unknown 'ring' or perhaps past a new moon. Many scientists dismissed these possibilities because other explanations for the reduced particle count were available. However, while Voyager 1 was in the process of searching for new moonlets – a process which revealed the small worlds Thebe, Metis and Adrastea – it found a tenuous ring! Voyager 2's whistle-stop tour four months later yielded further information about the ring's composition, which has since been verified by Galileo and ground-based infrared observations.

Unlike the rings of Saturn, discussed in the next chapter, the Jovian ring system is intrinsically dark and probably consists of very small, rocky grains. In fact, some of the ring particles seem to display a distinctly reddish hue, which is similar to the

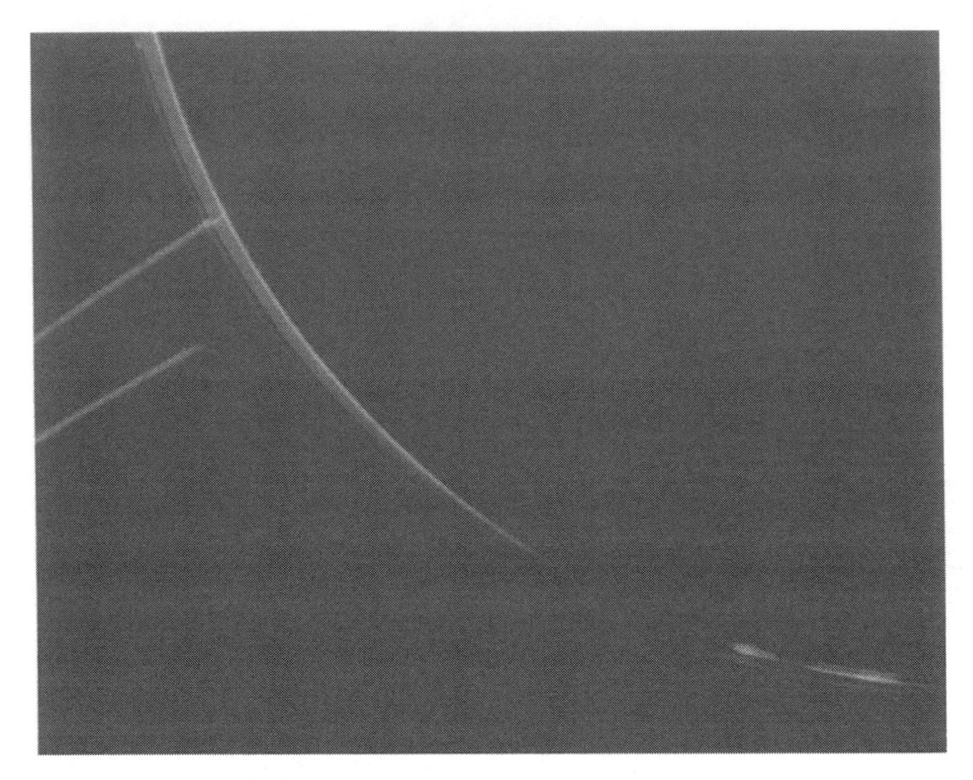

Jupiter's narrow ring system, as viewed by Voyager 2. Note that one section of the ring is in the planet's shadow.

surface colour of the small nearby moonlets. It was initially suggested that the ring material was what remained of an unfortunate moon ripped apart by tidal stresses, and that the small moonlets Metis and Amalthea – the latter of which had been discovered by Edward Barnard of Lick Observatory in 1892 – could succumb to a similar fate. The fact that the moonlets are reddish implied a relationship between them and the rings. Some scientists have argued that since sulphur dioxide from Io's volcanic plumes pervades the Jovian magnetosphere, the moonlets may be coated with it.

Although the closest Voyager pictures of irregularly-shaped Amalthea were from a distance of 420,000 km, they nevertheless revealed this distinct reddish hue. When Galileo inspected it in greater detail, it revealed a surprising 50-km-long 'streak' dubbed 'Ida', which may be the crest of a ridge or perhaps a ray of 'clean' ejecta thrown up by an impact.

Other features noted on Amalthea include a large crater called Gaia, near its south pole. Thebe and Metis were also resolved with a resolution of just 2 km per pixel, which prompted Galileo science planning and operations chief Duane Bindschadler to point out: "For the first time, we can start to resolve surface features on them. We see things that are distinctly impact craters." On Thebe, which has an equatorial diameter of only 60 km, an enormous, 40-km-wide crater known as Zethus was found.

By showing that the Jovian rings are actually sustained by these small moonlets and that dust blasted from their surfaces provides a reservoir of material, Galileo confirmed that they are parts of a single system. In view of the tiny size of the particles, the rings are probably quite young. They start at about 92,000 km from Jupiter's centre and extend outwards to around 250,000 km. When one takes into account that the planet's radius is over 70,000 km, the rings are very close to the surface.

In addition to the 'main' ring, which images from both Voyagers revealed to be 7,000 km wide and approximately 30 km thick, a broad 'halo' of tenuous dusty material was found inside it, extending out of the planet's equatorial plane. Such halos have not been seen thus far in any other ring system. Interactions with Jupiter's magnetic field increase its breadth to roughly 20,000 km.

Beyond the halo and main ring is a so-called 'gossamer' ring, which is the thinnest and least-visible of the trio. Intriguingly, it does not seem to 'end'! Rather, it peters out from the main ring until it fades into the background darkness, 180,000 km above Jupiter's cloud-tops. Although Voyager data on the gossamer ring is limited, it has been supplemented by Galileo, which alerted scientists to two main components: one inside the orbit of (and caused by) Amalthea, the other sustained by Thebe. In fact, all four moonlets shed material, but Amalthea and Thebe define boundaries in the structure.

Galileo showed that such moonlets – some of them embedded *inside* rings – are primary sources of material for these narrow bands. "We can see the gossamer-bound dust coming off Amalthea and Thebe," said ring specialist Joe Burns of

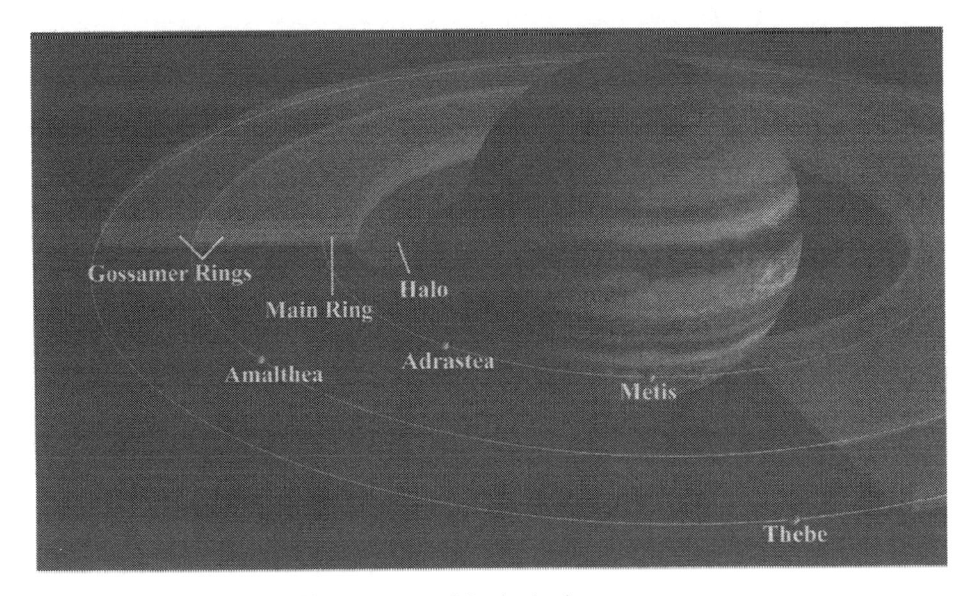

The structure of Jupiter's ring system.

Cornell University. A vital clue in reaching this conclusion was that, like the gossamer ring itself, both moonlets are slightly inclined to Jupiter's equatorial plane. As interplanetary meteoroids are drawn in by Jupiter's gravitational attraction, they are accelerated to such phenomenal speeds that they penetrate deeply into the moonlets' surfaces and vaporise themselves, blasting debris away with enough velocity to escape the moonlet.

In effect, were it not for these small moonlets orbiting so close to Jupiter, none of the rings would exist. "We now believe that the main ring comes from Adrastea and Metis," said Burns, pointing out that, like the main ring itself, both moonlets are very close to the planet's equatorial plane. As for the innermost halo ring, current thinking is that the magnetic field close to Jupiter draws out electrically-charged dust from the disk. The halo, therefore, consists of particles drawn from the main ring's innermost edge.

WATERWORLD

It seems unlikely, having journeyed thus far into the realm of Jove and through data from the Voyagers, Pioneers, Galileo and HST, that life or the tenets of life could have thrived in this ferocious patch of space. The effect of Jupiter's radiation on the Pioneers has already been mentioned and, indeed, over the course of its extended eight-year mission even Galileo has absorbed more than four times as much radiation as it was designed to withstand. On several occasions, most recently in January 2002, it was temporarily knocked into precautionary safe mode by the giant planet's fierce radiation belts.

Yet, hard as it may seem in the court of this unforgiving planet, life may once have evolved here, or at least on one of Jupiter's many moons. Europa, one of the four large worlds discovered by Galileo Galilei in 1610, has long been known to possess a water-ice crust, which aroused speculation that beneath this frozen veneer may lie an ocean of liquid water up to 100 km deep, perhaps capable of sustaining primitive life. Although the Galileo spacecraft's pictures and magnetometer data have offered persuasive evidence for a subsurface ocean, a lander or radar-equipped orbiter would provide conclusive answers.

Such missions are on NASA's drawing boards, albeit only in the planning stages and probably a decade away from reaching fruition. If such an ocean does exist, Europa will be the only other place in the Solar System, save Earth, to possess a substantial body of liquid water. In fact, to avoid a slim chance that, at the end of its mission, Galileo could accidentally hit Europa, NASA directed it to make a controlled, destructive dive into Jupiter's atmosphere late in September 2003. By so doing, Galileo will help to keep the intriguing moon – and perhaps extant life – pristine for future explorers.

Some scientists, however, have argued that the case for a subsurface ocean is already proved. On 3 January 2000, Galileo swept 343 km past Europa and provided a chance to test the hypothesis that a salty ocean was responsible for generating a magnetic field within the moon. The axis of Jupiter's magnetic field is inclined with respect to its rotational axis, so the ambient field in Europa's location periodically reverses polarity as the magnetospheric 'equator' sweeps past the slowly-orbiting moon. If Europa's weak field was produced by electrical currents induced by this changing field, its magnetic poles ought to be near the equator and ought to migrate.

Galileo revealed that Europa's north pole was changing alignment. "In fact, it is reversing direction entirely, every five and a half hours," said Margaret Kivelson of the University of California at Los Angeles. This period was precisely as expected by considering that the moon was travelling partway around its 3.5-day orbit as Jupiter, and its magnetosphere, rotated. Other than seeing a cryovolcanic flow underway, this is our best circumstantial evidence for a subsurface ocean. "Currents could flow in partially-melted ice beneath Europa's surface, but that makes little sense," continued Kivelson. "Europa is hotter toward its interior, so it is more likely the ice would melt completely. In addition, as you get deeper towards the interior, the strength of the current-generated magnetic field at the surface would decrease."

THE RIGHT INGREDIENTS FOR LIFE?

The fluid had to be close to the surface. It would appear from the magnetometer data, therefore, that there may be an ocean – perhaps as deep as 100 km – beneath Europa's surface. What about life? "There are three main criteria to consider when you are looking for the possibility of life outside the Earth: the presence of water, organic compounds and adequate heat," says Ronald Greeley of Arizona State University. "Europa obviously has substantial water-ice and organic compounds are known to be prevalent in the Solar System. The big question mark is how much heat is generated in the interior?"

In fact, the missing factor – internal warmth – could come from the radioactive decay of rocky materials inside Europa or tidal heating with Jupiter, Ganymede and Io. Certainly, Galileo pictures from December 1997 revealed three pieces of evidence to suggest that, far from being frozen and cold, Europa may indeed be warm and slushy just beneath its icy crust and even hotter at greater depths. This evidence included: (1) a strangely shallow impact crater, (2) 'chunky', textured surface types, similar to icebergs rising from and frozen in a jumbled matrix and (3) gaps where new crust seems to have formed between continent-sized ice plates.

The crater in question, known as Pwyll, probably formed quite recently (between 10 and 100 million years ago) by a meteorite impact and, judging by the large amount of dark debris cluttered around its rim, pulled up a lot of deep-buried material. Yet its shallow nature and a high series of mountain peaks nearby could mean that the subsurface ice was warm enough to collapse and fill in the deep impact hole.

A subsurface ocean sufficiently warm to be 'slushy' in nature might also explain the origins of an area littered with fractured and rotated blocks of crust, dubbed 'chaos terrain'. The Galileo images revealed rough, 'swirling' material between the fractured chunks, which may have been suspended in slush that froze at the very low surface temperatures. There also appears to be activity like terrestrial plate tectonics going on. Galileo clearly revealed dark, wedge-shaped gaps where the ice split apart.

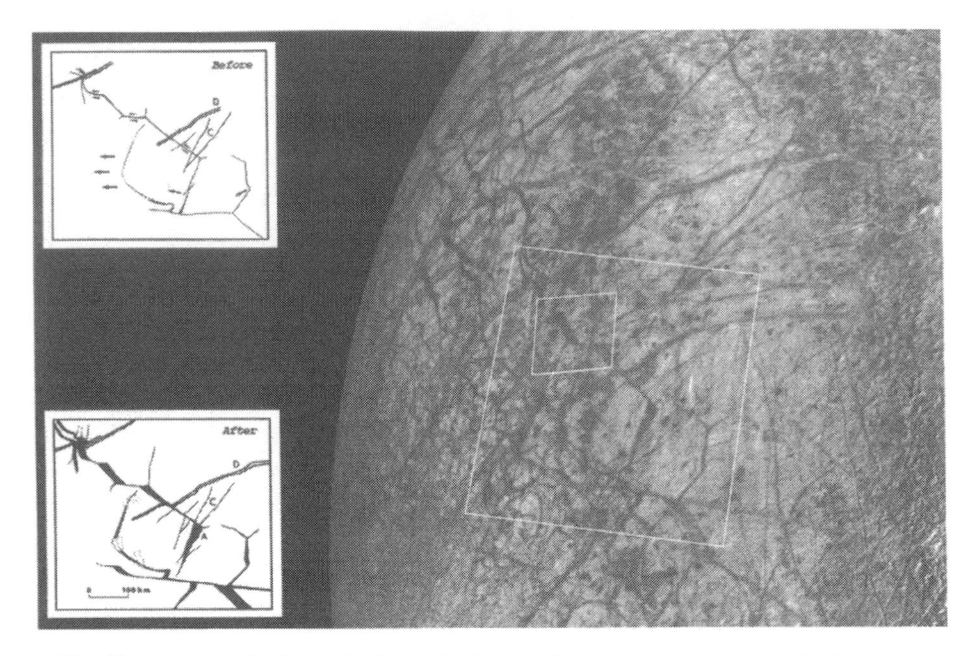

This Voyager mosaic shows the intensely fractured terrain around the anti-Jovian point of Europa. The outer box corresponds to the area represented by the inserts, which depict how one particular set of fractures formed. Notice how the wedge-shaped spreading zones work with 'strike-slip' faults to open rhomboidal 'gaps' which are in turn sealed by slush that oozes up and freezes. Several chains can be seen in the image. (Inserts courtesy of Paul Schenk of the Lunar and Planetary Institute, Houston.)

Although there is evidence of spreading, there are no signs of 'subduction' on Europa, so it is not plate tectonics in the terrestrial sense.

The new crust welling up beneath the Europan ice plates was most probably slushy ice or even liquid water that has since frozen and fractured. "In some areas, the ice is broken up into large pieces that have shifted away from one another, but obviously fit together like a jigsaw puzzle," says Greeley. "This shows the ice crust has been or still is lubricated from below by warm ice or maybe even liquid water." Such results provide tantalising clues that Europa may harbour environmental niches warm and wet enough to host bacterial life.

A BRIGHT YOUNG SURFACE

Europa's water-ice crust, which may be just 5 km or as much as 150 km thick, is the main reason it has one of the brightest surfaces in the Solar System. This brightness also suggests that its present surface is quite young – perhaps just 30 million years old in places – although its exact age remains uncertain. Indeed, some geologists speculate that it may be undergoing active resurfacing. One peculiarity is that, although its leading hemisphere seems to be shiny water-ice, the trailing side is distinctly darker and redder, perhaps due to charged-particle impacts from Jupiter's rapidly rotating magnetosphere modifying the surface chemistry.

Much of what we know about Europa comes from Galileo's multiple encounters with the moon. The Pioneer pictures were of insufficient resolution to reveal much, and even the Voyager images are only "of snapshot quality", but the remarkably sharp terminator indicated that, in general, the range of elevation is confined within a few hundred metres. The Voyagers saw Europa as a smooth white globe, reminiscent of a string-wrapped baseball, but Galileo took surface resolution down to a few metres. This allowed geologists to analyse surface composition, temperature and physical structure from ultraviolet to infrared wavelengths.

Particularly striking is the complex network of crack-like ridges, which criss-cross Europa's entire surface. "Ridges are visible at all resolutions," explains Robert Sullivan of Arizona State University. "Closely paired ridges are most common. With higher resolution, ridges seen previously as singular features are revealed to be double." Some of these features may have formed through crustal stresses: as two plates pulled apart, warm slushy material from below pushed up and froze to form a

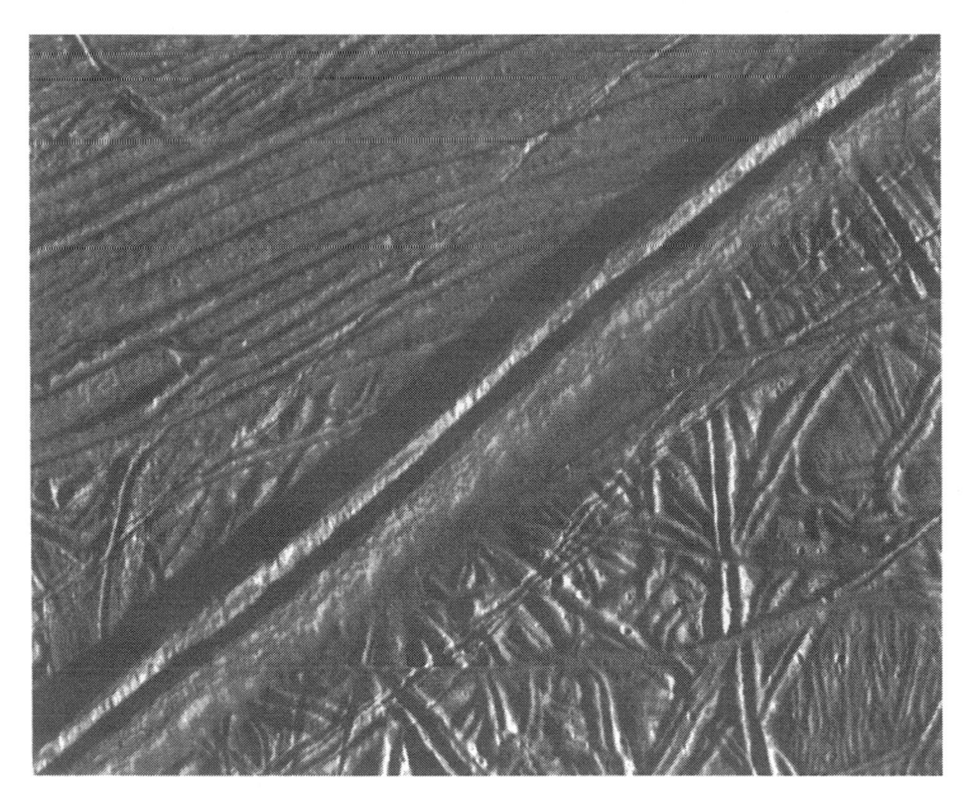

High-resolution image of triple bands on Europa.

ridge. Others might have formed by compression, whereby two plates pushing against each other caused the material at the centre to crumple into a ridge formation.

When the Voyagers flew past Europa in 1979, they photographed astonishing linear features, ranging from scalloped ridges to long dark stripes crossing the surface in complex patterns. More recent observations have revealed that the classification of many of these features was actually affected by the lighting conditions and viewing geometry at the time, combined with the resolution limitations of the Voyagers themselves. High-quality Galileo pictures, for example, have shown that many features first thought to be 'individual' are actually part of much wider networks of ridges and grooves stretching across the surface.

'TRIPLE BANDS'

Most intriguing and enigmatic of all are Europa's 'triple bands', which the Voyagers revealed to consist of two parallel dark zones separated, Mohican-like, by a bright central stripe. Typically, the central part is depressed and the flanking dark areas are raised. Although each triple band shares this characteristic, each one is very different from the next in terms of its appearance, brightness and size. Some range in width from just a few kilometres to more than 20 km, while others have been measured to run to lengths in excess of 1,000 km across the surface.

Before Galileo's arrival, most geologists thought the triple bands formed through straightforward tectonic processes. In the 'block-faulting' model, it was argued that the surface experienced extensional stress, which caused it to drop down between parallel sets of fault lines. The sunken valley, or 'graben', formed by this process was then flooded by an icy slush rich in non-ice components, which would explain the dark stripes. Other suggestions involved crustal fracturing which released hydrated silicates like clays onto the surface; the water they contained then separated to leave relatively pure ice in the central bright band.

When Galileo reached Europa, however, its high-resolution imagery apparently ruled out both possibilities because the outer boundaries of the dark stripes are not sharp, as would be characteristic of these processes. One alternative explanation involves an extreme form of cryovolcanism, in which geysers eject carbon dioxide or other volatiles explosively through the icy crust. If such an eruption were to occur along a curvilinear fracture, it could lead to a continuous line of geysers, akin to the 'curtains of fire' seen in Hawaii. Indeed, splotchy patches can be seen along some of these curvilinear features.

After leaving the fracture, the non-ice components would fall to the surface and form deposits around it. Thicker concentrations would accumulate along the fracture, grading to more diffuse edges further out. Eventually, the cryovolcanic source would become depleted and the explosive phase would end. Alternatively, as upwelling currents brought heat to the surface, the icy crust might sublimate away, leaving a dark deposit. This would be most concentrated where the heat was greatest and diminish with increasing distance from it. In both models, cleaner subsurface ice might then squeeze through the fractured crust to form a central ridge.

If cold geysers are responsible for the creation of triple bands on Europa, they might also be responsible for pushing up some of the other unusual surface features, such as dome-shaped lumps found here and there. Moreover, if cryovolcanism has happened – or may still be happening – on the frozen world, it might help to explain strange flow-like patterns, some up to 20 km long, which appear to cut across or through ridge formations. The height of the 'fronts' of these features, up to 100 m in some cases, has suggested to many geologists that they were originally very viscous and sluggish flows.

No source vents are apparent in any of these flow-like features and it seems likely that they represent regions of localised heating. The paucity of craters on Europa has already been mentioned and their conspicuous absence on the flows strongly suggests they are so young that geyser eruptions could still be ongoing! Although Galileo, when its primary mission ended in December 1997, began an extended period of operations that included multiple Europa flybys to look for erupting geysers, it seems that scientists will have to wait for future robotic explorers for the issue to be resolved.

Whether the triple bands form through cryovolcanism or by some other means, their degradation over time – including increases in the brightness of the dark stripes – has aroused debate. Three main possibilities have been put forward in recent years: (1) the non-ice components are extracted by erosion, (2) the darker material sinks into the crust because it absorbs more solar energy than its brighter neighbouring ice or (3) the material becomes 'masked' by a deposit of frosts.

LACK OF CRATERS ON EUROPA

Although the number of craters on Europa is lower than that on Ganymede or Callisto, Galileo's high-resolution images revealed more of them than the Voyagers, allowing tentative estimates to be made about the surface's age. The 30-km-wide Manann'an crater, named, appropriately, after an ancient Irish sea god, was caused by an impact which punched through Europa's icy crust and sprayed subsurface material several hundred kilometres across the terrain. The 26-km-wide crater Pwyll, whose existence was first hinted by Voyager pictures in 1979, is surrounded by bright 'rays' of ejecta running more than 1,000 km from its main basin.

Both impacts appear to have been sufficiently violent to excavate dark subsurface material and spread it out like a 'skirt' around their rims. Significantly, the Pwyll impact appears to have occurred on a particularly thin patch of crust; Galileo images showed that it lay at the same level as the surrounding terrain, suggesting that it immediately filled with slushy material. It also displays large iceberg-like chunks of material that stick up from its floor. If Pwyll *did* form recently, this offers persuasive evidence that liquid water or an ice-water slush underlies parts of the crust.

Furthermore, several blotches, nicknamed 'maculae' (Latin for 'dark spots'), were identified by the Voyagers and although their precise nature is still unknown, they are probably the result of upwellings rather than impacts. Two maculae, called Thera and Thrace, were first suspected to be similar to the chaotic terrain in which icebergs were trapped in a very dark fluid matrix.

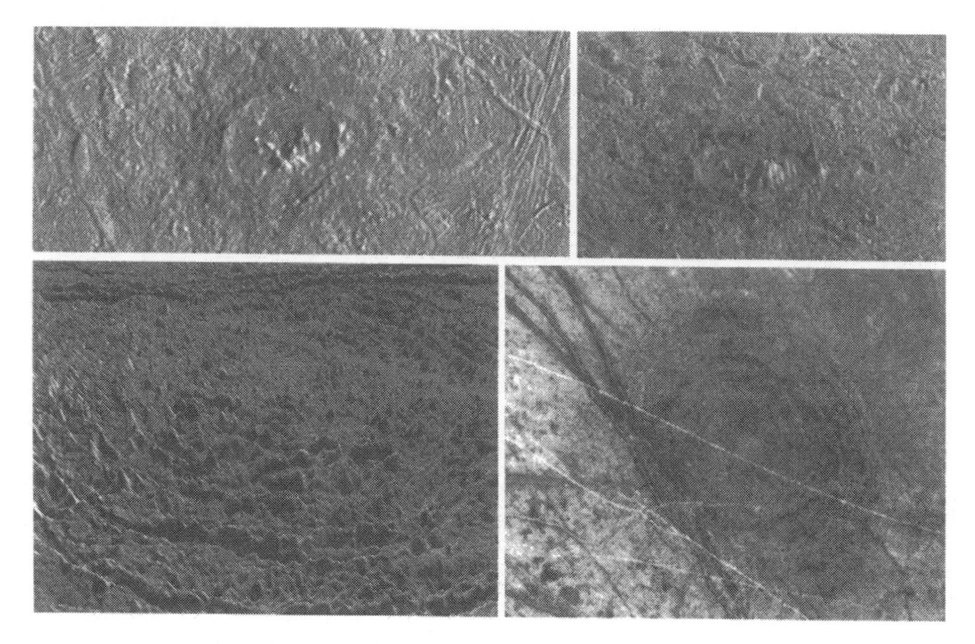

Top: Three-dimensional model of the structure of the 27-km-diameter crater Pwyll showing the rim and central peak complex, based on stereo and topographic analyses of Galileo images by Paul Schenk at Houston's Lunar and Planetary Institute, released in May 2002. Vertical exaggeration is roughly 50 times normal. *Bottom*: The 41-km-diameter multi-ring impact feature Tyre on Europa, as photographed by Galileo in April 1997. The perspective view shows concentric ridges and fractures. The rim of the actual crater, which has collapsed leaving a smooth plain surrounded by these concentric fractures, is located just inside the innermost concentric ring. Vertical exaggeration is roughly 50 times normal. Note the concentric rings are superimposed on older linear features and are transected by more recent features.

Galileo found that Thera, which measures about 80 km across – a little smaller than Thrace – lies at a slightly lower level than its neighbouring terrain. Curved fractures around its periphery suggest that its formation may have involved some form of collapse. Thrace, on the other hand, appears to stand at or slightly above the surrounding plain. Recent observations show that it has been flooded by an upwelling of salty fluid.

Two other features, originally thought to be maculae but now reclassified as 'relaxed' craters, are Callanish and Tyre. The former is 100 km across and includes a series of concentric fractures, including several trough-like grabens. Tyre is even bigger at 200 km wide, and many of its concentric rings are superimposed on sets of older, linear markings. Moreover, Tyre has itself been overprinted with triple bands and several narrow bright bands a few hundred metres wide.

Clearly, Tyre had been tectonically modified both before *and* after the feature was fully formed. In general, however, Europa's craters and maculae are difficult indicators of its surface age. A simple extrapolation of the number of craters found per unit area on our Moon to a similar-sized unit area on Europa is unacceptable, due to insufficient reliable evidence about impacting bodies in the outer Solar System. Nevertheless, when Comet Shoemaker–Levy 9 hit Jupiter in July 1994, it allowed dynamicists to estimate the current rate of collisions with the Jovian system and extrapolate backwards in time.

This pointed to frequent impacts and suggested that Europa's surface is only a few million years old. Another approach, however, generates a much older Europa: 3 billion years older, in fact! This uses the cratering of battle-scarred Ganymede and Callisto. It assumes that their largest extant basins are a remnant of the final stage of heavy bombardment in the Solar System's infancy, when our Moon's youngest basins formed. Ganymede's youngest surface can then be matched with our Moon to establish an impact rate over time, then adjusted because Europa's close proximity to Jupiter makes it more susceptible to impacts.

The age debate continues to rage, thus far unresolved. "There are too many unknowns," says Michael Carr of the US Geological Survey. "Europa's relatively smooth regions are most likely caused by a different cratering rate for Jupiter and Earth. We believe that both Earth's Moon and Ganymede have huge craters that are 3.8 billion years old. But when we compare the number of smaller craters superimposed on these large ones, Ganymede has far fewer than Earth's Moon. This means the cratering rate at Jupiter is less than in the Earth–Moon system."

THICKNESS OF EUROPA'S CRUST

Computer models of impact craters by two planetary scientists from the University of Arizona's Lunar and Planetary Laboratory did, however, succeed in providing a lower limit for the thickness of Europa's icy crust. Elizabeth Turtle and Elisabetta Pierazzo reported in *Science* in November 2001 that their efforts to simulate impacts powerful enough to produce the central peaks seen in craters like Pwyll led them to believe that the crust was at least 3–4 km thick.

"We have geological evidence from Earth and the Moon that shows the material that collapses up into the central peak is material that was previously buried, but has been uplifted and broken up," says Turtle. "Central peaks are deep bedrock that has been uplifted. What we're seeing on Europa appear to be standard central peaks. Since central peaks are deep material that's been uplifted, that means these impacts could not have penetrated through Europan ice to water. Water would not have been able to form and maintain a central peak."

Since Gerard Kuiper and others showed that Europa's crust has a predominantly water-ice composition, the evidence for a possible subsurface ocean has grown steadily more substantial. One of the most important recent indicators of the existence of a salty, conducting ocean came in December 1996, when Galileo detected hints of a magnetic field around the moon. Each time the spacecraft flew past Europa, it was possible to divulge information about its gravity field and internal geophysical structure by very precisely tracking changes in Galileo's position.

Data returned by the spacecraft on this particular occasion returned what

Margaret Kivelson has called "a substantial magnetic signature". Based on these results, Kivelson and her colleagues concluded that Europa has a magnetic field a quarter as strong as Ganymede's, leading geophysicists to wonder how it is generated. One theory is that the outermost water-ice shell, 100–200 km thick – the lower part of which is fluid – surrounds a core of metal and rock, but the favourite envisages a three-layered Europa with a 1,250-km metallic core encased firstly in a rocky mantle and then a 150-km-thick outer crust.

According to this theory, which received support from Galileo data gathered in December 1996 and February 1997, Europa was entirely molten at some stage early in its evolution. During this process, dense metallic elements sank inward to form its core and lighter compounds floated to the surface, leaving intermediate-density materials in the middle; a process called 'thermal differentiation'. These lighter compounds consisted mostly of water, which formed a global ocean whose exterior froze into the icy crust we see today.

DIFFERENT TERRAIN TYPES

Clearly, Europa is an important member of the Sun's family. Its physical characteristics, says Ronald Greeley, mark it out as a 'transitional' world between the rocky moons of the inner Solar System and their predominantly ice-rich counterparts in the outermost portion of the Sun's empire. The Voyagers saw three main surface types on Europa – dubbed 'bright' plains, 'mottled' terrain and 'disrupted' regions – which have since been subdivided further by Galileo observations.

Each of these three surface types has shown itself to be overprinted with criss-crossing ridges and linear markings, some of which stretch thousands of kilometres across the Europan landscape. During one of its two flybys of Earth, *en route* to Jupiter, Galileo had calibrated its imaging equipment by mapping deposits of ice in Antarctica. When it reached Europa and turned its multi-spectral gaze to a far more alien ice field, it revealed significant differences between the two, perhaps a result of variations in ice grain size or composition.

Of the three surface types, the bright plains predominate and provide a source from which the others are derived. Galileo's high-resolution pictures have shown that in some regions the plains consist of multiple sets of ridges and grooves as small as just a few tens of metres wide. In conditions of low-angle sunlight, the spacecraft was also able to photograph enormous ice plateaux up to 10 km wide and many tens of metres tall. These, together with numerous pits and depressions dotted here and there, suggest that the plains formed when the icy crust was repeatedly deformed by tectonic processes.

Mottled terrain, as its name implies, includes an irregular patchwork of dark zones which grade haphazardly into the bright plains. Current thinking is that these peculiar patches – which range in size from 50 to 500 km across – form in response to tectonic stresses. One region in particular, known as Conamara Chaos, consists of bright plains that have apparently been broken up into city-block-sized 'icebergs'

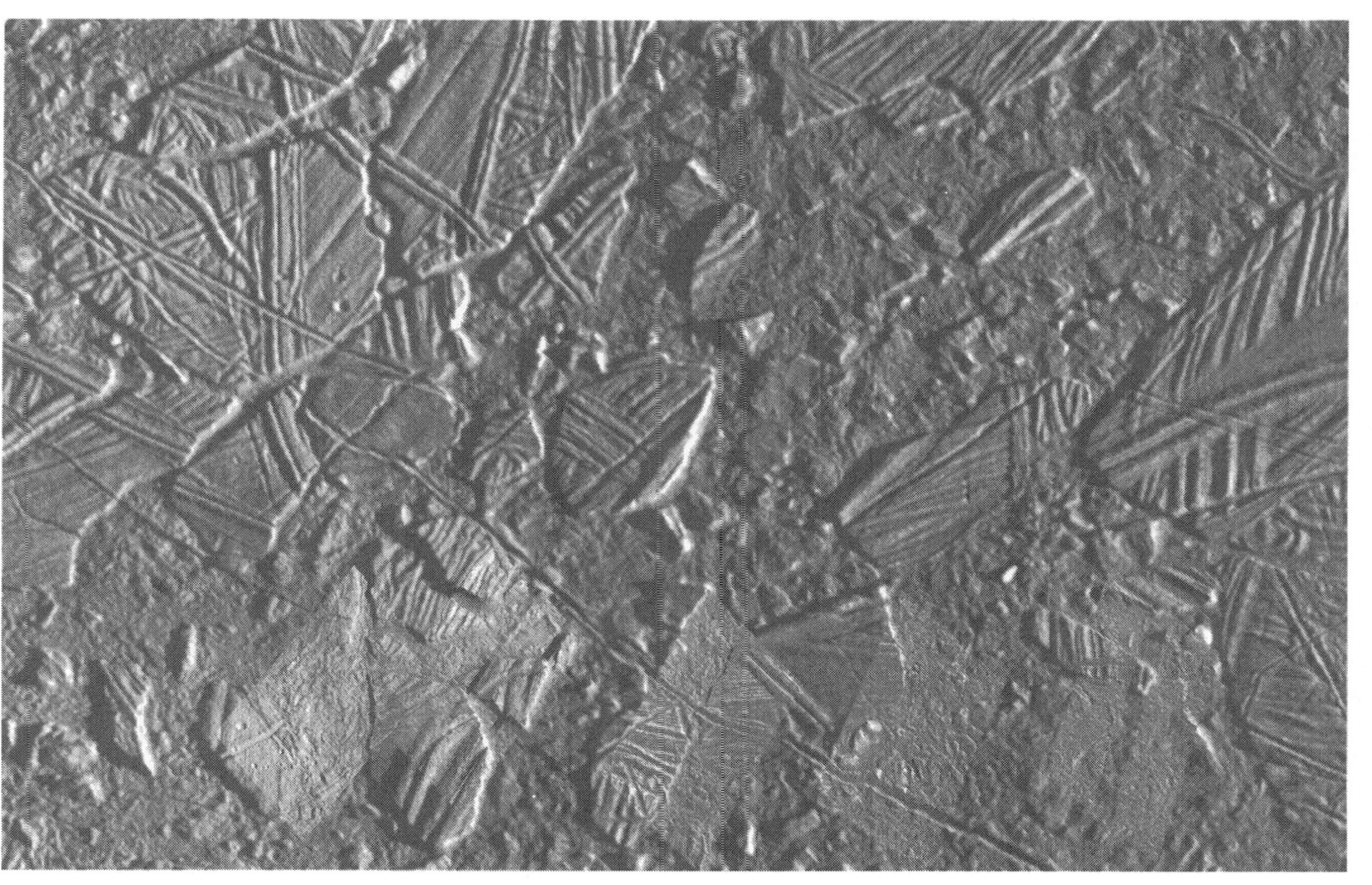

The Conamara Chaos region of Europa as recorded by the Galileo spacecraft. Several high-resolution frames have been superimposed on the main image.

poking out of the chaotic terrain. These icebergs are just a few kilometres across and stand 25–200 m above the tortured landscape. Many of them appear to be tilted, as if partially submerged into the chaotic terrain.

This low-lying chaotic material is visibly darker than the fragmented surface, pointing to a frozen salty fluid composition. Globally, these observations suggest that the Europan crust features an ice-laden exterior which covers a large fraction of some other, darker material. If Conamara Chaos is representative of all mottled terrain on Europa, it seems that they form at the expense of the bright plains. Some other surface features close to Conamara Chaos have actually yielded tantalising clues about the physical processes responsible for disrupting the icy crust.

It seems from these regions, which display gentle, up-bowed surfaces with central fractures and domes surrounded by moat-like depressions, that the crust has been pushed upwards by some force exerted from below. Some geophysicists have argued that this force may be attributable to buoyant, low-density masses (called 'diapirs') rising towards the surface, convective upwelling from deep inside Europa or even slushy intrusions of water-ice deposited by cryovolcanism. The latter is a topic of some debate: although Galileo images have shown no actively erupting ice volcanoes or geysers, they have revealed surface flows probably created by them.

"This is the first time we've seen actual ice flows on any of the moons of Jupiter," said Ronald Greeley after the release of Galileo's initial pictures in December 1996. "These flows, as well as the dark scarring on some of Europa's cracks and ridges, appear to be remnants of ice volcanoes or geysers." When these observations and speculations are taken into account, it seems likely that the mottled areas are parts of crust that have been modified by localised disruption. As this disruption occurs, it exposes a subsurface layer containing salts, clays or other dark materials.

Whether or not this is a local phenomenon or a global one is unknown and will presumably have to await future visits to be resolved. It is clear, however, that there are significant differences between the hemisphere that faces Jupiter and that which faces away from it. Voyager images showed that the latter is characterised by enormous, wedge-shaped dark bands, each more than 100 km long. These regions appear to represent places where the crust was stretched apart and, indeed, some of them are quite similar in structure to sea-floor spreading centres on Earth.

Many geophysicists think the wedges of Europa are the result of repeated stretching and ripping-apart of the crust, after which the opened fractures were filled with darker subsurface material. Such observations reinforce the argument made about the mottled terrain: that there is a darker material lurking just beneath the crust. Furthermore, Galileo maps of these patches of surface have shown that, although these wedges can be fitted back together like a jigsaw puzzle, there is still about 8 per cent of the original surface 'missing'. This suggests that parts of the original crust have been consumed by the separation process.

EUROPA'S TENUOUS ATMOSPHERE

Not only does Europa have an interesting, enigmatic surface, but HST images reported in *Nature* in February 1995 also revealed traces of a tenuous molecular oxygen atmosphere. This marked it out as only the fourth world in our Solar System, besides Venus, Mars and Earth, to possess traces of molecular oxygen in its atmosphere. Yet this gaseous veil is incredibly thin. "Europa's oxygen atmosphere is so tenuous that its surface pressure is barely one hundred billionth that of the Earth," said Doyle Hall of Johns Hopkins University in Baltimore, Maryland.

HST's observations were later confirmed by Galileo in 1996. As the icy surface is exposed to fierce sunlight and hit by dust and charged particles from Jupiter's magnetosphere, the frozen surface water-ice produces water vapour and gaseous water molecules. These then undergo a series of chemical reactions to form molecular hydrogen and oxygen; although the lighter hydrogen escapes into space, the sluggish oxygen lingers to form an atmosphere that may persist up to 200 km above Europa's surface. Nevertheless, because it 'leaks' away, this 'atmosphere' is better described as an 'exosphere'.

Much of our attention to Jupiter's moons thus far has centred on Io and Europa – two of the most visually stunning, perplexing worlds in our Solar System. Yet Jupiter has a vast number of moons – one larger than the planet Mercury, others so small that new ones are found every year as the resolution of Earth-based instruments improves. Eleven were found in 2002 alone. In fact, all four gas giants are adding so many 'new' moons to their personal tallies that the author can only say that, at present, Jupiter has more than five dozen attendants. Undoubtedly, this will change.

A PLANET-SIZED MOON

It seems fitting, however, that the King of the Planets should hold the King of the Moons captive in his retinue. Measurements from both Voyagers confirmed that, contrary to previous estimates, which bestowed the honour on Saturn's large moon Titan, Ganymede is actually the biggest planetary satellite in the Solar System. Previously, ground-based spectroscopic observations of Titan had, through necessity, been forced to include its substantial atmosphere when calculating its diameter, and thus produced an inaccurate result. By employing their on-board radios during occultations, the Voyagers pegged Ganymede's diameter at 5,276 km; Titan is a little smaller at 5,150 km.

In fact, as already noted, Ganymede is bigger than Mercury, although only half as massive. Present scientific models and gravity data acquired during Galileo's visits to Ganymede suggest that it has an iron core – part of which may still be molten – about the size of our Moon. This might then be surrounded by a thick covering of ice roughly equal in mass.

BRIGHT AND DARK TERRAIN

The Voyagers showed Ganymede to possess two distinct types of terrain: dark, heavily-cratered regions and brighter patches, overlaid with mysterious networks of ridges and furrows. This suggested that its entire icy crust had been under significant tension from violent tectonic activity on a pretty much global scale. Ganymede's strange, two-tone 'patchwork' had also been hinted at in Pioneer images taken more than five years earlier. Clearly, crater densities in the dark regions suggest that they are much older than their brighter counterparts, although both are ancient. Overall, the cratering points to an approximate surface age of almost 4 billion years.

If such estimates are correct, Ganymede's dark terrain at least displays one of the oldest surfaces in the Solar System. Most planetary scientists think that the dark terrain contains a higher percentage of 'rocky' material than the presumably more ice-rich brighter regions. However, there are subtle variations in the density of craters found in different patches of dark terrain, implying that some portions may have been reshaped by subsequent events. Arcing across vast expanses of the dark terrain, too, are long furrows, each up to 10 km wide, which trace out enormous concentric circles, like bull's eyes.

Planetary scientists Bill McKinnon, of Washington University in St. Louis, and Jay Melosh, of the University of Arizona, have suggested that these furrows are the consequence of violent impacts sustained in Ganymede's infancy. They think that, towards the end of the moon's accretion period, a mobile layer of warm ice lay just a few kilometres beneath the crust. The impacts punched straight through the crust, causing the slushy ice to seep inward and fill the newly excavated crater basin. During this tumultuous phase, giant slabs of cold surface material were dragged towards the centre, breaking along faults to create the furrows.

The Voyagers noticed that this dark layer is occasionally punctuated by bright, low-relief splotches, nicknamed 'palimpsests' after reused bits of medieval parchment in which earlier writing is incompletely erased. Like old text still partially visible on parchment, hints of Ganymede's ancient furrows are sometimes discernible beneath the palimpsests. These features were probably also created by the after-effects of asteroid or cometary impacts: after punching through the thin crust, they spewed out slushy ice that partially covered the older furrows. The surface relief has relaxed to the extent that these features have become bright, level plains.

When Galileo sped past Ganymede in June 1996, its more advanced imaging equipment showed the dark terrain – particularly an area known as Galileo Regio – in stark relief. The spacecraft's stereo pictures enabled scientists to create three-dimensional maps of the surface, which revealed the tall furrow rims, crater rims and isolated hills to be 'bright', while low-lying regions such as craters and furrow floors are considerably darker. Streaks of dark material were also seen to have spilled down the slopes of the furrows into valleys and surface hollows.

Although Jupiter and Ganymede reside 5 AU – three-quarters of a billion kilometres – from the Sun, this is nevertheless close enough for ice to sublimate into space. This means that slopes of dark terrain that are tilted slightly towards the Sun tend to rapidly lose their ice cover, leaving a thin layer of dark, rocky material. As a

result, because 'darker' objects absorb more heat than 'brighter' ones, the sublimation-darkened areas become even warmer, speeding the process along. Brighter, and colder, slopes, however, retain their ice, which helps to explain the segregation of dark terrain into a checkerboard of bright and dark splotches.

Until Galileo's arrival, it was speculated that the dark terrain was probably composed primarily of rocky materials. Data gathered since 1996 by the spacecraft's NIMS instrument has revealed it to possess a clay-rich veneer, perhaps only a few metres deep, overlying an icier substrate. Downslope movement of material and the sublimation of ice to space, mentioned above, helped to 'concentrate' this dark outer layer, which was stirred into the interior by asteroid or cometary impacts. The latter were probably also responsible for excavating 'clean' ice from several kilometres inside Ganymede to form the palimpsests.

The boundaries between dark and bright terrain types can be quite abrupt and remarkably straight. Galileo's infrared data showed that the bright regions are predominantly composed of water-ice. These are criss-crossed by sets of parallel ridges and valleys, nicknamed 'grooved terrain', tens of kilometres long and hundreds wide. Judging from the way these 'lanes' cross-cut one another, geologists suspect they formed over a long period of time. The grooves themselves seem to have been shaped by tectonic stretching of Ganymede's surface, which broke the brittle crust along inward-sloping faults to leave the long, sunken grabens that we see today.

This kind of faulting may also cause mountain-forming slabs of crust to rotate and tilt, similar to a stack of books that topples over as soon as its bookends are removed. Such tectonic phenomena is quite different from that seen on Europa, where long sections of crust seem to have been yanked completely apart, with icy subsurface material seeping up in between them. In fact, Ganymede shows little evidence that older craters or furrows were sliced and entirely separated across lanes of grooved terrain. Rather, it seems the moon's 'stretched' dark terrain is somehow converted into bright terrain!

CRYOVOLCANISM?

In order to understand how this happens, geologists have firstly attempted to explain the differing brightness of the grooved and smooth terrain. The most likely cause of this brightness is cryovolcanic activity, which covered the ancient dark surface with an icy veneer. Such a theory is supported by the presence of craters in bright regions, some of which exhibit dark halos around them; perhaps the result of impacts which excavated ancient rocky material from beneath the present surface. It is quite possible that this cryovolcanic activity is caused by the extension and faulting processes that produce Ganymede's ridges and grooves.

If this was the case, it would not be dissimilar from the way faulting and volcanism occurs together on Earth, within rift zones in eastern Africa. A key problem, however, is how the cryovolcanic material would reach Ganymede's surface. A watery magma would be denser than the surrounding ice and should conceivably sink rather than rise.

However, dissolved salts and subsurface gases might be sufficient to drive water upward, allowing it to flow onto the moon's surface. On the other hand, warm, buoyant ice could simply rise to the surface and 'erupt' across the terrain.

The Voyager pictures of Ganymede's surface led many scientists to expect to find evidence of icy volcanism when Galileo, with its higher-resolution battery of imaging gear, arrived towards the end of 1995. Certainly, Galileo revealed much more intricate detail of surface faulting, particularly in the Uruk Sulcus region, but saw very few 'smooth' areas. If the cryovolcanism idea is valid, it might be supposed that the Uruk region at least was initially brightened by icy outpourings, then torn beyond recognition by massive faulting.

There are also a number of plank-like patches in other regions, such as the border of Nippur Sulcus, which seem to have been smoothed by cryovolcanic activity. Other regions of Ganymede display strange 'scalloped' depressions – icy calderas – that may be local sources for fluid eruptions. Still, the relative paucity of identifiable cryovolcanic features suggests that faulting is the key action in shaping the grooved terrain. Furthermore, the sheer ubiquity of this grooved terrain – and the assumption that it is caused by crustal stretching – leads many geophysicists to suspect that the entire moon expanded in its adolescence.

Cryovolcanic activity during this period of Ganymede's youth points to an interior warmed to near the melting point of ice. The concentration of instances of surface stretching, volcanism and 'strike-slip' faulting such as on Tiamat Sulcus – where blocks of crust lurch past one another, like along the San Andreas fault-line in California – strongly suggests that buoyant slabs of warm material rose through the moon's cold interior towards the surface. The fact that the grooved terrain apparently post-dates the dark surface and most of the palimpsests implies that this fierce activity took place after Ganymede had fully accreted and its brittle crust thickened.

CRATERS ON GANYMEDE

The large concentration of craters, particularly in the expanses of dark terrain, has already been mentioned. These scars range in size from small, bowl-shaped depressions to enormous basins and palimpsests. Some craters are encircled by low 'pedestals', probably sheets of icy debris that was splashed out by impacts. Others, such as En-zu in the Uruk Sulcus region, are surrounded by dark rays, perhaps caused by the remains of an ancient, dark projectile that splattered across the surface. Still others have dark floors, where rocky dust and debris may have collected over time.

Younger, fresher craters, by contrast, display extremely bright rays running many hundreds of kilometres from their impact centres. These are particularly striking in 'global' views of Ganymede taken by both the Voyagers and Galileo. In time, of course, the brightness of all rays fades as their icy material mixes into the surface through meteoroid and charged-particle bombardment.

Many of the mid-size and large craters – those with diameters between 20 and 30

The enormous Gilgamesh impact basin on Ganymede, with surrounding ejecta, compiled from a mosaic of Voyager images.

km – also possess peculiar central 'pits'. Past arguments have suggested that these pits were formed by the initial crater-forming impact and then 'frozen' permanently in place. Others have countered that they may have been created by the impact of unusually slow-moving meteoroids. Another strange feature, exclusive to craters more than 60 km across, is a broad central dome.

Some planetary scientists, including Paul Schenk, have speculated that these were produced by deeply-penetrating impacts that threw subsurface ice into the crater's centre, where it rapidly cooled. These 'pit-and-dome' features tend to be quite shallow with respect to their diameters and their floors are often bowed upwards. It is possible that the initial impact warmed the surface so much that the crater floor 'rebounded' slowly, like a baking cake that bounces back when prodded with a finger.

The largest and youngest of Ganymede's impact basins is a monster named Gilgamesh, after the ancient Mesopotamian hero whose unsuccessful search for the secret of everlasting life spawned the original story of the biblical Flood. It has a 150-km-wide central depression, which forms a smooth plain surrounded by hummocky ejecta and disrupted terrain. The outermost ring, marked by a scarp, has a radius of 275 km. Overall, however, Gilgamesh's ring structure is quite subdued.

Although the crater's own age is unknown, it has clearly achieved greater longevity than its ill-fated mythological namesake. Whatever the age of Ganymede's surface, whether ancient or relatively geologically 'young', it nevertheless remains an intriguing world that warrants further examination.

A THIN ATMOSPHERE

Although, as mentioned earlier in this chapter, ice does sublimate into space, the overall process is much too slow to create a substantial atmospheric shield. Early in its mission, however, Galileo revealed that Ganymede does possess a magnetosphere, which forms a 'bubble' inside Jupiter's own magnetosphere that helps to protect the moon from high-energy charged particle impacts. Nevertheless, ultraviolet sunlight and this impacted energy is enough to tear water molecules apart, readily allowing the lighter hydrogen to escape into space, leaving behind oxygen atoms. Recently, Earth-based observers John Spencer of Lowell Observatory and Wendy Calvin from the US Geological Survey identified the spectral signature of molecular oxygen and HST has found ozone in a thin gaseous veneer surrounding Ganymede.

Attempts to determine the structure of the atmosphere have also been aided substantially by Galileo's observations of Ganymede's glowing aurorae since December 2000. Galileo has also found hydrogen, but did not see oxygen, suggesting that the bulk of this stays and accumulates on the moon's icy surface. Indeed, Ganymede does possess a thin dusting of frost near its poles, probably composed of water-ice tinged with frozen carbon dioxide, and in some regions this may be several metres deep. Some of this frost may, in fact, have migrated poleward after sublimation from the warmer equatorial regions.

OCEANS ON GANYMEDE AND CALLISTO TOO?

One of the more important factors in the formation of Ganymede's polar caps might be the bombardment of the surface by energetic particles. Galileo's discovery of the moon's intrinsic magnetosphere implies that charged particles coursing through Jupiter's own magnetosphere are funnelled onto Ganymede's polar regions. Their impacts would splatter and redistribute water-ice molecules and 'sputter' liberated atoms out of the surficial layer.

Ganymede is the only Jovian moon to have a 'dipole' magnetic field of its own. This intrinsic field undoubtedly derives from its iron core, making it not dissimilar to the Earth in this respect. It is this field which forms the moon's magnetosphere. In addition, Galileo found that the Jovian magnetosphere induces a 'secondary' field in Ganymede in the same manner as it does in Europa, which, according to Margaret Kivelson, is "highly suggestive" that it too has a subsurface ocean. "It would need to be something more electrically conducive than solid ice." She thinks a melted layer several kilometres thick, beginning about 200 km beneath the surface, might be as salty as Earth's oceans.

However, according to geophysicist Thomas McCord of the University of Hawaii at Honolulu, the Galileo infrared data acquired throughout the late 1990s and into this century does not indicate whether or not an ocean persists on Ganymede today. Natural radioactive decay of elements in the moon's rocky interior should provide sufficient heat to maintain a stable layer of liquid water between two icy layers approximately 150–200 km beneath the surface. This marks a significant difference from the hypothetical Europan ocean, which is most likely caused by internal flexing from the tidal effects of Jupiter's enormous gravity.

Before 1995, it was generally thought that both Ganymede and Callisto consisted of predominantly rocky cores, encapsulated in large mantles of water or water-ice and topped with icy crusts. Galileo results have since suggested that Callisto has a more uniform internal composition, while Ganymede is a three-layered structure with a rocky iron core surrounded by a rocky silicate mantle and an outer icy shell.

Callisto is another Jovian moon now thought to possess an ocean of some sort. It is a world of mystery regarded by some as having one of the most ancient surfaces in the Solar System. In fact, many planetary scientists have called Callisto and Ganymede a pair of siblings separated at birth that went on to lead very different lives. They shared similar infancies, forming in close proximity within the enormous mass of gas and dust that yielded the Jovian family, and were steadily showered over countless millennia by rogue impacts that increased their size and mass.

SIBLINGS WITH VERY DIFFERENT LIFE STORIES?

Far from the Sun, both moons were probably capable of accumulating and retaining large stocks of rock and ice during their formative years. Estimates have been put forward to suggest that both moons had a mixture of about 60 per cent rock and 40 per cent ice in their youth. Earth-based measurements had already revealed the infrared signature of frozen water on their surfaces, but it was not until the arrival of Galileo that the true differences between the pair were identified. Explanations for these differences, however, are still scarce.

In late 1997, ultraviolet data from Galileo recorded hydrogen atoms escaping from Callisto, implying that, like Ganymede, it too has oxygen locked up in its ice and rocks. Unlike Ganymede, however, it seems that sunlight striking Callisto's

Callisto.

rock-hard surface ice is the principal mechanism for separating the hydrogen and oxygen atoms.

"Because it is further away from Jupiter, Callisto does not interact as strongly as Ganymede with the charged particles in the planet's magnetosphere," said planetary scientist Charles Barth of the University of Colorado at Boulder. "Instead, we believe it is the ultraviolet solar radiation that is knocking the hydrogen atoms out of the ice on Callisto." Present measurements suggest that Ganymede's surface contains about 50 per cent ice, compared to just 20 per cent in the case of Callisto.

In spite of these similarities, scientists remain divided over why two sibling moons, so much akin in terms of origin, size, density and composition, should have taken such diverse evolutionary paths. Certainly, the gravity-field data from Galileo during its encounters with both moons points to changes in their internal structures. Whereas Ganymede was once warm enough for its icy component to separate fully from its rock and metal and, perhaps, turn the interior into an enormous melting pot, Callisto could not have melted throughout and was never hot enough to form a metallic core.

Galileo's findings suggest that because Callisto is the furthest-out of the four large moons, it was never subjected to the same gravitational stresses as its innermost cousins and therefore never experienced enough heating to form different internal layers. "Callisto had a much more sedate, predictable and peaceful history than the other Galilean moons," said planetary scientist John Anderson of JPL. It seems probable that Callisto has no core, but has a homogeneous structure, consisting of 60 per cent 'rock' and 40 per cent compressed ice.

Although there may well have been some shallow melting and a slow sinking of rocky chunks through warm ice, Callisto's internal rock and ice were never completely separated. Instead, the moon's interior may grade slowly from an icier surface to a rockier core and, when compared with Ganymede, it had a much cooler and less tumultuous history. Another difference is that Callisto has no intrinsic magnetic field, although magnetic disturbances have been noted in its vicinity. Presumably, like Io, Callisto has a magnetic field induced by Jupiter's own enormous magnetosphere in a global, subsurface ocean.

If this is true, Io may be the only one of the four Galilean moons not to possess traces of a subsurface ocean, because its volatiles long since boiled off. Certainly, Galileo data implies that Callisto may have a global, salty layer of water with a depth of 10–100 km. "But it's not a smoking gun," says Torrence Johnson, "and a lot more evidence needs to be uncovered before we will know for sure whether Callisto has a subsurface ocean." In view of its incompletely differentiated interior, however, the mechanism responsible for creating and sustaining such an ocean is unknown.

Despite the exciting possibilities of microbial life on Europa, the likelihood of anything similar on Callisto is very low. If an ocean does exist there, it probably resides much further below the surface and would be trapped between two layers of ice, rather than sitting on top of a 'warm' rocky layer, as most theoretical models suggest for Europa.

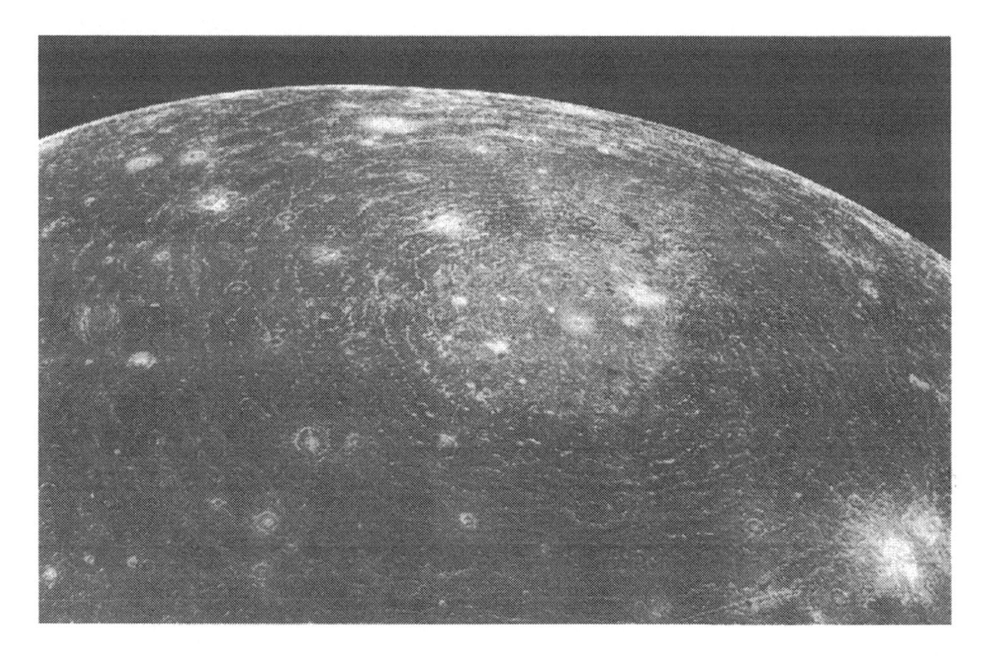

The enormous, multi-ringed Valhalla impact basin. Notice the linear chain of small craters (lower right) where the fragments of a comet rained down.

AN ANCIENT, CRATER-SATURATED WORLD

Models of a weakly differentiated Callisto seem to be reinforced by its stark, debris-laden surface. It has often been nicknamed the 'ugly duckling' of the four large Galileans. Its battered surface of ice and rock makes it the most heavily cratered moon in the Solar System, indicating it is geologically 'dead'. "The craters are the visible record of what sizes of comets and other objects have pelted Jupiter and its moons with what frequency over the past 4 billion years," says Johnson.

Its largest impact crater is Valhalla, thought to be several billion years old. It consists of a smooth central plain, 600 km across, and a series of rings spaced 20–100 km apart, which extend to a radius of 2,000 km. The bright central part of this ancient feature is roughly comparable to Colorado in terms of size, while its enormous bull's eye of concentric rings dominates an entire hemisphere and could span the entire contiguous United States. It can perhaps best be compared to the vast, multi-ringed Orientale basin on the far side of the Moon.

Most other, younger craters on Callisto are bright or dark spots, whose shape gradually changes into more diffuse splotches over time. Like Ganymede's palimpsests, the bright central part of Valhalla and other impact features elsewhere could indicate the intrusion of 'clean' ice excavated from beneath the dirtier surface layers. The appearance of Valhalla's rings also changes with increasing radial distance from the crater and some bear a close resemblance to the sunken grabens on

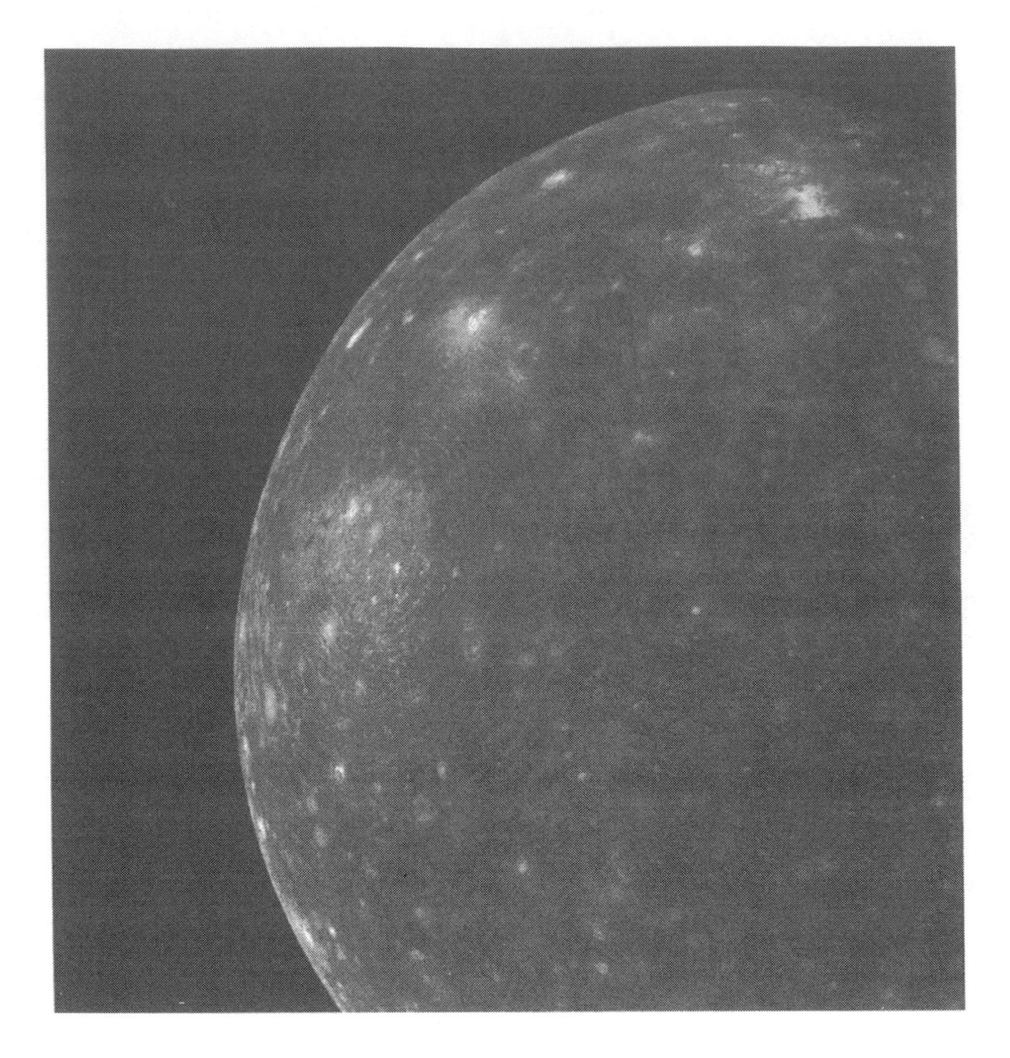

Vast tracts of dark, ancient, heavily cratered terrain cover Callisto. The prominent multi-ringed feature is Valhalla. (see colour section)

Ganymede. The formation of Valhalla itself, for example, probably triggered massive faults as the crust was dragged inward on top of a layer of warmer ductile ice.

There is no clear evidence that Callisto experienced the kind of volcanic or tectonic activity that shaped the faces of the other three Galileans. Although patches of material along the bases of Valhalla's scarps, photographed by both Voyagers during the course of 1979, hinted that sporadic episodes of cryovolcanism might have taken place, subsequent investigations by Galileo found no conclusive evidence; just blankets of smooth, dark material. This blanket is seen almost everywhere, punctuated only where bright crater rims poke through it.

Pictures from both Voyagers revealed Callisto's surface to have vast tracts of dark, heavily cratered terrain. On the eve of Galileo's arrival, most planetary scientists confidently expected that the spacecraft's improved resolution would reveal a myriad of small impact craters. However, very few 'small' craters were identified and those that were present were found to range from fresh to quite subdued. It seems from these findings that such craters are somehow 'filled in' and degraded over time. One possible explanation for this is that ejecta hurled out by impacts can soften or bury topographic relief, as on our Moon.

Not only is Callisto's surface more degraded than that of our Moon, but its composition is also very different. It seems that resurfacing by impact ejecta is more efficient on the ice-rich Callisto and the bright rims and central peaks of the larger craters provide clues to the bright ice lying under the dark surface. One possibility is that sublimation gradually robbed exposed crater rims of their ice, leaving dark rocky surface deposits. Numerous large pits may be remnants of the surface which collapsed after being severely undermined by the gradual loss of ice cover.

Some of the ice extracted through sublimation must escape into space, but the rest should be redeposited in bright, cold patches, especially at higher latitudes. This may explain the isolated patches of very bright material in several of Callisto's craters. Downslope movements can also redistribute dark rocky material across the surface and Galileo has observed an unusual chain of craters, each with a small, dark 'talus' at its base, which seems to have come from the erosion of the crater walls.

Infrared data from Galileo reveals that Callisto's dark terrain seems to consist of widespread hydrated minerals (clays) and also concentrations of sulphur dioxide, particularly in the hemisphere that faces forward as the moon orbits Jupiter. It seems likely that this sulphur dioxide is leaking from Io's torus and is being swept up by the outer moons, including Europa. As a result, some dynamicists have argued that the dark terrain may be related to micrometeorite impacts, which preferentially hit a moon's leading side. Clearly, Callisto's ancient surface is a unique textbook of a broad period of Jovian history, waiting to be read.

ONWARD TO SATURN

The Voyagers, of course, each had only a few weeks, with all 10 of their instruments working feverishly, to cram in as much information as possible before pressing on to Saturn. It was left to the astronomers, planetary scientists, geophysicists, geologists and atmospheric physicists back on Earth to sort through the reams of data and make sense of the jumbled story of Jupiter and its realm. It is a story that, even at the start of the twenty-first century and with the important insights offered by Galileo, Cassini and HST, continues to astound and perplex us.

One thing that was known with relative certainty in 1979, however, was that Jupiter's gravity would provide more than enough impetus to hurl both Voyagers more than three-quarters of a billion kilometres across space, towards Saturn, increasing their speed by over 50,000 km/h in the process and slinging them on different trajectories. Voyager 1 flew close to Jupiter to gain a slingshot for a 'fast' trajectory so that it could reach Saturn, inspect Titan and chart that planet's magnificent, frost-encrusted rings from close range. Voyager 2, on the other hand, kept its distance from Jupiter, to pursue a slower trajectory timed to fly past Saturn at a point that would automatically set it up for a Uranus rendezvous early in 1986.

4

Ringworld!

WORTH THE WAIT

Nineteen months after leaving the vicinity of Jupiter, on 12 November 1980 Voyager 1 sped 64,200 km over Saturn's cloud-tops and took more than 16,000 photographs of the Solar System's most visually stunning jewel. But, despite this triumph, a trajectory-correction burn of the spacecraft's attitude-control thrusters just five days earlier gave then-JPL Director Bruce Murray cause for concern. He knew that, in order for the burn to take place, Voyager 1 would have to be rotated on its gyros and its radio link with Earth temporarily broken.

"Isn't it risky to break communications so close to encounter?" he recalls asking at the final planning meeting held at JPL before the spacecraft hurtled past Saturn. Fortunately, the burn was successful; its purpose, said then-Voyager Project Manager Ray Peacock, was to improve the chances of getting a good look at Saturn's giant moon Titan. Atmospheric scientists were keen to develop vertical, temperature and pressure profiles of its thick canopy of gases and particulate haze, and the best way of doing that was to probe it with radio waves from the RSS instrument.

Late on 6 November, the Goldstone DSN antenna transmitted commands for the burn to Voyager 1 and, precisely on cue, the spacecraft obeyed by rotating a quarter of a turn away from Earth. Abruptly, as expected, radio communications were cut off as the high-gain antenna moved away from its lock on the home planet. Shortly afterwards, Voyager 1's thrusters hissed noiselessly for nearly 12 minutes to nudge the spacecraft gently 'sideways' and onto a path to shave past Titan. Next, with pinpoint accuracy, and to sighs of relief at JPL, it automatically realigned the antenna towards Earth, by now 1.6 billion km away.

Titan would be worth the wait. Voyager 1 swept within 3,800 km of it on the morning of 11 November, just a few hours before it reached its closest point to Saturn. It was at this stage that the spacecraft's radio signal performed its own magic: it gradually faded and for 13 minutes vanished entirely, before re-emerging on the other side of Titan, quivering, then returning to normal. This radio 'occultation' procedure helped to determine that the giant moon's atmosphere was much thicker and deeper than had previously been thought, and measured the diameter of the body within that atmosphere for the very first time.

A BLAND, ORANGE WORLD

In some ways, Titan is akin to the planet Venus, in that both have thick, opaque atmospheres. On the other hand, the Saturnian moon is also reminiscent of Mars, because its equatorial tilt causes pronounced seasons and movement of atmospheric gases and particles from one hemisphere to the other. Mars' atmosphere, however, is very thin and its seasons cause carbon dioxide to sublimate in one hemisphere and 'snow' in the other. Furthermore, the planet's atmospheric pressure is subject to changes as high as 30 per cent.

Unlike Mars, Titan's atmosphere is extremely thick, so the effects of seasonal change are much more subtle. Particulate haze in its upper atmosphere seems to be driven from the 'summer' hemisphere to the 'winter' one, which changes Titan's brightness quite dramatically. The presence of these particles, as well as gas, had already been determined from ground-based observations before the Voyagers even left Earth. The two spacecraft were able to discern darker splotches in the northern hemisphere compared to the southern one, although their imaging gear was unable to resolve any direct pictures of a solid surface beneath the murky canopy.

Nevertheless, several distinctive, detached regions of particulate haze were visible over the main opaque layer. They seemed to 'merge' with the main layer somewhere over Titan's north pole to form what scientists initially suspected to be a kind of dark 'hood'. It later turned out to be a very dark brown, circumpolar 'ring' of material. The southern hemisphere was found to be considerably brighter than its northern counterpart, possibly due to seasonal variations. At the time of the Voyager encounters, it was early spring in Titan's northern hemisphere and early autumn in the south.

CHEMICAL COCKTAIL

To both Voyagers, Titan was a featureless orange ball. The dominant atmospheric gas is nitrogen, much like Earth, although in larger relative quantities. There is 10 times more nitrogen on Titan than on Earth, resulting in a surface pressure one and a half times that at terrestrial sea level. This discovery proved at odds with pre-Voyager estimates for pressures at Titan's surface, which swung between 20 millibars at one extreme and 20 bars at the other. Nevertheless, the high-pressure, nitrogen-rich model for Titan's surface, advocated by Donald Hunten of the University of Arizona at Tucson, did prove consistent with observed temperatures and pressures elsewhere in the atmosphere, although the surface pressure is only 1.5 bars.

The only gas identified with certainty from Earth before the Voyagers arrived was methane, which had been detected by Gerard Kuiper in the 1940s and which actually turned to be a relatively minor player in the atmosphere, with an average abundance of just a few percent. It seems that high above Titan's surface, molecules of nitrogen and methane are broken apart constantly by bombardment from the Sun's energetic ultraviolet photons and Saturn's magnetospheric electrons. The resulting fragments of these molecules then recombine to form an impressive line-up of trace constituents, some of which were detectable using the Voyagers' IRIS instruments.

In fact, this thick atmosphere has been partly to blame for a long-held, yet erroneous, assumption that Titan was the largest planetary satellite in the Solar System. Even as late as 1979, when it was finally shown by the Voyagers that Jupiter's moon Ganymede is actually slightly bigger, Titan was placed first on the list. It later turned out that previous measurements of Titan's equatorial diameter had also included the size of its thick atmosphere. The actual equatorial diameter of the moon is 5,150 km, making it smaller than Ganymede, yet still larger than the planet Mercury.

Remarkably, in spite of its present appearance, in some respects Titan may closely resemble what Earth was like billions of years ago, long before life evolved here. Several factors have led atmospheric specialists to draw close parallels between Titan and many of the Solar System's terrestrial planets. As already mentioned, the nitrogen composition and pressure of the Titanian atmosphere are broadly similar to our own, but unlike every other 'substantial' planetary atmosphere in the Sun's realm. No other atmosphere so closely matches ours, in this respect at least.

However, that is where the similarities end. The trace gases in Earth's and Titan's atmospheres are poles apart, largely due to enormous temperature differences. Titan lies more than 1.6 billion km from the Sun, making it an extremely cold, icy world

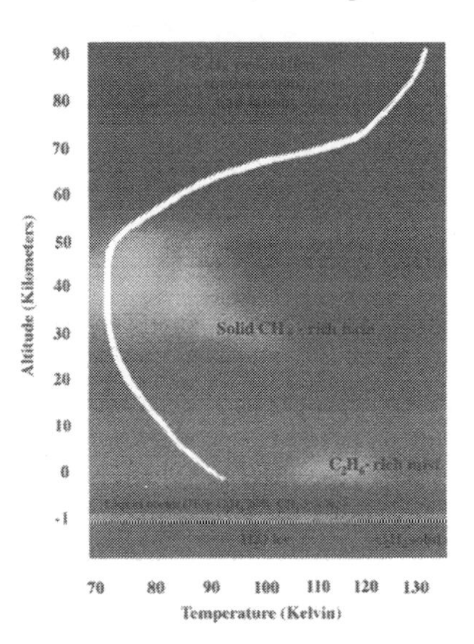

Structure and composition of Titan's atmosphere.

with an average surface temperature of just 92 K. As a result, many terrestrial trace gases are frozen solid, and instead there arises an unusual cocktail of carbon-based chemicals, including ethane, acetylene and carbon monoxide.

The presence of carbon monoxide, together with some carbon dioxide, in Titan's atmosphere has presented a special puzzle. The moon is so cold that water vapour pressures are extremely low and, consequently, water-ice cannot participate to any significant degree in atmospheric photochemistry with methane, hydrocarbons and nitriles. Mars, on the other hand, is sufficiently warm to drive water vapour into the atmosphere and convert methane and its byproducts into carbon dioxide. To explain the presence of carbon dioxide and monoxide in Titan's atmosphere, in view of the paucity of oxygen, there are two possibilities.

One is that the gases are primordial – perhaps trapped in the same ices from which the moon first formed – but were later released when bombardment from Saturn's magnetosphere broke carbon monoxide apart and left oxygen exposed. The creation of carbon dioxide could then ensue if one of these oxygen molecules encountered and reacted with a methane molecule. This would produce a hydroxyl radical, which would only have to combine with another carbon monoxide molecule to produce carbon dioxide. It is also feasible that carbon dioxide arrived inside impacting comets.

TITAN'S EVOLUTION

Some planetary scientists, including Toby Owen of the University of Hawaii, point out that the final composition a moon takes is not only dictated by how far its parent planet resides from the Sun, but also from *where* in relation to that planet it evolved. In this sense, Owen draws parallels between the infant Saturnian system and the primordial solar nebula. He points out that, in the case of Jupiter, moons closer to the parent planet tend to be much denser in composition than those farther away. However, at Saturn, the planet's reduced mass and smaller internal heat source make this effect much more subtle.

This, says Owen, also helps to explain why Saturn's moons, and even its ring particles, are exclusively icy in nature. Nevertheless, many of them still possess, within their ices, volatiles like methane, ammonia, nitrogen, carbon monoxide and argon, suggesting that the planet's centre was still quite hot early in its evolution. The cold, icy nature of the Saturnian system may actually explain why Titan possesses a thick atmosphere and others – like the four large Jovian moons, for example – do not, despite their similar densities and (presumably) physical compositions. Moreover, being closer to the Sun, the Jovian moons are warmer than their Saturnian counterparts.

Owen refers to laboratory simulations which have shown that crystallising water-ice loses its ability to 'trap' and retain gases at higher temperatures. In view of their closer proximity to the Sun, ices inside Ganymede and Callisto must have condensed at higher temperatures than would have been available in the Saturnian system. This would imply that the gaseous constituents of these Jovian moons would have been markedly depleted. Titan, on the other hand, formed from gas-rich ices know as 'clathrates' that condensed at much lower temperatures.

Heat generated during the moon's accretion would have been primarily responsible for releasing these gases. More gases might then have been contributed by cometary impacts, to form the first version of Titan's atmosphere. Such impacts would probably not have worked in the same way at the Jovian moons, however, because Jupiter's powerful gravity would have accelerated cometary impacts to such high speeds that they would have quickly driven any liberated gas into space. In Titan's case, additional heat could also have been generated by the decay of short-lived radioactive isotopes in its rocky component.

Like Earth and other terrestrial planets, Owen argues, Titan's interior gradually 'differentiated' to form a dense, rocky core, overlaid by a mantle of ice. This would have pushed along the development and thickening of the moon's atmosphere and, more intriguingly, the heat generated in Titan's interior and its effect on the icy mantle might have produced an ocean of liquid water somewhere beneath the crust. Intriguingly, such an ocean might still persist there! This possibility will be explored in greater depth later, but at present whatever surface Titan has is obscured by its thick atmosphere.

ATMOSPHERIC PHOTOCHEMISTRY

Although the Voyagers were not equipped with radar to penetrate this thick, smog-like haze, the Cassini spacecraft currently *en route* to Saturn certainly is, and our understanding of this strange world should change substantially over the next few years. Nevertheless, the Voyagers were able to work out how atmospheric temperatures varied with height, thanks to the RSS occultations and IRIS measurements. They found that temperatures actually decrease with altitude from the surface, until they reach the tropopause at an altitude of approximately 42 km. After this point, temperatures climb rapidly to 80° C higher than at the surface.

This relatively common 'temperature inversion', which was predicted by G.E. Danielson in 1973, can be attributed mainly to ultraviolet sunlight entering Titan's upper atmosphere, which adds to the complex photochemical processes going on. In contrast to terrestrial smog, which typically forms within a kilometre or so of the surface, the smog on Titan extends much higher, to an altitude of around 200 km for the main canopy, coupled with another more tenuous layer, 100 km deep. The bulk of this material seems to be a condensed form of many of the gases in the atmosphere. However, the reality may be much more complicated.

Ordinary condensation products should create white or grey aerosol layers, whereas those actually seen on Titan are a dirty orange. This suggests that some additional chemistry is turning simple molecules into more complex substances. Some of these end products include dark polymers, formed by hydrogen cyanide and acetylene, which probably contribute to the observed colour. Ground-based laboratory experiments have also mixed the main ingredients of Titan's atmosphere, using a range of energy sources, and produced organic compounds that closely match the observed dark colour.

Nevertheless, the precise composition of the thick smog is unknown, but it certainly cannot remain suspended above the surface indefinitely. The Voyager observations suggest that most of the airborne particles are anywhere between 0.2 and 1.0 micron in diameter. As these particles increase in size, they precipitate out of the atmosphere to deposit a kind of 'snow' on Titan's surface. The consequences could be quite remarkable and have led to a number of intriguing possibilities. For example, methane 'lakes' could form on the surface, together with low-hanging clouds of methane crystals.

A GLIMPSE OF SURFACE FEATURES

It is also quite possible that ethane – a trace constituent in Titan's atmosphere – could form clouds and hazes, as well as depositing liquid onto the surface. How much of this liquid ethane there might be, if any, is not known, with some scientists even speculating early on that there could be a global ocean several kilometres deep. This is now thought unlikely, however, particularly in light of Earth-based radar observations of Titan made by Duane Muhleman of Caltech in 1990, which revealed a bright 'spot' on the moon's leading hemisphere. The feature stretched one-sixth of the way around the equator. At the very least, this observation confirmed that the surface is not homogeneous.

Since this pioneering work, near-infrared HST measurements have allowed scientists to crudely map Titan's surface. Infrared radiation can pass through hazes

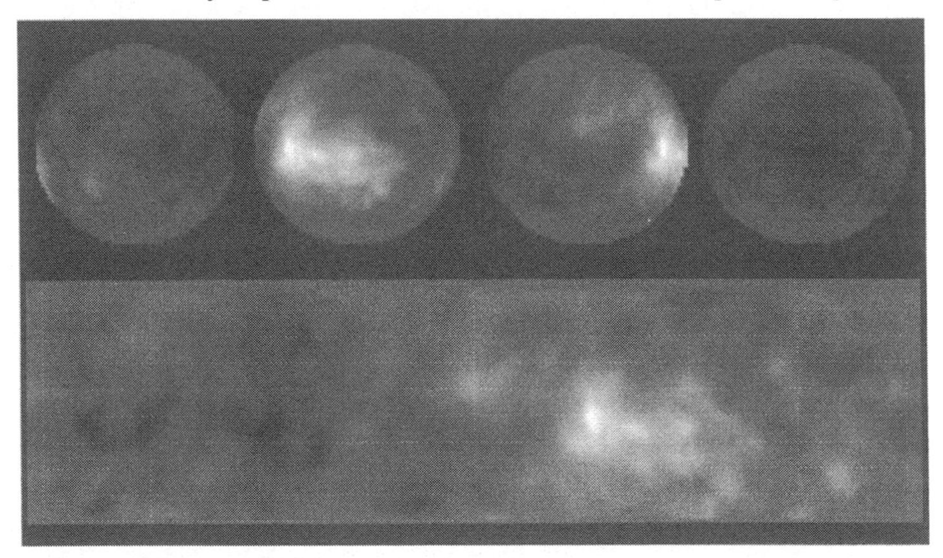

Infrared images of Titan's surface, obtained by the Hubble Space Telescope in 1994, showing a bright, Australia-sized 'continent' on the leading hemisphere.

made up of small particles and, in late 1994 pictures taken by HST revealed a bright 'continent' 4,000 km across, about the size of Australia. Centred at latitude 10°S and longitude 110°W, the object seems to be the same as the radar-bright spot observed by Muhleman's team. At present, little more is known of its nature with any certainty, but the radar brightness implies either that its terrain is rough in nature or that there is a difference in chemical composition.

The HST data was acquired over two weeks from 4 to 18 October 1994, marking almost one complete revolution of Titan around Saturn and hence one axial rotation. Led by Peter Smith of the University of Arizona's Lunar and Planetary Laboratory, a team used WFPC-2 at near-infrared wavelengths (between 0.85 and 1.05 microns) to probe the atmosphere. They were hoping to see tropospheric clouds in an effort to track their motions because, said Smith, there was no guarantee of being able to resolve surface features.

"When I wrote my proposal," he recalled, "I said that we intended to map the surface features, but the reviewers said: 'That's impossible.' Of course, we saw no clouds, and we mapped the surface!" In the near-infrared range, Titan's haze is sufficiently 'transparent' to allow surface features to be crudely mapped according to reflectivity. Much of the surface was mapped, except for the poles, which were out of HST's view, resolving features as small as just 580 km across.

Although the radar brightness suggested that the large spot is rough terrain or a compositional difference, Smith's team were quick to stress at the time that they were unsure if it represented a 'continent', an ocean, an impact crater or something else, although many scientists believe it to be elevated terrain. All that could be determined with certainty was that there could not be a totally global ocean, because there were at least some traces of a solid surface. More recent ground-based observations, made in July 1999 with the Keck telescope on Mauna Kea in Hawaii, have improved the picture still further, by providing tantalising hints of what may be a series of dark, hydrocarbon 'seas', possibly occupying crater floors.

The pictures, taken by a group of scientists from Lawrence Livermore National Laboratory and the University of California, led astrophysicist Bruce Macintosh from Livermore to speculate that the dark patch may be "a big body of liquid", or possibly some form of solid organic material. If liquid, it may contain many of the constituents known to exist in Titan's atmosphere. If solid, it could be methane-ice or snow. Other observations from the Canada–France–Hawaii Telescope on Mauna Kea noted several bright spots on Titan, which led Athena Coustenis of the Meudon Observatory in Paris to speculate that they "could be a plateau with peaks".

It is not unreasonable to suppose that the bright patches could represent a 'frost' of methane-ice on the summits of several tall mountain peaks. Several scientists have disputed this possibility, but Ralph Lorenz of the University of Arizona – a "big fan of mountains on Titan" – thinks the peaks may appear bright because they stimulate rainfall that has washed them 'clean' of a mantling deposit of darker material. Smith also believes them to be a very large range of ice mountains which could be produced by wet air blown from a hypothetical methane ocean. The air, says Smith, would freeze out and produce clouds hanging over the mountains.

Ongoing HST and Keck observations, says Martin Tomasko of the University of

Arizona at Tucson, will provide context in which to interpret the results from the Huygens spacecraft as it drops into Titan's atmosphere in January 2005.

HYDROCARBON 'SEAS'?

These findings lend some credence to ideas that bodies of liquid volatiles like ethane and, perhaps, propane condense on to the surface as seas, lakes or ponds. Perhaps even 'rivers' of liquid hydrocarbons flow sluggishly over Titan's water-ice bedrock. One can envisage many of the solid landforms poking out of the strange seas being drenched in these aerosols which, on Earth, would be highly flammable, but not so in Titan's virtually oxygen-free reducing atmosphere. Wind-blown drifts of particles might also gather in depressions or hollows, turning them into organic-rich swamps. There may even be the Titanian equivalent of a hydrologic cycle, with ethane substituting for water.

Evaporation from the seas and lakes, if they exist, followed by condensation in the atmosphere, could cause rainfall. Raindrops would fall with the grace of snowflakes on Titan. This might lead to rivers sculpting the alien landscape into weird and wonderful shapes, just as wind and water play such a vital role in shaping terrestrial geological formations. A human observer sailing or drifting in this slow-moving liquid would undoubtedly behold a dusky, cold world. In fact, Titan's atmosphere is so dense that the environment close to the surface would probably experience very little temperature change, even across seasons.

Saturn and a distant Sun peek through the clouds in artist Don Dixon's concept of Titan's surface.

This dismal picture is, however, brightened by the possibility that the lower part of the atmosphere may be quite clear, unless you were stuck in an ethane fog bank, because most of the aerosol layers are concentrated at higher latitudes. Yet the brightness of the Sun, filtered through the dense smog, would cast a dull orange glow over the strange landscape – a thousand times fainter than terrestrial daylight, and roughly comparable to a full Moon on Earth. It is also possible that the seas or lakes may even have waves churning through them, although scientists will have to await Huygens' arrival for data on winds near Titan's surface.

As the resolution of mapping by HST and ground-based telescopes improved, it began to seem ever more unlikely that a global ocean exists. Infrared 'speckle interferometry' from the Keck telescope in 1999 revealed a complex surface, comprising bright areas that might represent ice-and-rock 'continents' and a large number of extremely dark areas, each several hundred kilometres across. This latter deposit has been described by Bruce Macintosh as "one of the darkest things in the Solar System" and may be part of a thick blanket of hydrocarbons, perhaps with the consistency of a semi-solid sludge. Later observations, however, pointed to the global predominance of water-ice and perhaps even a complete *lack* of hydrocarbons!

It would seem, therefore, that research has undergone a turnaround since a global ocean was first proposed in light of Voyager results, to such an extent that many scientists now doubt that anything more substantial than shallow 'seas' inside craters actually exist. However, this raises the question of where Titan gets its supply of methane to replenish its atmosphere. Perhaps its icy lithosphere is porous and methane is stored in fluid form underground, and is slowly diffusing into the atmosphere. What is needed are *in situ* measurements. Hopefully, in the spring of 2005, Huygens will tell us more.

WEATHER

Titan is probably too small to support the sort of complex, large-scale wind patterns seen on Earth, such as hurricanes, but it clearly receives intense solar heating around its equator, which in turn forces some kind of circulation. Like Venus, Titan's atmospheric circulation system has a single 'Hadley cell' on either side of the equator. Warmed air rises near the equator, flows at high altitude towards the poles, cools and sinks to the surface, before returning to the equator. This simplistic circulation process is possible on worlds like Titan and Venus, because they rotate so slowly. On Earth, however, the Coriolis effect disrupts such a system by inducing swirling airflows and producing tropical, temperate and polar components.

Not until comparatively recently was it possible to identify the first tropospheric methane clouds on Titan. In September 1995, a team at the UK Infrared Telescope (UKIRT) in Hawaii watched a thick methane cloud at an altitude of about 15 km, which gradually rotated over the limb into view and ultimately covered 10 per cent of the moon's disk. It was strongly suggestive of a developing hurricane system, but further observations were not possible, and not surprisingly, when the site was inspected again two years later the cloud had vanished. Since then, however, smaller clouds, covering no more than 1 per cent of Titan's disk, have been seen and used to make inferences about tropospheric wind speeds.

Recent ground-based computer simulations of atmospheric-circulation models carried out by Pascal Rannou at the University of Paris, reported by *Nature* in August 2002, have theoretically tracked the formation, evolution and movement of smog particles in Titan's atmosphere. The nature of this atmosphere has for some years led astrobiologists to speculate that the moon might be similar to Earth in its youth and may even harbour clues about how life forms.

The work by Rannou and his team showed that smog particles stay intact as they move from the equator to the poles, which helps to build the global opaque layer up to 400 km above the surface. At the poles, the particles descend, which helps to explain the lower-altitude 'hood' photographed by the Voyagers. In the same edition of *Nature*, Robert Samuelson, a University of Maryland astronomer, wrote: "The smog and organic gases on Titan share a common characteristic. Both are born at altitudes higher than where they die."

Thus far, Voyager 1 and Voyager 2 – the latter of which flew within 907,000 km of the moon on 27 August 1981 – have provided our only close-up glimpse of Titan. More information on the composition of its thick atmosphere, smog and the tantalising possibility of finding 'pre-biotic' molecules are optimistically anticipated when Huygens lands there in January 2005. Titan is a mystery among the Saturnian moons, not least due to the fact that it is so much larger than any of its companions.

LIFE ON TITAN?

When the European-built Infrared Space Observatory (ISO) was launched towards the end of 1995, it undertook detailed spectroscopic studies of Titan's atmosphere. Not only did this work help to verify and refine the chemical abundances reported by the Voyagers, but its wider wavelength coverage allowed atmospheric specialists to probe deeper into the gaseous envelope. It established the presence of carbon dioxide and also water vapour, which helped scientists to better understand Titan's atmospheric chemistry.

Where that water vapour came from is not known for certain, because Titan's surface is much too cold for it to be released from ices. It is possible that the water comes from the rain of interplanetary dust, which in turn contains grains released by comets. This has prompted Athena Coustenis to draw parallels between the present situation on Titan and what conditions may have been like on the young Earth, before life evolved here. "We are seeing a mix of elaborate organic molecules closely resembling the chemical soup out of which life emerged."

It was Harold Urey, the Nobel Prize-winning chemist, who suggested that Earth formed with a hydrogen-rich gaseous envelope drawn directly from the primordial solar nebula. Although molecular hydrogen would very quickly have been lost into space, prodigious amounts would have remained in the form of methane, ammonia and water vapour. In an environment such as this, the free oxygen would have been 'reduced' by being bound up chemically. The reactions which ultimately gave rise to life could not have taken place in an oxygenated environment, because the extremely reactive oxygen would have readily broken down the organic molecules.

When Voyager 1 flew past Titan in November 1980, its IRIS instrument measured the abundances of molecular hydrogen in its troposphere as being 0.2 per cent, which indicates a reducing atmosphere in which there can be no free oxygen or oxygen-rich compounds. Clearly, all the oxygen Titan received from the solar nebula is stored in the rocks in its interior and its water-ice crust. In this sense, the atmosphere is thought to be akin to the oxygen-free envelope that surrounded the infant Earth. This has for several years raised the tantalising question: could life develop on Titan?

One of the IRIS' most important discoveries was the presence of hydrogen cyanide, which is formed by the dissociation by ultraviolet sunlight of nitrogen and methane molecules in Titan's upper atmosphere. Its significance is that it plays a crucial role in the chemical synthesis of amino acids and marks Titan out as the only world in the Solar System, besides Earth, to demonstrate the 'building blocks' of complex organic chemistry in such abundance. Although some scientists are tempted to think of this moon of Saturn as a deep-frozen version of Earth before life evolved, it is probably too cold for even bacteria to survive.

Nevertheless, in 1992 physicists Carl Sagan and Reid Thompson of Cornell University suggested that meteorite impacts could melt Titan's crust and create temporary pools of liquid water. Although the atmosphere lacks oxygen, it was calculated that for several million years organics in the crust around an impact site might be able to react with the oxygen in liquid water. Given enough time, simple amino acids might develop. If it could be determined that self-replicating molecules exist on Titan, the discovery would have profound implications for humanity. It would, as Philip Morrison of Massachusetts Institute of Technology (MIT) commented, "transform life from the status of a miracle to that of a statistic".

In 1997, Ralph Lorenz and his colleagues postulated that indigenous biology may develop on Titan in about six billion years' time, at around the same time the ageing Sun will evolve into a 'red giant'. They argued that, for a short time at least, the increased solar heating would be absorbed by the moon's haze-laden stratosphere, causing it to inflate, but only raising the surface temperature to around 200 K, possibly melting the surficial ice. This prompted Arthur C. Clarke to speculate that Titan might be the final refuge for our descendants, fleeing the destruction of the inner Solar System, who may look to the outer planets and their moons for their new, albeit temporary, home.

RHEA

The next-biggest moon after Titan is Rhea, which is, at 1,530 km across, less than a third the size of its planet-like sibling. The differences do not stop there. Unlike Titan, Rhea has no appreciable atmosphere and is literally saturated with craters. The magnitude of the bombardment necessary to produce these craters indicates that

Crater-saturated surface of Rhea.

more than one episode of particularly violent impacts took place, and that the later impacts may have been unique to the Saturnian system. They probably represent something more than the ongoing trickle of comets and asteroids suffered elsewhere in the Solar System.

The centre of Rhea's disk, particularly north of the equator, is very heavily cratered and if this is the result solely of meteoritic or cometary impacts, it suggests that the moon took a considerable battering in its youth. However, the majority of Rhea's craters are unique compared to those on the other Saturnian moons. Mostly, they are irregular and several are actually polygonal in shape, with very few 'flattened' ones. It would appear that Rhea's low gravity and relatively small size led to rapid cooling and freezing and helped to preserve the original, 'sharp' crater forms.

Most of the craters are quite small – there are very few 'large' basins – and some form enormous rings or 'chains' across the surface. The Voyager pictures strongly hint at two main cratering epochs, dubbed Populations I and II, in which Rhea's surface was bombarded, although less severely in Population II. Between the two epochs, there may have been an episode of 'resurfacing', in which dark subsurface material was pushed upwards to cover the craters from the first bombardment. Localised bright patches seen in some craters could point to relatively fresh ice excavated from beneath Rhea's surface.

Other indications of this subsurface (or 'endogenic') activity come from several strange linear 'troughs', such as the Kun Lun Chasma, which the Voyagers showed to run clearly through crater fields laid down by the Population I bombardment, but not the second. The troughs might arise from thermal expansion as Rhea heated up after its initial accretion or from changes in the moon's volume as its interior underwent differentiation into a series of layers of differing chemical and mineralogical composition. Alternatively, they could simply be impact fractures.

At present, unfortunately, scientists' understanding of Rhea is limited, because it was incompletely surveyed by the Voyagers. Their pictures – particularly those of Voyager 1, which flew within 73,980 km of the moon – present most of the 'leading' and Saturn-facing hemispheres in astonishing detail, but the trailing one is very poorly resolved. In fact, the images of the leading hemisphere typically reveal details as small as 20 km across, compared to 50 km for its trailing side. Hopefully, the more advanced imaging equipment on board Cassini should rectify the situation when it surveys Rhea sometime after July 2004.

HYPERION

In general, cratering on many Saturnian moons has drawn scientists to suspect that two epochs of bombardment did occur. Population I is usually attributed to an initial assault of cometary and other debris, probably producing some of the larger craters visible on some moons, but the source of the Population II impacts is somewhat harder to determine. However, it is quite likely that something indigenous to the Saturnian system was responsible, and one possible 'smoking gun' is a strange little world, next door to Titan, called Hyperion.

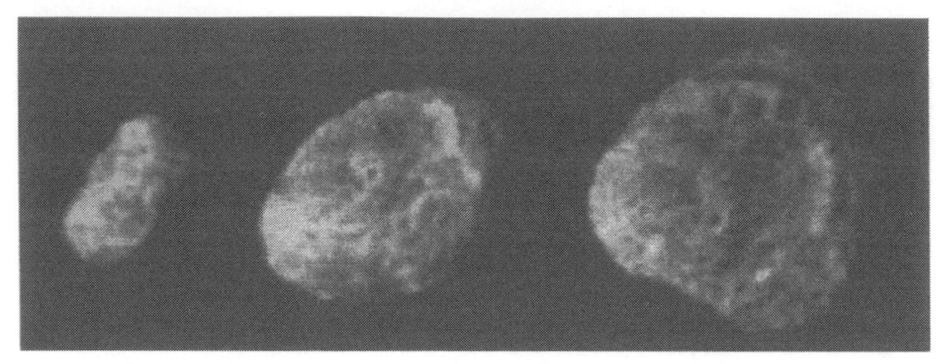

Hyperion, as seen by Voyager 2 in August 1981.

This moon circles its parent planet just beyond the orbit of Titan. It is irregularly shaped, looking a bit like an elongated potato, and measures 360 km along its long axis. Such an unusual shape is not thought to occur naturally; rather, Hyperion probably represents the shattered result of a number of devastating collisions sometime in its history, which have left the battered world we see today. Moreover, its strange shape causes it to tumble chaotically. One possibility is that whatever freak impact broke up and created Hyperion also sprayed debris across the Saturnian system which, as Population II, battered the inner moons.

Initially, Hyperion's rotation rate was a mystery because the Voyager pictures were unable to identify its spin axis, although a periodicity of about 13 days was inferred from photometric variations. In 1983, however, it was realised that the moon has a 'chaotic' rotational state. It is locked in a 'resonant' relationship with Titan: for every three orbits of Saturn it makes, its larger cousin makes four. Hyperion's orbit is also slightly eccentric and resultant perturbations seem to have set it rotating on several axes, in a motion much like a pendulum.

Despite its long axis being comparable to the diameter of the moon Mimas, Hyperion clearly has not reshaped itself and must therefore have significant structural strength, perhaps due to its density being enhanced by virtue of having once been part of a larger object. Unfortunately, its exact density has yet to be determined. Certainly, its icy surface suggests that its composition is the same as most other Saturnian moons. Also, if the 10-km-high ridges – as well as impact craters up to 120 km across – observed by the Voyagers are relics of large structures on the surface of the as-yet-hypothetical 'big Hyperion', their sheer size alone points to the Population II bombardment as the possible cause.

TETHYS AND MIMAS: VICTIMS OF POPULATION II?

Two of the moons that may have received some of the devastating Population II debris include enigmatic Tethys and Mimas, both of which boast the largest craters in proportion to their own size ever seen in the Solar System. The pair are roughly spherical and composed mainly of water-ice. Voyager 1 was first to photograph a

Looking like a cyclops from classical lore, Mimas and its vast Herschel crater emerges from the darkness to reveal itself to Voyager 1.

crater on Mimas that was so enormous that the impact that caused it must have very nearly torn the low-density moon apart! The crater, known as Herschel, is about 130 km across – a third of Mimas' own equatorial diameter – and some 5–10 km deep.

At Herschel's centre is a huge mountain, jutting 6 km up into space – almost as high as Mount Everest. This may have been produced from a 'rebound' of the crater floor under Mimas' extremely weak gravity shortly after impact. The crater floor includes a lot of detail: it is adjoined by a long, winding valley and numerous other, smaller craters. In fact, very few craters greater than 50 km across have been found on Mimas. This might imply that it may have a younger surface on which later processes destroyed ancient, larger craters.

Furthermore, Herschel itself displays very few smaller craters superimposed on top of it, suggesting that it may well be much younger than neighbouring areas. In general, most of Mimas' craters are very roughly bowl-shaped and much deeper than instances of comparable size on our Moon or the large Jovian moons. This is probably due to its low density – Mimas is only 1.2 times as dense as water – which allowed impacts to penetrate much more deeply into the crust.

Numerous valleys, in addition to those adjoining Herschel, were detected in Voyager 1's pictures, some of which extend as much as 90 km in length by up to 10 km wide. The most conspicuous of these winding grooves tend to head in northwesterly or west-northwesterly directions. Some of them are remarkably straight and may have formed over underlying crustal fractures, whereas the more irregular and meandering ones could represent chains of coalesced craters. There have been a few suggestions that the grooves were produced as a result of the Herschel impact or perhaps by internal tidal activity.

Clearly, as with all of Saturn's moons, Mimas is almost exclusively icy in nature. Ground-based infrared observations, made before the Voyager encounters, suggested that water-ice on Mimas is almost pure. Yet, at Saturnian distances from the Sun, it is frozen so solid that it can probably absorb the punishing hypervelocity impacts that have dented and chipped its surface repeatedly over billions of years. As Mimas contains very little in the way of naturally occurring radioactive elements such as uranium, thorium or potassium-40, any heat derived from their decay is modest.

The closest pictures of Mimas came from Voyager 1, which flew within 88,400 km of the moon. Most of its surface was photographed, although the north polar regions were inaccessible. Like all of Saturn's moons, except Phoebe and Hyperion, Mimas has a 'synchronous' rotation, always keeping the same hemisphere turned towards its parent planet.

In some ways, Tethys is like a scaled-up model of Mimas, in that it too has a large impact crater. However, this crater – called Odysseus – is, at 400 km across, four times larger than Herschel and actually wider than the equatorial diameter of Mimas! Yet, because Tethys is more than twice the size of Mimas, the size of Odysseus is proportionately about the same: covering about a third of the moon's total diameter. In further contrast, Odysseus' floor returned to approximately the original shape of the surface, probably due to Tethys' stronger gravity and relatively fluid water-ice.

It is possible that a slightly warmer layer beneath Odysseus caused the interior to become flattened. Like Herschel, it has a central peak – in fact, it is the largest crater with a well-developed peak yet found in the Solar System – and its present layout implies that the floor may have rebounded above the subdued rim. The interior surface of Odysseus is less well preserved than that of Herschel, suggesting that it is an ancient feature. Its rim has been softened by a series of 'terraces' and the central peak has degraded into a ring-like structure.

The oldest surface type on Tethys is a hilly cratered terrain that is characterised by very rugged topography. It is heavily cratered, but many of the larger scars are degraded. A notable landmark, after Odysseus, is a vast tectonic fracture known as Ithaca Chasma, which covers at least three-quarters of the total circumference of Tethys, and probably runs all the way around. It may be related in some way to Odysseus, although it is not actually connected to the giant crater. In fact, if the crater's central peak is considered to be a 'pole', the chasm tends to trace the great circle of the associated 'equator'.

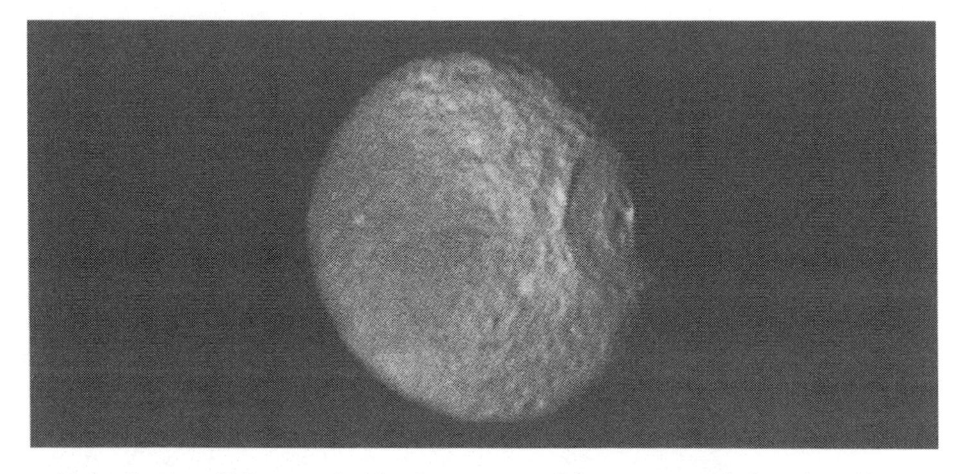

Twice the size of Mimas and with a huge crater – Odysseus – roughly four times bigger than Herschel, Tethys is another strange world in Saturn's retinue.

Ithaca Chasma, seen in this Voyager 2 image extending from centre-left to upper-right.

Voyager 1 did not fly particularly close to this moon, yet was able to see the enormous canyon; its twin, however, flew past at less than 93,000 km and mapped most of Tethys' surface. It is possible that, in its early stages of evolution, Tethys consisted of a liquid water globe, surrounded by a thin solid crust. If this happened, and if both features formed at the same time, the force of the Odysseus impact may have temporarily deformed Tethys into an ellipsoidal shape, inducing tensional stresses that produced the huge canyon as a surface crack. That Odysseus and Ithaca Chasma are approximately the same age was hinted from Voyager pictures comparing craters within the canyon to those on neighbouring terrain.

Why, however, was only one crack produced, rather than many? Answers to this conundrum are scarce, but Ithaca Chasma is certainly not a crater chain, yet does hint at some sort of catastrophic event on Tethys in its remote past. The canyon has been mapped all the way from the north pole to the equator and from there to the south pole and beyond, maintaining an average width of 100 km and a depth of 4–5 km. On each side of the trench there is a raised 'lip' of terrain, which typically reaches a height of about 500 m.

Tethys (1,050 km) and another moon, Dione (1,120 km), both have four times the surface area of Mimas. Voyager pictures of both worlds revealed greater diversity in terms of their geological evolution. Both have terrain ranging from ancient and crater-saturated to younger and more modestly scarred. However, Tethys is somewhat different in nature from Dione, due mainly to its very low density, which is close to that of water. In fact, Tethys seems to be made of almost pure ice and, of all the Saturnian moons, is only surpassed in brightness by Enceladus.

ENCELADUS

Enceladus reflects nearly all of the sunlight that strikes it. It has a geometric albedo of 95 per cent, brighter than a field of fresh snow, and its high reflectivity gives it a very low surface temperature of only 73 K. Although Voyager 1 did not approach

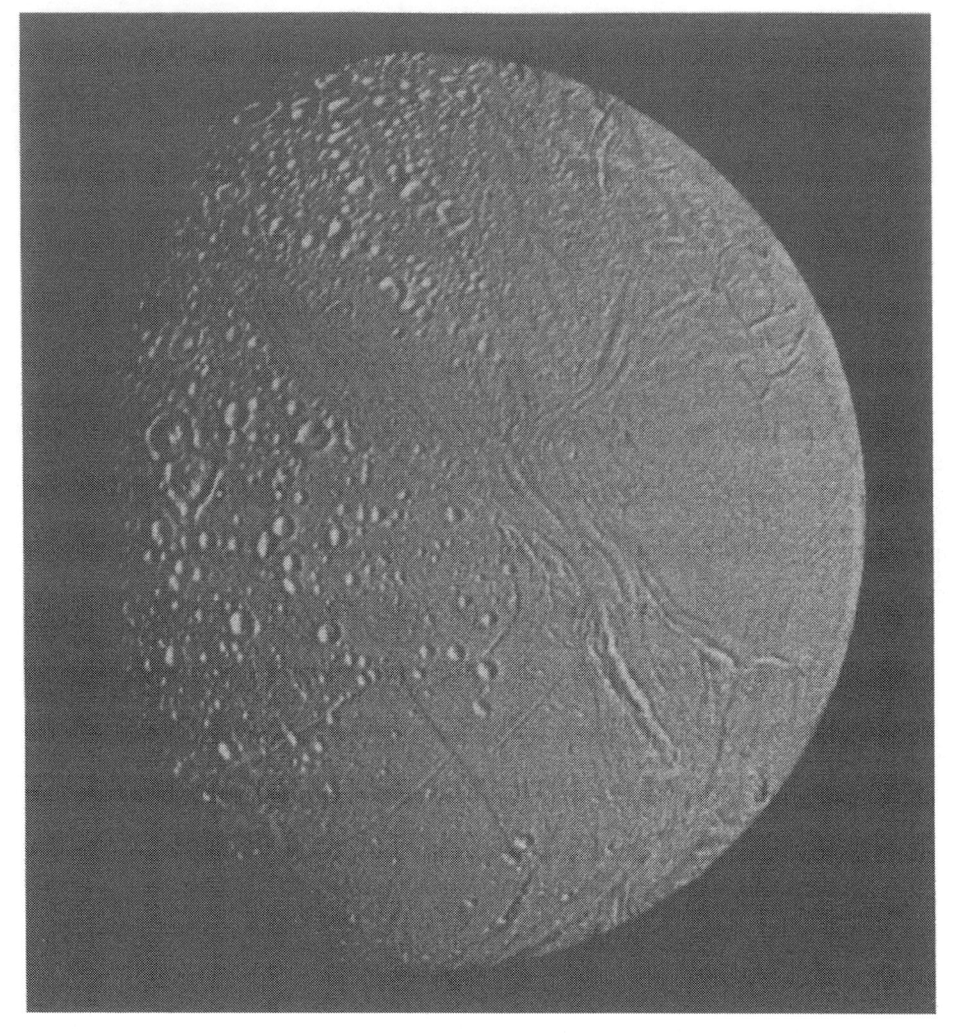

Enceladus.

the moon closely, its mate was able to provide imagery of the northern portion of its trailing hemisphere at a resolution of several kilometres per pixel.

Enceladus' reflectivity has led to some exciting possibilities, not least of which is that it may have been volcanically active until fairly recently. Its high reflectivity gives it the brightest surface of any moon in the Solar System. This indicates a fresh icy surface, which may have been the result of a recent deposit of frost particles from a cryovolcanic source. Enceladus' low gravity means that even a single-vent, geyser-like eruption could easily have coated the entire surface.

Marred by faults and valleys, it has by far the most active surface of any of the Saturnian moons studied to date. The Voyagers photographed cratered, 'ridged' and smooth types of terrain. The cratered type is thought to be the oldest terrain on Enceladus. It possesses no 'large' craters, but numerous smaller ones with diameters ranging from 10 to 30 km. Most scientists assume that these were produced when Enceladus was bombarded by debris in orbit around Saturn.

The cratered terrain can itself be further subdivided. In some areas, craters, 10–20 km across, show evidence of collapse and those that have central peaks display very gentle, rounded mountains. Other areas show similar-sized craters, but they are in a much better state of preservation and are much deeper, suggesting that their thermal histories and chemical composition are somewhat different. Studies of the crater forms concluded that Enceladus' lithosphere is a mixture of ammonia-ice and water-ice.

In larger areas, the early craters have been 'replaced' by plains, whose paucity of impacts implies either a very low cratering rate or the relative youth of the surface. In some places, these plains are criss-crossed by a multitude of long grooves, thought to be subdued grabens formed as subsurface ice froze, expanded and 'cracked' the brittle outer shell. There are also long, winding valleys and ridges rising up to a kilometre high, which seem to cut across older crater fields. These features cover most of the middle of Enceladus' visible disk and may be fluids that oozed from fractures and solidified in place.

Although the very ancient parts of its surface were, predictably, pitted with dozens of craters, their conspicuous absense in other areas strongly betrayed an age of less than a few hundred million years for the youngest regions. Clearly, there has been a significant period of resurfacing, fairly late in Enceladus' history

It would also seem that some areas of the surface are still undergoing tectonic change, as shown by areas that are covered by ridged plains with no evidence of craters. Still other areas showed patterns of criss-crossing linear fault lines. It seems likely that the heating needed to produce tectonic modification on this scale comes partly from tidal bulging caused by a tug-of-war between Saturn and close-by Dione, which perturbs Enceladus in its orbit and keeps its interior warm. In this respect at least, Enceladus can be likened to the Jovian moon Io.

DIONE

Dione is unusual in several ways. Its diameter is only slightly greater than Tethys, but its brightness distinctly less, so that from Earth it appears no brighter than Tethys. The trailing hemisphere is comparatively dark, while the brightest features on its leading face have almost twice the albedo. Its density, at 1.4 times that of water, is greater than for any of the other icy moons and it has been suggested that Dione may affect the radio emissions from both Saturn and the surface of Enceladus, although the latter idea rests on fragmentary evidence.

The surface features of Dione differ from those of the other moons. It was surveyed by Voyager 1 and less effectively by Voyager 2. A considerable portion of the entire surface has been studied, although there are gaps, and some features have been recorded in each polar region. Dione has one small co-orbital moon, called

Dione.

Helene, which moves ahead of it, and Tethys was also found to have two diminutive companions: Telesto and Calypso.

The predominant terrain is a rugged 'highland' with several craters up to 100 km across. In general, these craters are shallower than those on Tethys and most of them have terraced walls and central peaks. Scarps up to 100 km long were found to run

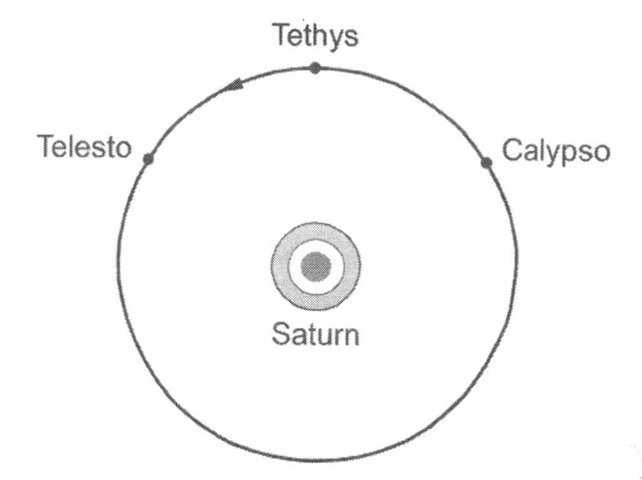

Diagram showing the orbital positions of Telesto and Calypso in relation to Tethys.

across this terrain. In addition to a less ruggedly cratered terrain, defined as 'cratered plains', there are also 'smooth plains' that show some troughs, and although these are typically less than 100 km in length, a few exceed 500 km having, in most cases, a 'pit' at each end. There may, in fact, be a global network of fractures.

The most striking feature on Dione is elliptical Amata, which lies at the centre of a system of bright, wispy features that divide up the dark trailing hemisphere. These appear to be associated with narrow linear troughs and ridges that are extensions of bright lines. They are possibly the result of internally generated stresses such as those that might have been produced by the freezing of the interior.

The bright material, probably water-ice, may have come from inside Dione. Amata has a diameter of about 240 km and its nature is unclear: it may be an irregular crater or it may be more in the nature of a basin. Images so far have not been clear enough to tell, although it is probably not a 'ray crater' like those seen on our Moon. The wispy features are thought to be faults or fractures and there can be little doubt that Amata is related to their formation, but they remain mysterious.

Few craters on Dione measure above 30–40 km across. The exceptional, largest ones – up to 200 km in diameter – are much shallower than those of Tethys, but some have pronounced central peaks. Craters tend to be most populous on the vast plain that covers most of Dione's observable façade and this is generally assumed to be its most ancient extant terrain. The brighter areas were probably resurfaced with a layer of material thick enough to conceal craters that had been formed there.

One possible implication of this is that an ancient internal heat source – perhaps some material undergoing radioactive decay – kept Dione at least moderately active until quite late in its evolution. This would appear to be consistent with the moon's density, which many geochemists consider to be 'intermediate' between Enceladus on one hand and Mimas or Tethys on the other.

NEW MOONS

Clearly, Saturn is a world of many moons. Before the Voyagers even arrived, 11 had been spotted from ground-based observations or analysis of images from the Pioneer 11 spacecraft, but numerous others were found during and after the 1980–1981 encounters. Three were found by Voyager 1, three more from subsequent Earth-based work and a further three 'possibles' from an interpretation of imaging data after the Voyager 2 flypast. Two of the new arrivals, Prometheus and Pandora, were actually found to 'shepherd' Saturn's F ring, lending support to theories that nearby moons help to orchestrate the flow of ring material.

Two others, detected by ground-based observers, were Epimetheus and Janus, which actually share the same orbit 91,000 km from Saturn. They orbit about 50 km apart and, when they pass close to each other, they exchange orbits! Still others share the orbits of Dione and Tethys, either 'leading' or 'trailing' their larger hosts, and have been dubbed 'the Trojans' for these moons.

The name 'Trojan' derives from asteroids that circle planets at so-called 'lagrange points', where the gravitational attraction of the Sun and a planet are balanced. Trojans either precede or trail the planet in its orbit. Jupiter has the most Trojans; those preceding the giant planet are named for the Greek heroes of the Trojan War, while those trailing it honour the Trojan heroes. When similar, 'co-orbital' moons were found at Saturn, the name stuck and they too became known as Trojans, although only one – Helene, named after Helen of Troy – is directly linked to the famous conflict. There may, indeed, be many more Trojans. Several dark objects were picked up in Voyager images, but were never spotted in more than one photograph, and as such their existence remains to be confirmed.

Still more have been found in recent years. Astronomers in Chile, Hawaii and

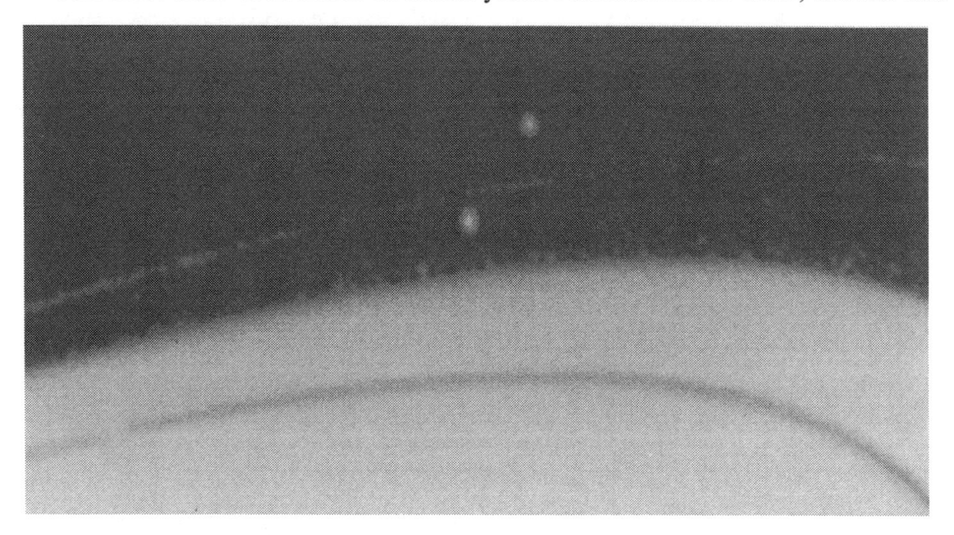

This image shows two tiny moons – Prometheus and Pandora – that help to 'shepherd' material in Saturn's F ring and keep it in place.

Arizona announced in October 2000 that they had found four new moons, which at the time pushed Saturn's tally to 22. In doing so, it pipped Uranus, which at the time was thought to have 21 moons. This author will make no attempt to create a 'league table' of planets with the most moons because, as 2003 has shown us, new discoveries come thick and fast. At present, however, it would seem that Jupiter is top of the league with, at the end of April 2003, at least 60 small worlds in orbit around it.

THE RINGS

Undoubtedly, with the possibility of shepherd moons close to Saturn's rings, still more, smaller objects may await discovery. A survey of the planet's rings is one of the key goals for Cassini, as it was an important objective for the Voyagers. Since the Italian scientist Galileo Galilei first spotted them in 1610 and later wrote "I do not know what to say in a case so surprising, so unlooked for and so novel", they have presented an aura of mystery and wonder, as well as making Saturn undoubtedly the most stunning world in the Solar System.

Galileo never found out what caused Saturn's strange – and ever-changing – appearance in his telescope eyepiece, but did come up with some interesting suggestions. Their true nature, as a flat disk of particulate material encircling the planet, was unknown in Galileo's day and the great scientist pondered for years whether they were separate celestial bodies or even two enormous 'arms' or 'handles' stretching towards Saturn. Exactly what was responsible for holding these arms in orbit around the planet constantly perplexed him.

It was only in 1659 that the Dutch astronomer Christiaan Huygens correctly inferred that they were in fact a disk-like structure around Saturn. Nevertheless, for more than three centuries, little else was known. Why was Saturn the only planet with rings, scientists wondered, and what kept them in orbit around the planet? Obviously, with the benefit of hindsight, we know that gravity is responsible for keeping them there, but it was not until 1687 – nearly half a century after Galileo's death – that English scientist Isaac Newton finished his *Principia* with the first universal theory of gravity.

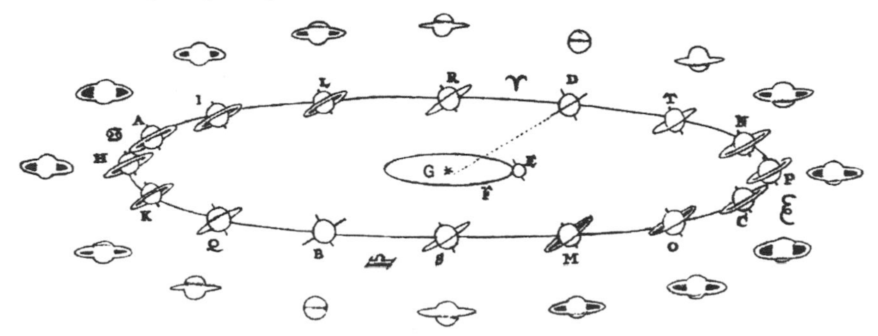

Taken from Christiaan Huygens' *Systema Saturnium* (1659), the true disk-like nature of the strange appendages pondered by Galileo is realised.

Of course, we have known for several decades that not one, but four, worlds possess ring systems. The planet Uranus was found to have a dusty set in 1977 and the Voyagers successfully located tenuous ring-like systems around Jupiter and Neptune in 1979 and 1989 respectively. The question now seems to be: why do only the gas giants apparently have rings? They may, in fact, not be limited to the gas giants and some dynamicists have argued that it is perfectly reasonable that Earth and Mars may have had rings in the distant past or may develop them in the future.

It was not until the middle of the nineteenth century that theoretical calculations by James Maxwell inferred that Saturn's rings were probably composed "of an indefinite number of unconnected particles". This was supported by James Keeler's observations about 50 years later, which revealed that sunlight reflected off the rings was Doppler-shifted in such a way that the ring particles must occupy separate orbits around their giant host, and the innermost ones moved faster than the outermost ones.

At present, ring dynamicists believe that these strange structures begin when a swarm of particles, loosely held around a planet, is flattened into a thin 'disk' in a specific plane, usually around its equator. If the swarm is quite large, particles will collide with one another on a regular basis, which circularises their orbits and reduces the chances of renegade ones escaping from the disk. Saturn's rings, as can be seen through even a modest telescope, are nearly opaque – they must contain trillions of particles – and this 'collisional dynamics' process helps to maintain their average thickness at just a few hundred metres.

This is remarkably thin in comparison to their 282,000-km breadth and, in this case, can be likened to a sheet of tissue paper spread across a football field. In fact, their total extent is about three-quarters of the distance from Earth to the Moon. Eventually, under the gravitational influence of Saturn, the innermost particles are pulled closer to their parent planet, while the slower-moving outermost ones fall further away. This causes the disk to spread out, and, in Saturn's case, their diameter is today nearly three times that of the giant planet itself.

Joe Burns of Cornell University has likened such ring systems to protoplanetary disks similar to the one that probably encircled our Sun during its infancy. However, unlike a protoplanetary disk, Saturn's rings do not and could not harbour large objects evolving within them. The reason for this is that the largest particles tend to lie close to the planet, within its 'Roche limit'. Anything lying inside this limit probably could not remain intact and large, low-density bodies composed of ices would be ripped to shreds by tidal forces induced by the planet's gravitational pull.

Admittedly, some large, solid objects *can* survive within a planet's Roche limit, but only if they are of higher density, and even then they can only achieve a certain size before the planet's tidal forces break them up. Over time, no matter what their size, these particles would continuously crash against others and this 'collisional grinding' would gradually reduce them to the size of houses, cars or even sand. Ultimately, after countless millennia of relentless collisional grinding, even the largest intact particles would be no bigger than Central London. This idea, when carried further, seems to fit most of the observed facts.

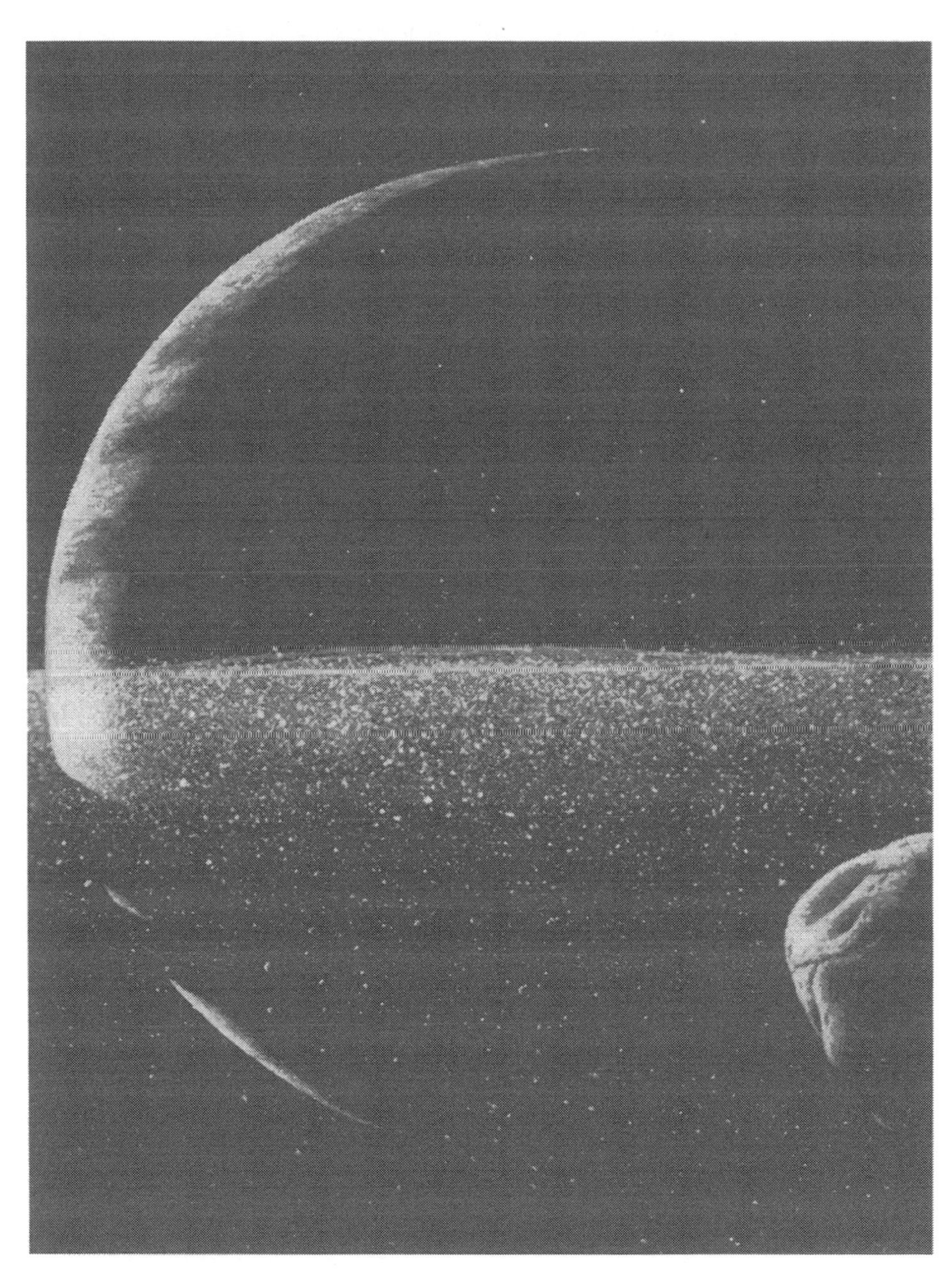

William K. Hartmann's concept of 'dynamical ephemeral bodies' within the rings of Saturn from a vantage point in the outer part of the rings. These bodies are thought to originate in particles that aggregate into larger lumps before being sheared apart by dynamic forces acting inside the ring plane. Note that some of the material is in an inclined plane, due to the presence of larger bodies.

For example, some scientists have postulated that, if tidal forces within the Roche limit are so intense, they might not even have allowed a large object to develop in the first place. Therefore, how could it be destroyed? Any 'smaller' material that existed close to the planet might, similarly, have been dragged into its extended atmosphere. The result, Burns says, would be an 'empty' region of space immediately above the planet's surface, a ring system beyond it and a series of large moons exterior to that. This is similar to what can be seen currently at Saturn.

Others have argued that objects forming very close to a planet's Roche limit neither fully coalesce nor are totally destroyed. Rather, clusters of particles group together for short periods of time, before being torn apart by tidal forces when they have grown large enough; in other words, perhaps Saturn's ring particles change their characteristics over timespans as short as just a few weeks! Certainly, observations from both Voyagers strongly suggested that, in the F ring at least, bright clumps of material formed quickly, before gradually dissipating.

The presence of small 'shepherding' moons has, since the late 1970s, been held up as an explanation for why the rings appear as they do and what keeps them in place. In fact, when two small shepherd moons – Cordelia and Ophelia – were found close to Uranus in January 1986, the theory seemed vindicated. They also helped to create unusual 'arcs' around Neptune and probably influence the structure of the tenuous Jovian ring. Such objects, which are frequently embedded *within* rings, often act as sources or sinks for new particles, as well as accounting for the differing brightness of some of the rings.

The seven known rings of Saturn are named alphabetically in the order that they were discovered, from A to G. Of these, the brightest rings are the A, B and C bands, which were discovered long before the first spacecraft ventured towards Saturn. Others have since been found during the Pioneer 11 and Voyager encounters with the giant planet. In July 2004, when Cassini enters orbit around Saturn, it will actually fly through a gap between the F and G rings, although as a safety measure its cameras and most of its scientific instruments will be turned off during the passage.

CHARACTERISTICS OF THE RINGS

Even when viewed from Earth, Saturn's rings show several openings, the largest of which – the Cassini Division, a gap 4,700 km wide – was first spotted by Giovanni Cassini in 1675. It was initially believed that such gaps are formed through a process called 'resonance', in which particles are confined to specific regions of the rings by the gravitational influence of nearby shepherd moons or larger particles. An example of this resonance has already been seen at Jupiter: in which Io is perturbed in its orbit by Europa and Ganymede.

Particles at the edge of the Cassini Division, similarly, may be affected by Saturn's moon Mimas, and those at the edge of the planet's bright A ring are controlled by the moon Atlas and might also be perturbed by Janus and Epimetheus. Some dynamicists have gone so far as to argue that such resonance patterns might even be sufficient to heat the interiors of two other moons: Enceladus at Saturn and Miranda

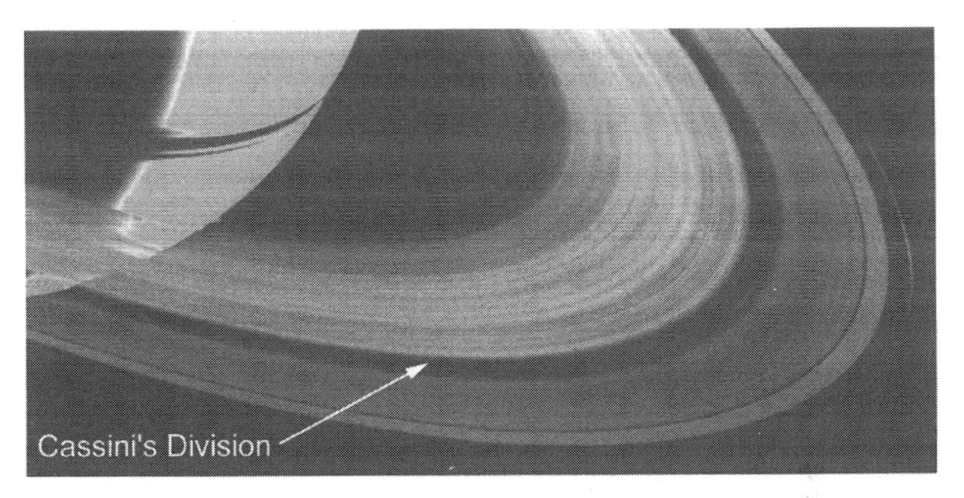

The Cassini Division.

at Uranus. The Voyager results, however, showed that the reality is probably not quite so simple. Both spacecraft picked out significant numbers of 'ringlets', even in the supposedly empty Cassini Division.

An interesting possibility had been proposed in 1979 by astrophysicists Peter Goldreich of Caltech and Scott Tremaine of Princeton University. They suggested that 'spiral density waves', thought to operate in galaxies, might originate at resonant locations in circumplanetary disks. Here, the gravitational attraction of a moon embedded in the ring plane would change the orbits of other ring particles, making them elliptical. In a heavily populated disk like that of Saturn, the effect is that particles 'bunch up' close to resonances. These bunches then pull on the rest of the disk, causing a spiral-shaped density wave to propagate outwards.

Certainly, both Voyagers detected dozens of these spiral-density tracks, especially close to the A ring, and found a few instances where inclined moons actually 'pulled' ring material up out of the ring plane and caused a rippling effect. They also supported the idea of shepherd moons being primarily responsible for constraining the radial extent of the rings. Voyager 1 spotted two such worlds, subsequently named Prometheus and Pandora, astride the faint F ring.

COMPOSITION

Saturn's rings are much more elaborate, complex and – significantly – brighter than any of the other structures found thus far in our Solar System. It is unsurprising, therefore, that they were discovered so far in advance of the ring systems around Uranus, Jupiter and Neptune. The Saturnian material is highly reflective, a key reason being that, like most of the planet's moons, the main parts of the rings are water-ice. This was ascertained in the 1970s, thanks to ground-based infrared observations, although there is probably also a degree of contamination and discoloration.

In some cases, there is a distinct reddening of particles in visible light, perhaps caused by trace impurities, micrometeoroid debris or chemical modifications caused by solar or magnetospheric radiation. More recently, infrared and microwave observations have persuasively argued that the ring particles' interiors are probably also composed of water-ice. Joe Burns considers it unlikely that the interiors are metallic, pointing out that, if present theories as to the formation of the solar nebula are accurate, water-ice should be the dominant solid condensate in the outer Solar System.

Furthermore, the density of Saturn's icy moons and most of the ring particles are close to 1 g/cm^3. This suggests that their nature may be something akin to dirty snowballs, a little like cometary material in this respect. However, in other areas of the system – particularly in the C ring and around the edges of the Cassini Division – particles are generally more neutral in colour (with less reddening), as well as being larger and darker than those from the brighter A and B rings.

As their names imply, the A and B rings were first to be discovered, around 1675, and are the brightest and most easily recognisable of the system. A third, C, ring, lying just inside these two, was added in 1850, but the greatest leap in our understanding of them had to wait until spacecraft went there. In November 1980, Voyager 1 detected a few dusty, narrowly-spaced bands of material, dubbed the D ring, just interior to the C ring and extending about halfway towards Saturn's cloud-tops. This ring, however, is so tenuous that it is barely detectable from Earth.

Other bands found by the Voyagers included the E ring, which lies outside the 'classical' system, and even to the two spacecraft revealed itself as little more than a slight concentration of debris. Nevertheless, it was dense enough to be visible to ground-based astronomers and HST in August 1995, when Saturn's rings appeared edge-on to Earth. This provided a unique opportunity not only to determine that the E ring extends as far as Rhea's orbit and broadens to perhaps 30,000 km at its outermost edge, but also to work out the average thickness of the entire structure.

Even when viewed edge-on, Saturn's main rings are just barely visible and bright enough to suggest that they are no more than 1.4 km thick. However, other dynamicists argue that their actual average thickness is closer to a few tens of metres, suggesting that the inclined F ring makes it extend above the plane of the main system. Other contributors to this 'illusion' of a thick ring may include localised dense patches, caused by their vertical ripples, or perhaps even mini-'atmospheres' surrounding some of the larger particles.

The Voyagers were not the first emissaries from Earth to visit Saturn. They were beaten to it by the irrepressible Pioneer 11 spacecraft, which hurtled past the giant planet in September 1979 and found the slender F ring. Described by Burns as "a contorted tangle of narrow strands", Pioneer 11 data revealed that charged particles were intermittently absent from the F ring, suggesting that unseen shepherd moons might lie there. This was confirmed the following year, when Voyager 1's cameras found Prometheus and Pandora.

This second visitor also spotted yet another band, labelled G, although this was even more narrow and tenuous than the E ring. Despite the differences in thickness and colour of all seven Saturnian rings, their composition in general is poles apart from those of the other three gas giants. The disks around Uranus, Jupiter and Neptune are much darker in colour, presumably due to the dominant presence of silicate- or carbon-rich material in their composition. Ring particle sizes are also different from planet to planet.

RING PARTICLE SIZES

Several ingenious attempts have been made to estimate the average size of Saturn's ring particles. As they enter the giant planet's shadow, the rates at which they cool imply that, if they are 'solid', they should be at least a couple of centimetres in diameter and probably much more. In fact, the bulk of the particles in the rings must be greater than a few centimetres in size because the whole structure is remarkably reflective to ground-based radar transmitting at frequencies most sensitive to particles larger than around 4 cm across.

Furthermore, radio signals transmitted by both Voyagers' RSS instruments suggested that particles in the A and C rings range from as little as a centimetre to 5–10 m in diameter. The high-resolution pictures returned by the two spacecraft also revealed that the rings are much more finely divided than had been expected. Rings lay *within* other rings – more than a thousand were counted in the images alone – but the RSS data suggested that smaller 'ringlets', less than a metre wide, might also exist. The ostensibly empty Cassini Division, for example, is thought to contain perhaps a hundred ringlets.

It was during the observations of the rings that another of Voyager 2's instruments was found to be very useful. The PPS, which had failed to work on Voyager 1 – again justifying the 'redundancy' concept of sending up two identical spacecraft – was able to measure the intensity and polarisation of light from the star Delta Scorpii as it peeked through the rings. This enabled scientists to discern structures smaller than about 300 m wide and showed that comparatively few clear gaps exist *anywhere* in the rings. Data from the PPS also helped them to work out the composition of the ring particles.

In fact, the Voyager results unveiled a range of particle sizes almost as broad as the rings themselves: from minuscule grains to car-sized boulders and monsters as big as houses, all of which sped around the planet on million-kilometre-long racetracks. It would also seem that, in the majority of cases, inter-ring 'divisions' are not at all absent of particles, but merely show relative absences of material when compared to adjacent, more populous regions. The A and B bands are especially bright, thanks to the close spacing of their ringlets, yet neither is fully opaque.

INTRICATE STRUCTURE IN THE RINGS

Close examination of the rings revealed some unexpected structures, such as 'kinks' and 'spokes', as well as very thin and very broad diffuse bands, some of which are almost invisible to ground-based observers. The Voyagers showed that the

Spoke-like structure in Saturn's rings.

gravitational effect of nearby moons does indeed directly influence much of this intricate structure. Their results also strongly suggested that the varying distances of these shepherd moons, and the changing intensity of the gravitational effects from some rings, could help to explain the origin of the kinks and spokes.

Both spacecraft recorded time-lapse movies of the rings, which revealed that these unusual structures seemed to form and dissipate over time. Exactly how Saturn, or its tiny shepherd moons, achieves this remarkable feat remains unknown, even two decades after the Voyager encounters, although the levitation of dust particles 'above' the rings by electrostatic charging is one persuasive possibility for the spokes. It will be left to Cassini's detailed, four-year survey to provide more clues.

However, it would appear that this small-scale structure is almost certainly

COLOUR SECTION

PLATE 1 – This picture of the Earth and Moon in a single frame, the first of its kind ever taken by a spacecraft, was recorded 18 September 1977, by NASA's Voyager 1 when it was 11.66 million km from Earth. The Moon is at the top of the picture and beyond the Earth as viewed by Voyager. In the picture are eastern Asia, the western Pacific Ocean and part of the Arctic. Voyager 1 was directly above Mt Everest (on the night side of the planet at 25 degrees north latitude) when the picture was taken. The photo was made from three images taken through colour filters, then processed by the Image Processing Lab at Jet Propulsion Laboratory (JPL). Because the Earth is many times brighter than the Moon, the Moon was artificially brightened by a factor of 3 relative to the Earth by computer enhancement so that both bodies would show clearly in the prints. Voyager 1 was launched 5 September 1977 and Voyager 2 on 20 August 1977.

PLATE 2 – (*Upper left*) Dramatically surreal view of the Great Red Spot (*at right*) and its tumultuous neighbourhood, as seen by Voyager 1. (*Upper right*) Europa and Ganymede, both from Voyager 2. (*Lower left*) Close-up of Ganymede's variegated terrain. (*Lower right*) The icy, grooved surface of Europa.

PLATE 3 – (*Upper left*) Blood-red Pele, as seen by the Galileo spacecraft in June 1996. (*Upper right*) This picture, which, says Mike Hanlon, "nearly caused Linda Morabito to fall off her chair", shows for the first time the strange bluish plume of the volcano Pele emerging from behind the limb of Io. (*Centre left*) Galileo pictures of Pele from June 1996 (*top*) and September 1997, showing the large grey pyroclastic deposit produced in the wake of the Pillan Patera eruption. The Pillan plume stretched 120 km above the surface. (*Lower left*) The Loki region. The large black area may be a lake of liquid rock and sulphur 200 km across. North-east of the lake is Loki's large, elongated fissure, producing a grey plume from its left end. (*Lower right*) An eruption of Loki – Io's most powerful volcano – as seen by Voyager 1 in March 1979. This image also reveals the unusual, pizza-like surface of Io itself.

PLATE 4 – Close-up image of Pele, in eruption.

PLATE 5 – (*Upper*) Io appears to drift against a backdrop of Jupiter's clouds in this picture taken by the Saturn-bound Cassini spacecraft in December 2000. (*Centre*) Close-up of the Conamara Chaos region on Europa. (*Lower*) A 70-km-wide region of the thin, disrupted ice crust in the Conamara Chaos region of Europa shows that the surface has been broken into blocks up to 13 km across which have been 'rafted' into new positions. This is not dissimilar to the disruption of pack ice on polar seas during springtime thaws on Earth. The size and geometry of the slabs suggests that their motion was enabled by water or soft ice not far below them.

PLATE 6 – (*Upper*) Dramatic shot from Galileo, showing the 'fire fountain' fissure eruption of Tvashtar Catena in February 2000. (*Lower*) Sometime in the future, an unmanned 'cryobot' like the one shown here may explore the hypothetical ocean of Europa. This artist hints at the possible existence of mineral-laden hot springs, perhaps similar to 'black smokers' known to exist deep beneath terrestrial oceans.

PLATE 7 – Voyager 2 mosaic of a crescent Europa, showing the long, dark linear features, mottled terrain, irregular dark maculae and an intensely fractured region near the anti-Jovian point. The sharply defined terminator indicates that there is very little surface relief.

PLATE 8 – (*Main*) Vast tracts of dark, ancient, heavily cratered terrain cover Callisto. The prominent milti-ringed feature is Valhalla. Although it is Callisto's answer to the huge Gilgamesh impact crater on Ganymede, Valhalla's morphology is actually much more subdued. (*Upper left and upper right*) These images of Callisto reveal an apparently ancient surface with little indication of resurfacing.

PLATE 9 – (*Upper*) A spectacular pre-encounter shot of Saturn by Voyager 2, with Tethys's shadow on the planet's southern hemisphere. (*Centre left*) Twice the size of Mimas and with a huge crater whose impact may very nearly have split it apart, Tethys is another strange world in Saturn's retinue. (*Centre*) Dione. (*Centre right*) This is perhaps the best picture of Titan obtainable by Voyager 2, revealing a world hidden under a thick orange atmosphere. Hopefully, the Cassini–Huygens mission will shed more light on this strange world sometime after 2004. (*Lower left*) Enceladus and the strange, two-toned moon Iapetus. (*Lower right*) Enhanced detail of Saturn's atmosphere.

PLATE 10 – (*Upper left*) Displaying a misleadingly quiescent façade in this Voyager 2 shot, planet Uranus is shown more or less as it would appear to the unaided eyes of a human observer. Although Uranus is often considered a near-twin of Neptune, the colours of the two planets are quite different: aquamarine in the former case and 'sky-blue' in the latter. (*Upper right*) This false-colour view of Uranus shows some banding which indicates that the planet was being viewed pole-on. (*Lower*) Part of an infrared 'movie' taken by the Hubble Space Telescope in August 1998 showing the 'wobbling' rings of Uranus, together with 10 of its moons. The atmosphere was surprisingly active, with about 20 clouds (nearly as many clouds on Uranus as the previous total in the history of modern observations) producing Jupiter- and Saturn-like latitudinal banding. Note that the planet is tipped on its side, with one pole facing towards us.

PLATE 11 – (*Upper*) In what has become perhaps the most famous picture of Neptune from Voyager 2, the giant planet is here revealed in all its splendour. Note the Great Dark Spot and white Scooter near the centre of Neptune's disk. (*Centre left*) Close-up shot of the Great Dark Spot and Scooter. (*Centre right*) Cirrus-like cloud streaks of frozen methane in Neptune's atmosphere. (*Lower left*) A pair of HST images, taken in June 1994, showing the conspicuous absence of both the Great Dark Spot and the Scooter. Courtesy of D. Crisp at JPL. (*Lower right*) Adaptive-optics image of Neptune, acquired by the Keck-II telescope on Mauna Kea in June 2000, showing an active atmosphere. Courtesy of Imke de Pater/UC Berkeley.

PLATE 12 – Photomosaic of Triton, based on Voyager 2 images, showing the distinctly pinkish south polar region and green cantaloupe terrain.

PLATE 13 – (*Upper*) Artist's concept of the Cassini spacecraft braking itself into orbit around Saturn in July 2004. (Left) An artist's impression of Cassini releasing the Huygens probe. Note that a certain licence has been taken concerning the range from Titan when this occurs. (*Upper right*) Artist's concept of Huygens descending through the Titanian atmosphere. (*Lower right*) Artist's concept of what Titan's surface may be like.

Plate 1

Plate 2

Plate 3

Plate 4

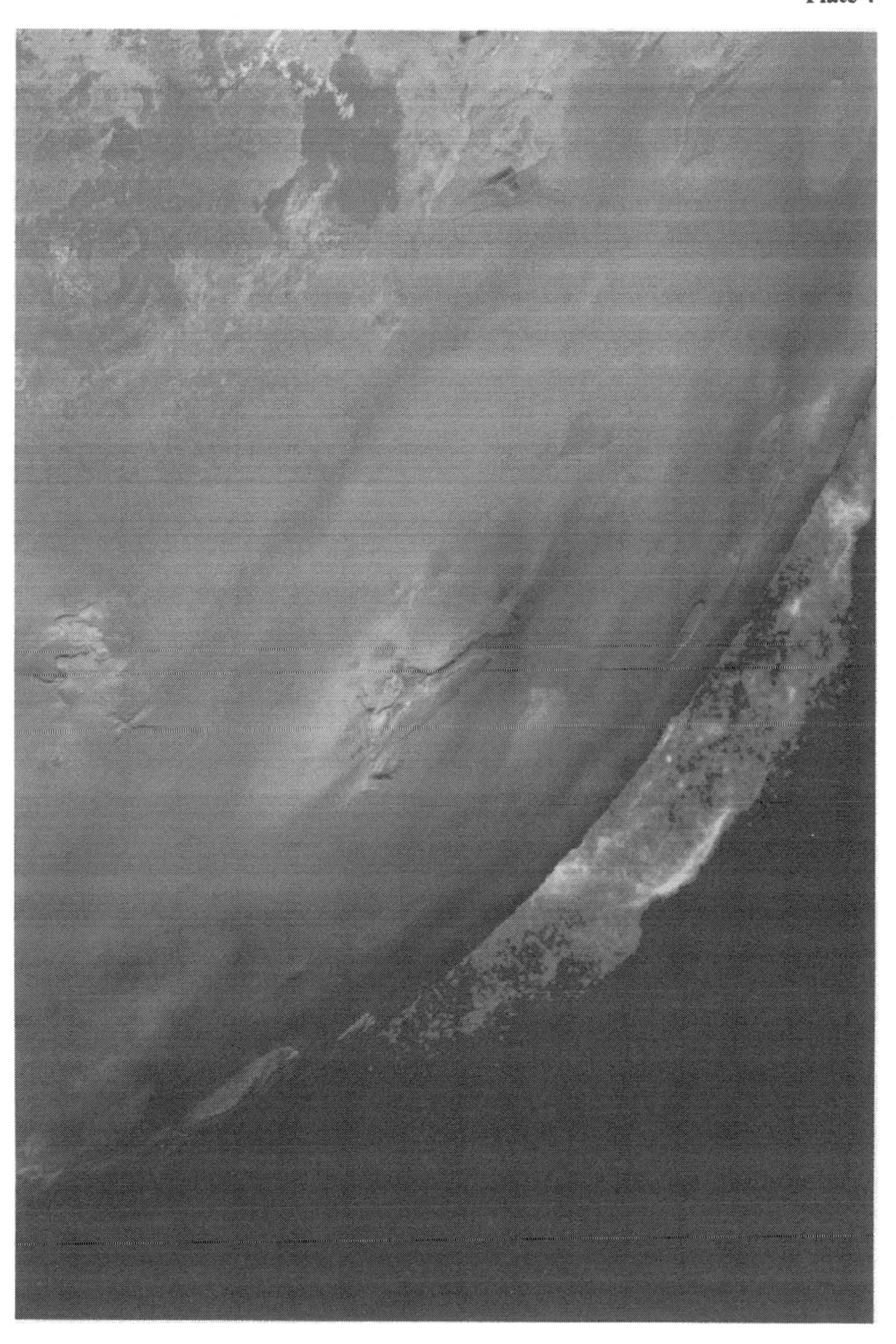

Plate 5

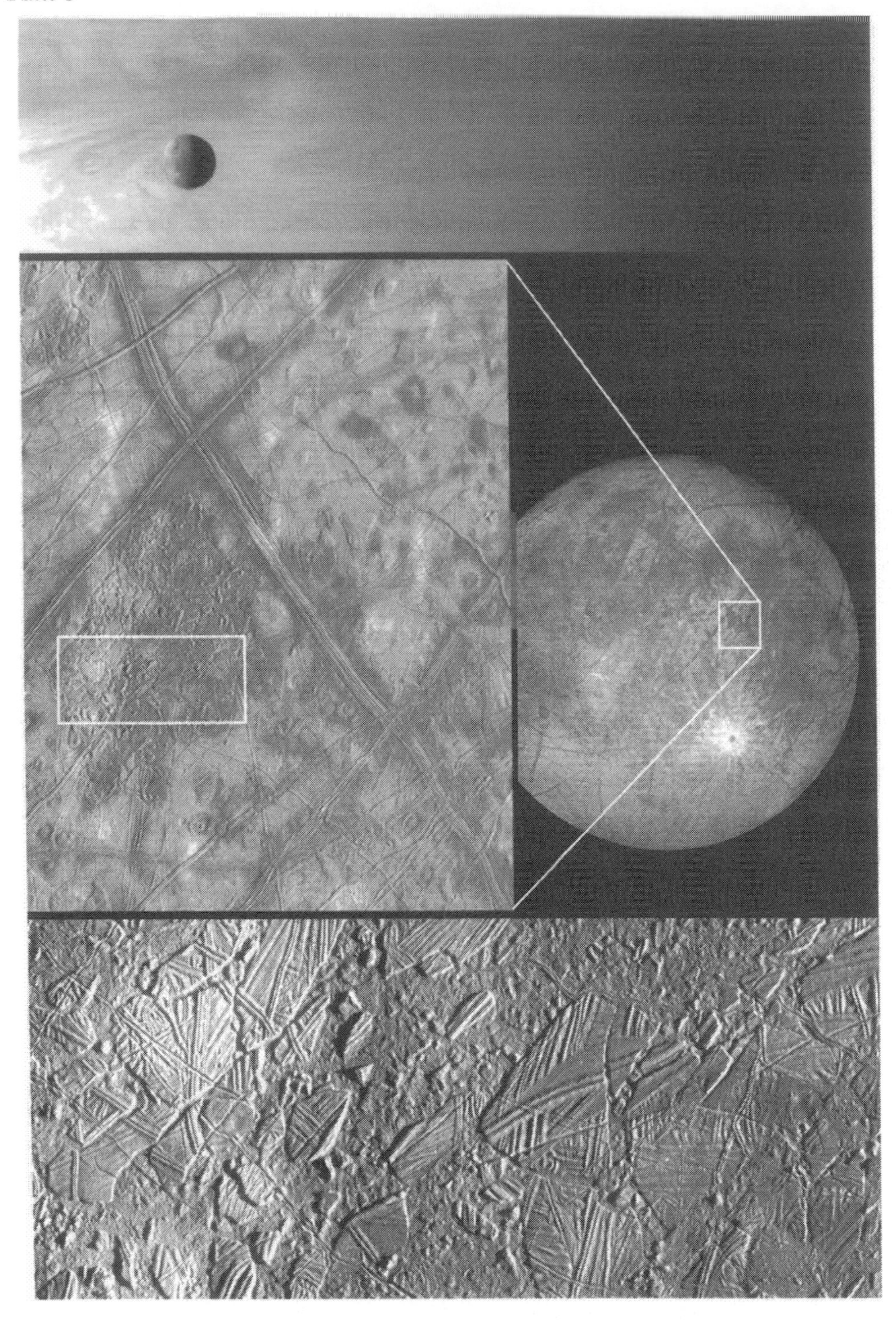

Plate 7

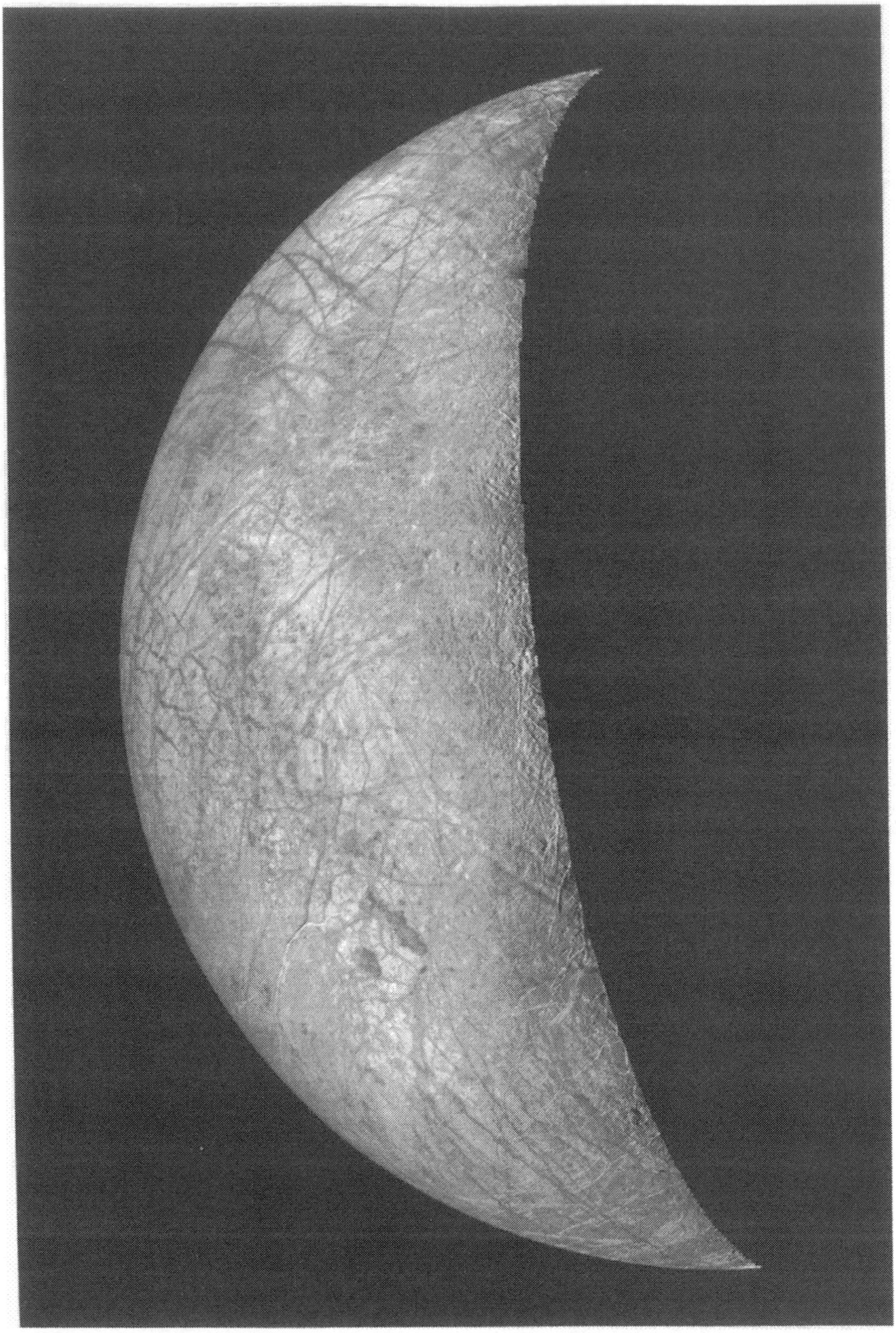

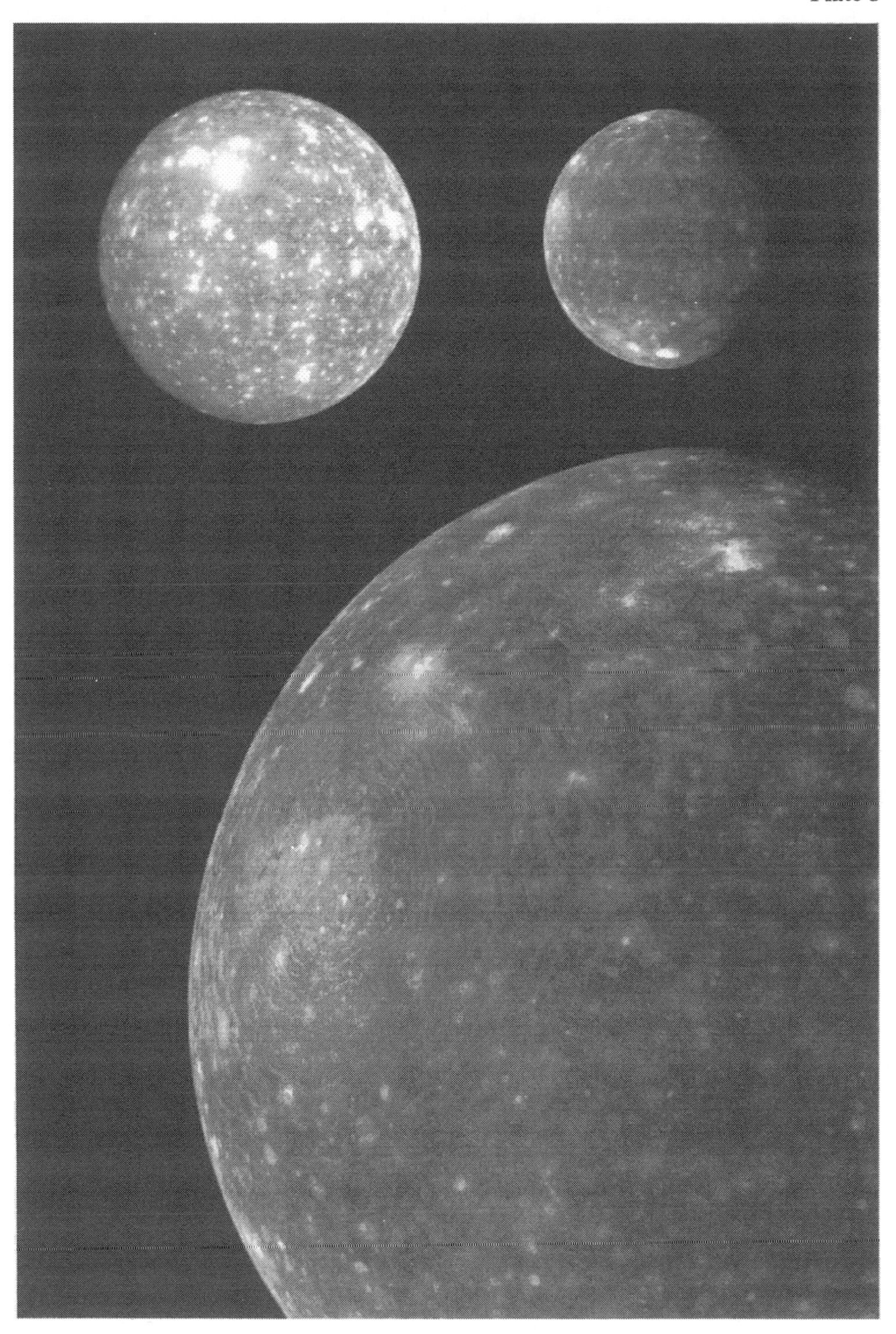

Plate 9

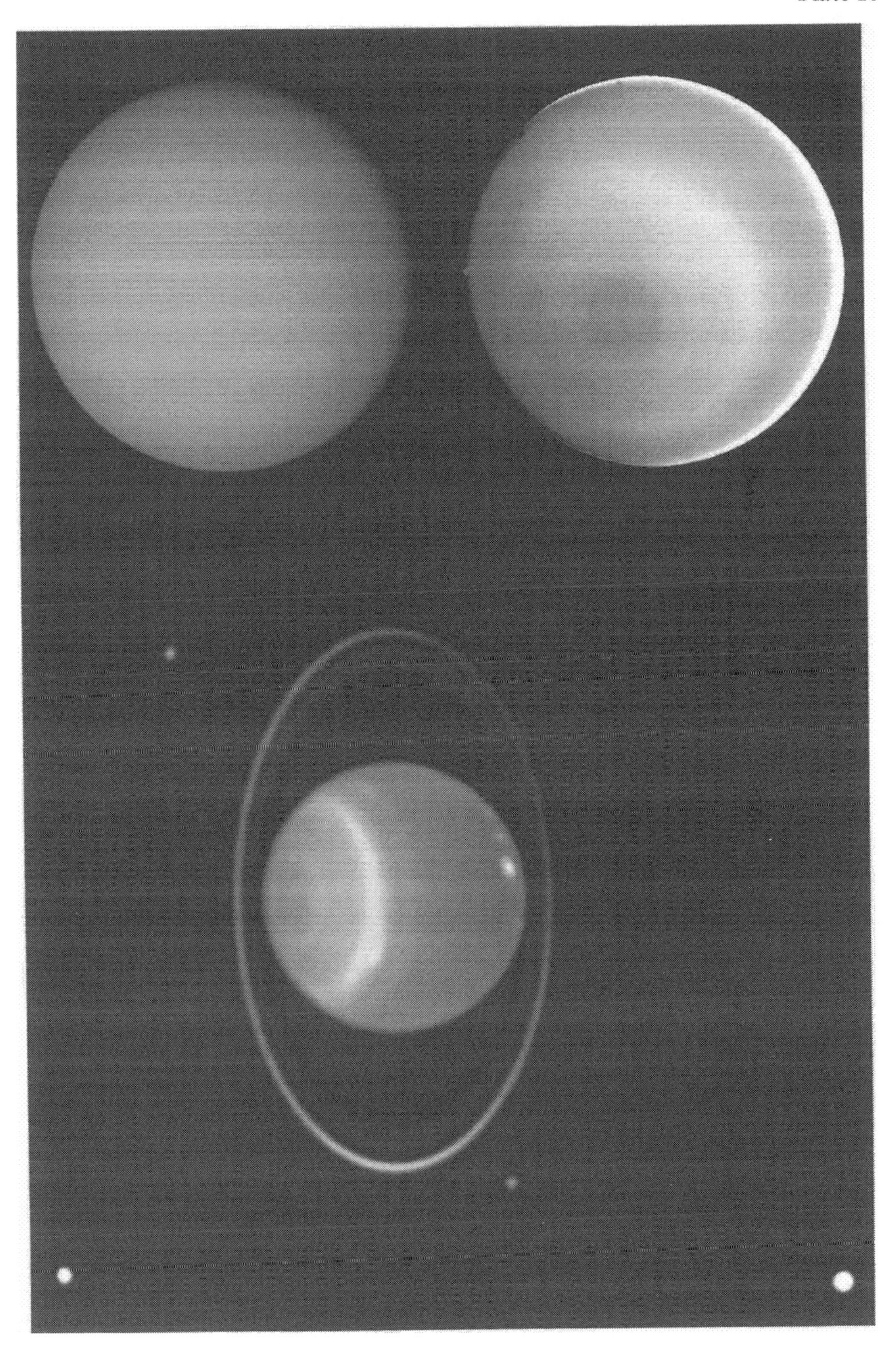

Plate 11

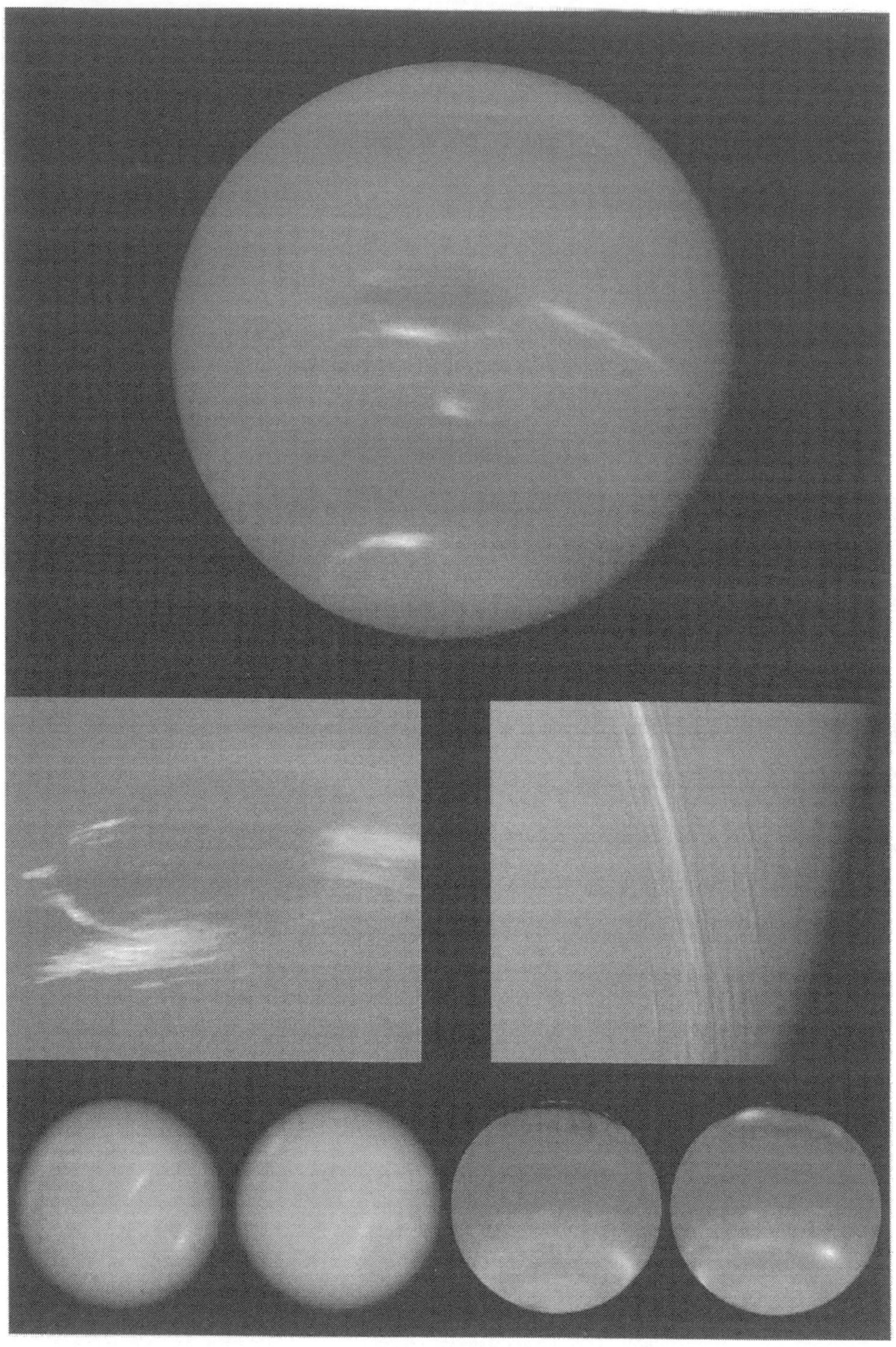

Plate 13

'transitory', while the larger features, like the Cassini and Encke Divisions, appear more permanent. The edges of these rings and other divisions where few clear gaps existed turned out to be so sharp that the ring must have been less than 200 m thick, and perhaps just 10 m thick in some patches. Another important discovery made by the Voyagers was that even the dimensions of the rings are not constant at all locations around the planet, with variations of 140–200 km recorded near the outermost edge of the B ring.

The majority of the spokes, particularly those in the classical B ring, are thought to have formed fairly recently and the speed at which they hurtle around Saturn seems to depend on their distance from the planet's centre. Some spokes actually appear to have been 'reprinted' over older ones. This phenomenon seemed to favour specific regions and was not confined solely to regions near Saturn's shadow. The increase in rotational period across the width of the B ring exceeds three hours; hence radial spokes should become distorted within minutes.

However, as the Voyagers watched, they revealed the spokes to persist for several *hours* as they rotated, became fainter and less well-defined with time and finally lost their integrity only just before re-entering the rings' shadow. On 25 October 1980, only two weeks and 24 million km from Saturn, Voyager 1 received a new command sequence from the ground to take 500-km-per-pixel shots every five minutes over Saturn's 10-hour rotation period. From this data, time-lapse movies of the spokes' rotation were produced. They revealed that, as their rotation matched Saturn's rotation, they were an artefact of the planet's magnetic field.

It was thus inferred that the spokes were charged particles being swept along over the B ring. The force that created the spokes was evidently at work where the planet's shadow fell onto the rings. Analysis concluded that the spoke particles were probably only a few microns wide and were 'elevated' away from the plane by either magnetic or electrostatic forces that were effective in darkness.

When Voyager 1 closed to within a few million kilometres of Saturn, it detected lightning-like radio bursts, apparently from the rings, which suggested that the material in the spokes was being 'charged up' in the shadow and the bursts were electrostatic discharges between the clouds of dust as it settled back towards the rings upon emerging into sunlight. Another possibility is that the radio bursts were due to atmospheric lightning in the super-rotating equatorial wind stream.

As the two Voyagers closed in on the giant planet, the spokes revealed themselves as dark lines against a bright backdrop of rings. However, as the spacecraft departed on the outward leg of the encounter, the spokes appeared much brighter than neighbouring areas, which led some scientists to suspect that the scattered material reflected sunlight more effectively in a 'forward' direction. This forward scattering is particularly characteristic of very fine, dust-sized particles and is our best interpretation so far as to their structure, at least until the arrival of Cassini.

THEORIES OF FORMATION

There are two plausible models for what causes planetary rings. The first model, developed by Edouard Roche, postulates that moons lying a certain distance from their parent planet will be torn apart by tidal forces. However, Roche's limit only truly explains ring systems located close to the planet; neglecting the outer reaches of, say, Saturn's rings. More recently, the destruction of Comet Shoemaker–Levy 9 when it hit Jupiter in July 1994 suggest that large ring systems could be produced by cometary debris that strayed inside the Roche limit.

Numerous efforts have been made to determine the number and packing frequency of craters on distant Saturnian moons in an attempt to work out how these might have contributed to the evolution of ring systems. The late Eugene Shoemaker argued that celestial bodies the size of Saturn's moon Mimas may have been broken apart on several occasions by large impacts, but their significant distance outside the Roche limit of the parent planet meant that they could coalesce into a single object again. Smaller, closer-in moons within the Roche limit, on the other hand, could not re-accumulate and their debris could form a permanent ring.

The second suggestion for the origin of the rings was proposed by Pierre Laplace and Immanuel Kant two centuries ago. They argued that Saturn's rings formed from the same circumplanetary nebula as the planet's moons, in a manner not dissimilar to the formation of the Solar System itself. In areas outside the Roche limit, the composition of ring particles may have depended heavily on their distance from the proto-planet. High temperatures near today's rings would have prevented the condensation of water for several million years.

This may explain why Saturn's rings are so bright and splendid in contrast to the darker, dustier rings of the other three giant planets. It is conceivable that Jupiter and Uranus lost their gaseous disks before cooling completely, which would have left behind only less-volatile materials like silicates from which to assemble their rings. In contrast, Saturn cooled much earlier, allowing water vapour to condense and eventually produce the brilliant rings that we see today.

It has long been assumed that planetary rings formed several billion years ago, although several lines of evidence strongly suggest that they are relatively young. Particles in Saturn's rings, in particular, are so bright and pristine that they seem unlikely to be much older than a few hundred million years. Darkening from cometary or micrometeoroid impacts might be expected to have taken its toll if they are much older than this, although Joe Burns points out that the number of these interplanetary intruders is dropping as more of them either hit planets or are ejected out of the Solar System.

Nevertheless, bombardment by Saturn's magnetospheric radiation should render the particles less 'clean' if they are more than a few hundred million years old. Moreover, the ring material absorbs charged particles circulating in the Saturnian magnetosphere; the 'cleanliness' of the particles implies that they are in a very benign environment in terms of radiation. In fact, when Voyager 1 flew 'below' the rings, it experienced the lowest level of ambient radiation since it had sat in the laboratory. It must also be borne in mind that some particles are so small – from car- and boulder-sized right down to sand-like fragments – that they would have been pulled into their parent planet's upper atmosphere and vaporised relatively quickly. The fact that some of them have not yet succumbed to this fate implies that they are quite young.

RINGS AROUND TERRESTRIAL PLANETS?

One key question that has been raised since the discovery that all four giant planets have ring systems is why there are no traces around the terrestrial worlds closer to the Sun. This may simply be a fluke at present, but there is no reason to believe that rings could not form around the terrestrial planets, nor that they have not done so in the distant past. In fact, computer models published in June 2002 suggested that temporary rings of debris might once have surrounded Earth.

According to Peter Fawcett of the University of New Mexico and Mark Boslough of the US Department of Energy's Sandia National Laboratories, a possible asteroid impact 35 million years ago might have kicked up a 'sea' of orbiting debris, creating a scaled-down version of Saturn's B ring around our own world. They argued that such a ring might form with a glancing blow in which an asteroid and the debris it carves from the planet ricochet into the atmosphere and thence into orbit. Over time, this debris would collapse into a plane matching Earth's equator.

Such a ring, said the pair, might have persisted for anywhere up to a million years and debris thrown into the atmosphere would have blotted-out the Sun and plunged Earth into a climatic 'cold spell' spanning 100,000 years or more. Certainly, geological records support the hypothesis by revealing a layer of melted meteoritic material thought to be associated with an asteroid impact around this time (in the Eocene epoch). "This cooling is longer than one would expect from a large impact alone, so we hypothesised that a temporary ring might have formed," said Fawcett.

"The jury is still out on this though." Of course, 35 million years ago there was no sign of even humanity's most distant hominid ancestors, but if the event were to occur today it is hard to say how easily visible the ring or rings would be to a human observer. Boslough, for his part, thinks that they may have been semi-transparent, much like Saturn's in this respect, and proposes that someone standing in their shadow would have seen a level of darkness on a par with twilight or a heavily overcast sky.

"I think the most spectacular view would be after sunset or before sunrise," he said, "when the sky is dark but parts of the sunlit ring would be brightly visible in the sky." It is quite possible that other planets may develop ring systems in the future; perhaps Earth, but also, according to Burns, maybe Mars, the orbit of whose moon Phobos is deteriorating. It is so close to its parent planet that it orbits within Mars' own 'day', so the tidal forces slow it and drag it closer. This may cause Phobos to impact Mars in several tens of millions of years' time, perhaps creating a temporary dusty ring, although this is purely speculative.

A LIGHTWEIGHT PLANET

Scientists' attention to the rings has not, however, distracted them from the sheer splendour of Saturn itself. With an equatorial diameter of almost 120,536 km, it is the second largest planet in the Solar System behind Jupiter, although in terms of density it ranks lowest. Its mass is only 30 per cent of that of Jupiter and it has often been said that if a large-enough bathtub could be found, Saturn would float on the water! Again, like Jupiter, it is primarily composed of hydrogen and helium, and also is believed to have a similar internal structure.

It probably consists of a rocky core about the size of Earth, around which circulates the enigmatic liquid metallic hydrogen layer first found at Jupiter and a topcoat of 'ordinary' molecular hydrogen. Various 'ices' are also present. Saturn's interior is thought to be as hot as 12,000 K at the core but, like Jupiter, the planet radiates more heat into space than it absorbs from received sunlight. Although it takes almost 30 years to complete a full circuit of the Sun, Saturn actually spins very rapidly: its 'day' lasts only 10 hours, which produces polar flattening and a bulging equator – the most pronounced 'oblateness' of all the planets.

The Voyagers revealed it to be a hazy, pale yellow world, with similar latitudinal banding to Jupiter, albeit much fainter and with less dramatic colour variations. Yet Saturn is by no means an inactive world. Although Voyager 1 noticed very little in the way of atmospheric markings, the cameras on its sister ship, which arrived nine months later and scraped just 41,000 km over the cloud-tops, uncovered long-lived oval storms and other subtle features.

ATMOSPHERIC STORMS

Nor were these merely transient or restricted to a specific time period, for HST has observed the ringed planet throughout the 1990s and found remarkably energetic storm patterns. Pictures from the telescope's WFPC-2, taken in December 1994, for example, revealed a white, arrowhead-shaped object near the equator. Such 'white storms' had been observed telescopically since the late nineteenth century. The 1994 storm turned out to be enormous – measuring some 12,700 km from its eastern to western extremities, or more than the diameter of Earth – and subsequent observations showed that its size and motion through the atmosphere remained remarkably constant through time.

By this point, of course, the notorious 'spherical aberration' suffered by HST shortly after launch, which prevented it from properly focusing its instruments, had been corrected by a Shuttle servicing mission the year before and the Saturn images turned out to be incredibly sharp. So sharp, in fact, that they even revealed the planet's prevailing winds forming a dark 'wedge' that ate into the westernmost edge of the giant storm. Moreover, comparisons with decade-old Voyager data enabled meteorologists to estimate what atmospheric conditions were like in the storm's neighbourhood.

They offer us a glimpse of a turbulent world. The close-range studies of Voyager 2

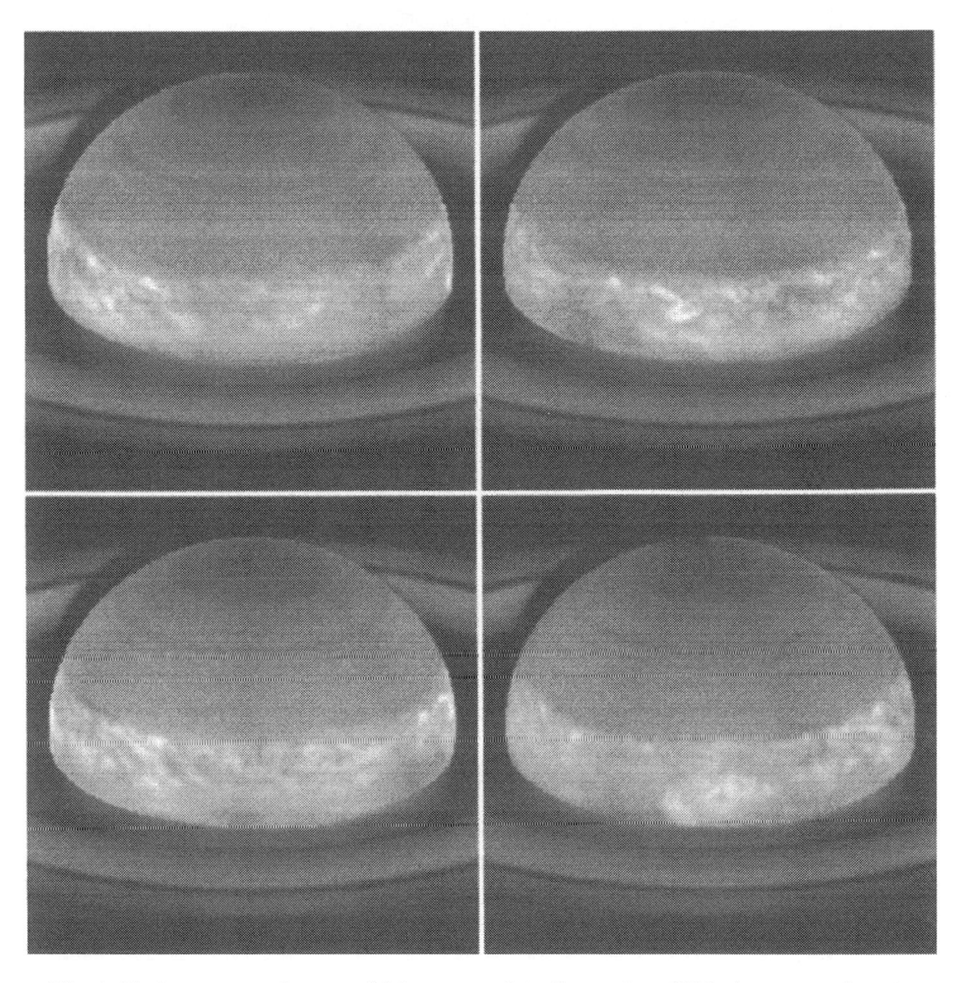

The 'white' storm on Saturn which appeared in September 1990, documented as it crossed the planet's disk in November of that year by the Hubble Space Telescope.

alone showed Saturn to be, at the time, the windiest place in the Solar System, with jetstreams flowing at up to 1,800 km/h near the equator. In contrast, the strongest hurricane-force winds on Earth go no higher than about 396 km/h. Other, gentler winds were recorded at higher latitudes, near the poles. In the case of the 1994 storm, 1,000 km/h winds seemed to decrease in speed around its northern rim, indicating that it was moving in an easterly direction.

Under the gaze of HST, clouds expanding north from the storm were gradually swept westwards by winds at higher latitudes. Strong winds close to the latitude of the dark wedge were also seen to blow over the storm's northern rim, creating a second disturbance which produced faint white clouds – probably made from ammonia-ice crystals – to the east of the system. The 1994 event was not the first such storm seen by HST; the telescope had observed a similar, though much larger,

A white, arrowhead-shaped storm, discovered in 1994, is shown against the backdrop of Saturn's clouds in this Hubble Space Telescope image.

phenomenon four years earlier. This storm, which first appeared in September 1990, came as quite a surprise.

It lay in Saturn's equatorial region, making it the first of its kind to be found since 1933. Discovered by amateur astronomers, the scientific community quickly realised that they were being handed what might be a once-in-a-lifetime opportunity to observe a rare equatorial storm on the planet. A White Spot Observing Team was quickly established and worked in conjunction with the HST team to spend several days observing the object in mid-November 1990, before Saturn approached conjunction with the Sun. Nevertheless, HST was able to record its progress in a 24-frame 'movie'.

The movie, the bulk of which was shot on 17 November, revealed that the storm actually ran completely around Saturn, although its characteristics at various locations differed quite markedly. In some areas, it looked like enormous masses of clouds, while in others it bore a closer resemblance to atmospheric turbulence. This data, combined with information archived from the Voyager encounters, allowed meteorologists to tentatively work out what might be going on in the giant planet's interior. The primarily easterly direction of the winds, for example, implied they were not confined to the upper levels of the atmosphere.

Rather, they probably extend to a depth of at least 2,000 km. Nevertheless, the Voyagers did identify winds that appeared to alternate from an easterly to a westerly direction at latitudes above 35 degrees. Not only that, but the Voyager 2 observations in particular noted a striking north–south symmetry that led some atmospheric specialists to speculate that winds might literally extend *right through* Saturn's interior!

AURORAE

The planet's somewhat chilly atmosphere – like Jupiter, a mix of predominantly hydrogen and helium, with compositional ratios similar to that of the primordial solar nebula – varies in temperature between 83 K and 140 K. This broad range, which both Voyagers showed were typically lower near to the north pole, may be purely seasonal. Like Jupiter, enormous aurorae were found in the polar regions, above 65 degrees latitude, and ultraviolet emissions of hydrogen were spotted by the UVS instruments in sunlit regions at mid-latitudes.

These phenomena puzzled some scientists, not least because upper-atmospheric bombardment by ions and electrons – thought to be the main cause of terrestrial aurorae – occurs mainly at higher latitudes. Aside from this problem, HST has presented a useful insight into the frequency and extent of Saturnian aurorae throughout the 1990s. One picture in particular, taken on 9 October 1995 at far-ultraviolet wavelengths, revealed unusually bright auroral displays in both the northern and southern hemispheres.

Notably, it picked out a luminous, circular 'band' centred on the north polar region, where a vast 'curtain' of light rose more than 2,000 km above the highest clouds. HST teams kept the telescope focused on the strange spectacle for two hours and watched as it changed dramatically in brightness and size. Similar phenomena, of similar intensity, were also seen two years later.

MAGNETIC FIELD

When Voyager 1 reached Saturn in November 1980, its results added to what had been learned from the Pioneer 11 spacecraft, which flew 22,000 km past the giant planet more than a year earlier. One of the most important findings from the Pioneer encounter was the unusual alignment of Saturn's magnetic field, which was found to be 'tilted' by less than a single degree relative to its rotation poles. Both Voyagers capitalised on this discovery by identifying several distinctive 'regions' within the field.

Notably, within 400,000 km of the planet's cloud-tops, they revealed a 'torus' of positively charged hydrogen and oxygen ions, probably caused by water-ice sputtered up from the surfaces of the moons Tethys and Dione. At the outermost edge of this torus, some of the ions were found to be accelerated to fantastic speeds and hit temperatures of 23 million K. Strong plasma-wave emissions – easily detectable to both Voyagers' PWS and PRA instruments – appeared to be associated with this 'inner' torus. Beyond this is a broad plasma 'sheet' stretching outward almost a million kilometres.

It was suggested at the time that the sheet is probably supplied with material by the combined atmospheres of Saturn and Titan, together with a neutral hydrogen torus – somewhere between 500,000 and 1.5 million km thick – that is known to trace the orbital path of the moon. Generally, Titan resides in the boundary (or 'magnetosheath') region of Saturn's magnetosphere, although sometimes its orbit does cause it to venture into the undisturbed solar wind when that is gusty. Voyager 1 found that Titan has no appreciable magnetic field of its own, which allows ions and electrons to interact directly with its upper atmosphere.

As a result of the high-velocity impacts with these charged particles, neutral gases in Titan's upper atmosphere quickly become ionised and swept off into space by

Saturn's magnetic field. The implication is that Titan must be gradually losing gases, and at the present rate of destruction all of the methane in its atmosphere should be lost within a few million years. This, of course, is only a tiny period of time in comparison to the 4.5-billion-year lifespan of the Solar System, so presumably there is something else at work – possibly cometary intruders or even volcanism – that replenishes the atmosphere.

Saturn's magnetic sphere of influence revealed itself to the Voyagers as considerably smaller than the Jovian field, although it still spanned more than 2 million km. However, this size is entirely determined by external pressures exerted on it by the solar wind. When Voyager 2 reached Saturn in the late summer of 1981, the wind pressure was particularly high and the planet's magnetic field extended only about 1.1 million km sunward. However, by the time the spacecraft left the Saturnian system, it had dropped and the field ballooned outwards in under six hours! It maintained this increased size for at least three days, by which time Voyager 2 had crossed the planet's magnetic field boundary and switched off its instruments in readiness for the trek out to Uranus.

In addition to their observations of the planet and its rings, the Voyagers returned considerable amounts of data and astonishing pictures of many Saturnian moons. As of late 2002, nearly three dozen moons were known to orbit the giant planet, and what we know of them stems almost entirely from those encounters more than 20 years ago. Undoubtedly, this will be updated significantly when Cassini arrives. The moons already known before the Voyager arrivals are, in order of distance from Saturn: Mimas, Enceladus, Tethys, Dione, Rhea, Titan, Hyperion, Iapetus and Phoebe.

'MID-SIZED' MOONS

Of these, the regular, 'mid-sized' ones, with equatorial diameters ranging from 400 to 1,600 km, occur in pairs that increase in size the further they lie from Saturn. Mimas and Enceladus are both roughly 500 km across, Tethys and Dione are approximately double this and Rhea and Iapetus are half as much again. Yet even big Rhea, the second-largest moon, is less than a third the diameter of planet-sized Titan.

Iapetus, the outermost mid-sized moon, was found by Cassini in 1671. It moves around its parent planet at a distance of around 3.6 million km in the relatively long period of 79 days and, at 1,440 km in diameter, is just smaller than Rhea. Even through his modest telescope, Cassini noticed something special about Iapetus. He watched it grow 'brighter', then 'dimmer', as it circled Saturn, and wrote "it seems that one part of [Iapetus'] surface is not so capable of reflecting to us the light of the Sun which maketh it visible, as the other part is".

Neither Voyager, unfortunately, approached Iapetus closely, but even their fairly low-resolution images unveiled a strange world, literally divided into rather bright and dark halves. The bright terrain seems to be very heavily cratered, much like Rhea is this respect, and covers most of the trailing hemisphere, together with portions of the northern polar region. The darker terrain has an albedo less than a tenth of the brighter areas and reflected so little light that very little detail was

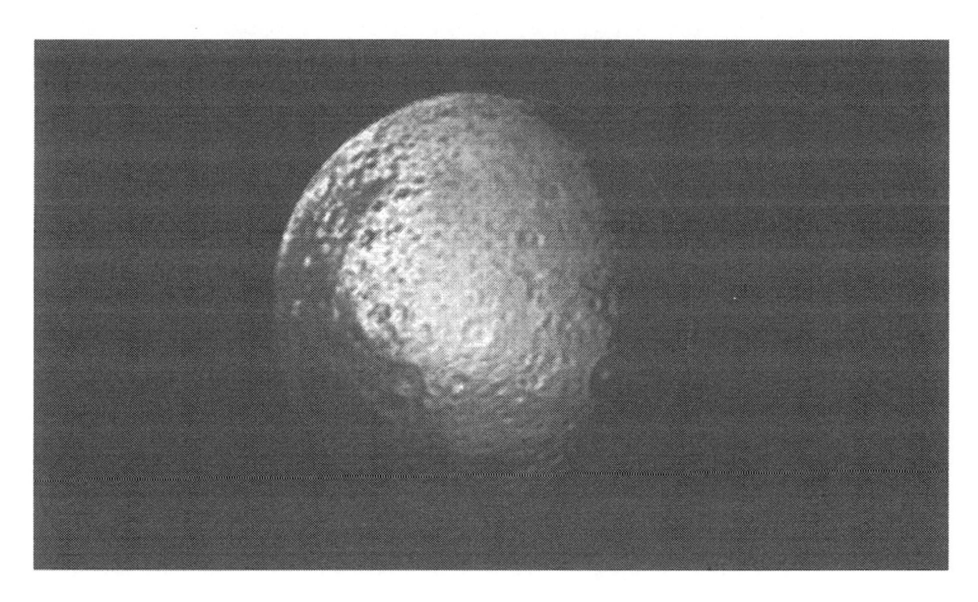

The strange, two-toned moon Iapetus.

discernible in any of the Voyager pictures. However, a recent re-analysis of the raw data has shown that what appear to be outlines are fairly large craters in the dark terrain. Significantly, the darkened zone is centred precisely on the leading hemisphere.

This has led some geophysicists to suspect that preferential bombardment of the leading hemisphere by dark material could be responsible for Iapetus' appearance, with internal activity – such as cryovolcanism – deemed unlikely. Moreover, the Voyagers photographed small white 'spots' in the dark portion near its border with the brighter region. This may suggest that the dark coating is quite thin and could easily be penetrated by impacts to reveal brighter subsurface material.

The demarcation between the dark and bright terrain is particularly intriguing. Voyager 2 flew within 909,000 km of the moon and its far-off pictures revealed a 'transitional' region 200–300 km wide. The dark–bright boundary of this region seems to be somewhat meandering in nature, with dark spots on the bright terrain and vice versa. In view of its distance from Saturn, Iapetus is certainly an icy world, and so, presumably, are the dark deposits that cover the brighter material. The sheer symmetry of the divide is so sharp, however, that the source may be traced to material impacting the moon.

PHOEBE: A CAPTURED ASTEROID?

Where might this material have come from? One popular contender is the small, very dark moon Phoebe, which lies a little further from Saturn than Iapetus and travels a retrograde orbit. Dark material from Phoebe, possibly ejected from some of its

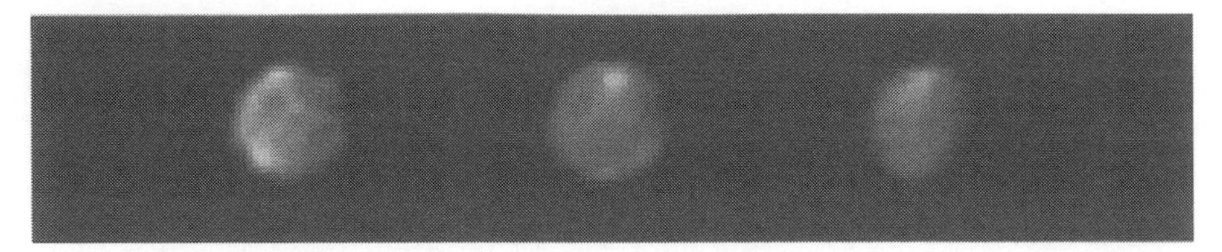

This roughly spherical blob – the moon Phoebe – was the first planetary moon to be discovered with the aid of photography in 1898. In view of its composition and likely origin, it may also be the first unmodified 'primitive' body in the Solar System to have been directly photographed by a spacecraft.

craters, could, it was theorised, have spiralled in towards Saturn and peppered Iapetus' leading face at speeds of more than 6 km/s. However, more recent spectroscopic research carried out by NASA astrophysicist Dale Cruikshank with the UKIRT on Mauna Kea has shown that, whereas Iapetus' leading hemisphere is dark-reddish in colour, the surface of Phoebe is more spectrally neutral.

An origin beyond the Saturnian system is strongly suspected. When Phoebe was discovered in 1898 – becoming the first planetary moon to be found with the aid of photography – its diameter was estimated at about 160 km. More significantly, its motion turned out to be retrograde, in that it orbits Saturn in a direction opposite to that of the majority of the planet's other moons. Moreover, unlike the rest of the Saturnian retinue, it lies in the plane of the ecliptic. This hints that Phoebe might be a captured asteroid. Unfortunately, neither Voyager flew any closer than 1.5 million km from Phoebe, so scientists have only a few long shots of its surface at present.

Despite the apparent mismatch in terms of surface colour with Iapetus, this does not mean that Phoebe was not somehow involved. It does, however, mean that if dust from Phoebe *was* responsible for part-contaminating Iapetus, there must also be something else at work that serves to intensify the darkness. Perhaps some as-yet-unknown ingredient in Iapetus' ice may account for the redness.

At present, dark-reddish material – which also characterises some comets and asteroids – is assumed to be organic, possibly comprising hydrogen, carbon, oxygen and nitrogen. It is conceivable that the early 'nebula' around Saturn, from which its moons would ultimately develop, may have been sufficiently cool at Iapetus' distance for methane to condense and be incorporated into its ices. When these were 'shock-heated' by impacting dust from Phoebe, the mix of volatile ices on Iapetus could be converted into reddish, tar-like compounds.

However, there are other scientists who argue persuasively that surface features picked out by Voyager 2 cast doubt on the idea that Phoebe is to blame for Iapetus' appearance. They point to a 'ring' of dark material, approximately 400 km across, situated close to the bright–dark boundary, which probably could not have been caused by material 'dusted' onto Iapetus' surface. Moreover, some craters in Iapetus' bright region have dark floors, pointing to a possible internal origin for the dark material.

It seems quite likely that this dark material has emerged from the interior, although in view of Iapetus' temperature, probably not by volcanic means. It could comprise ammonia, soft ice and a dark – maybe organic – ingredient. Scientists' knowledge of how thick the dark deposit may be varies from a thin 1–2-mm 'dusting' to several metres thick. It would appear that the dark material is either very young or extremely thick, because there are no visible examples of craters having penetrated the dark stuff and deposited bright ejecta on top of it.

Although Phoebe's surface is dark in colour, it is less 'red' than that of Iapetus. This makes it quite similar in appearance to a breed of asteroids known to inhabit the outer Solar System, and means that Voyager 2's images might make Phoebe the first relatively unmodified 'primitive' celestial body in the Solar System to have been directly photographed by a spacecraft. No surface features were discernible and even its rotation period is known only to be 'around nine hours'. In fact, the only definite characteristic of Phoebe, besides its colour, is that it is roughly spherical.

Many scientists suspect that it was captured by Saturn's gravity at some point in the distant past and has not changed much, geochemically or geophysically, since it first formed. It may have been ejected out of the inner Solar System by Jupiter's enormous gravitational field. Exactly when it was captured is unknown, but if the Cassini spacecraft finds that it is less heavily cratered than Saturn's innermost icy moons, it might be assumed that it arrived *after* the last major bombardment of Iapetus. Otherwise, we would expect Phoebe to have been disrupted or even destroyed as micrometeoroids pummelled its surface.

Some alternative arguments have proposed that Phoebe may be the ice-encrusted nucleus of an enormous comet, but its mass is too high for this to be a realistic possibility. An asteroidal origin has gained more support, not least following NASA astronomer Charles Kowal's discovery of a strange object called Chiron, which generally resides between the orbits of Saturn and Uranus, back in 1977. Kowal was searching for far-off comets with a Schmidt telescope at the Palomar Observatory in California when he first saw it.

In its orbit of the Sun, Chiron's path takes it as close as 1.3 billion km and as far as 2.8 billion km. Not much else is known about it, except that it may be rocky or icy in nature with an equatorial diameter of 100–400 km. More interestingly for Phoebe, it is believed that in 1664 BC, Chiron passed within 16 million km of Saturn, which is not much greater than the present mean distance between the planet and Phoebe, whose orbit is very elliptical. Some scientists, therefore, have gone so far as placing Phoebe and Chiron in the same 'class' of Solar System objects.

DIVERGENT PATHS

After leaving Saturn, both Voyagers hurtled off on wildly different routes to explore different parts of the Solar System. Voyager 1's path was bent northward, 'above' the level of the ecliptic plane, where it will eventually enter interstellar space. In doing so, it will depart the Solar System in the same direction that our Sun and its family are travelling through their local interstellar neighbourhood across the Milky Way in the direction of the constellation Hercules.

In the meantime, immediately after leaving Saturn, its cameras – and those of its sister ship, Voyager 2, at least for the time being – were turned off, although their UVS and fields and particles instruments continued to study the unique conditions of interplanetary space all around them. For Voyager 1, the remainder of its life would entail a long, lonely search for the edge of the Sun's realm. By mid-1981, however, NASA had reached the moment of truth when planning the future of its twin.

5

Bull's eye

TRIUMPH AND TRAGEDY

When Voyager 2 returned its first close-up pictures from Uranus in early 1986, they epitomised perfectly the extent of scientists' knowledge about the strange aquamarine world. They revealed a bland, featureless place, 2.9 billion km from Earth, with very little visible activity. Discovered two centuries earlier by a professional musician and amateur astronomer named William Herschel, Uranus had since remained not only on the fringes of the Solar System, but also on the fringes of human knowledge.

Its physical appearance, its moons – seen only as five distant, fuzzy specks in even the best pictures taken from Earth – and whatever bizarre conditions prevailed in its deep, cold atmosphere, were a virtual blank before 1986. It was left to Voyager 2 to shed new light on the world that Herschel had wanted to name in honour of King George III of England: "Georgium Sidus" or "George's Star".

The spacecraft's arrival came at a difficult time. As everyone who beheld the events of 1986 knows, a tragic twist of fate made Uranus NASA's only moment of glory in a year otherwise dominated by funerals, public disbelief, shock, then anger, and the painful process of rebuilding shattered dreams. Who can forget the horrific sight of the Space Shuttle Challenger being torn apart in a fireball above Cape Canaveral? Who can forget the ashen-faced flight controllers, staring aghast at their monitors? Who, indeed, can possibly forget the heart-rending scene of seven coffins, each draped in the Stars and Stripes, gently laid to rest? Such images indelibly burned their way into the public mind and turned many against NASA. It seemed in those dismal days that the exploration of space was over. Voyager 2's success not only proved otherwise, but helped to give the United States the courage and will to continue.

A TEMPTING OPPORTUNITY

Yet, had it not been for the concerted efforts of scientists at JPL more than a decade earlier, Uranus might not have happened at all. When the Grand Tour was proposed by Gary Flandro in 1965, both NASA and Congress had predictably balked at its steep price tag. Despite words of protest from some scientists, NASA opted for the simpler, cheaper MJS venture instead. However, many trajectory specialists were already aware that by carefully planning the route to be taken by one of the spacecraft after leaving Saturn, it should be possible to use the planet's gravity to propel it naturally to Uranus.

"At that time, even MJS looked good to most scientists," says Ellis Miner of JPL. "Those closest to the project knew from the start that the possibility of Uranus and Neptune flybys existed and expected NASA to approve that extension, assuming the spacecraft were still fully functional. The specifics were not laid out until after the Voyager 1 Saturn flyby, but each of the instrument providers did their best to prepare their instruments to be able to observe these more distant planets, although they had specific instructions not to add hardware costs purely for the sake of Uranus and Neptune observations."

The efforts of the planners paid dividends in mid-1976, a year before the missions began, when NASA conditionally approved Uranus, provided that the tasks planned for both Voyagers at Saturn were met. At about the same time, the agency also set aside funds to build a special Modified IRIS to overcome the predicted poor sensitivity of this instrument at Uranus. It was felt that the much lower intensity of reflected sunlight at such immense distances would render a standard IRIS ineffective.

"MIRIS was not much different from IRIS," says Miner, "except that the wavelength range was split into two segments and the response of the far infrared portion was considerably higher than that of IRIS." Although the modified variant never actually flew because of problems uncovered during pre-flight testing, its development at least showed that even before the Voyagers were launched, NASA was beginning to officially take an interest in the idea of a mission to Uranus.

By the end of August 1981, as Voyager 2 finished a spectacular, whirlwind tour of Saturn, an ecstatic NASA agreed that the tough little spacecraft could probably survive the 1.4-billion-km trek to Uranus with all of its instruments in good working order. An extension was duly authorised and funds provided for the Voyager Uranus Interstellar Mission (VUIM), which began on 1 October with the first of five planned thruster firings to set sail for the distant world.

According to Miner, however, Uranus funding was considerably lower than for either Jupiter or Saturn and the inevitable consequences were layoffs within the Voyager workforce. "The project staffing levels were lowered to about 50 per cent of those extant at Saturn," he says, "ostensibly with the idea that we could take twice as long to prepare for Uranus. The science team sizes were not reduced after Saturn, but their work level was considerably less than before Jupiter and Saturn."

AN UNKNOWN PLANET

As a result, when a hundred scientists met in Pasadena in early 1984 to discuss the current level of knowledge about Uranus and plan a comprehensive series of observations of the giant planet, they were welcomed by a somewhat leaner Voyager staff than had been present three years earlier. Furthermore, there remained many technical hurdles to overcome before a detailed survey of Uranus could even be contemplated.

One of the more pressing of these problems was Voyager 2's instrument-laden scan platform, which had experienced trouble during its last few days at Saturn, probably due to a loss of lubricant. It soon became clear that gears needed to move the platform backwards and forwards tended to grind together and seize if used at the fastest slew rate. After an extensive series of ground tests with 86 full-size replicas of the misbehaving device, engineers concluded that alternately operating the gears and giving them time to cool down should free the stuck platform.

Tests of the correctional procedures were radioed firstly to Voyager 1, which could safely be used as a test-bed since it had no more scheduled imaging targets, after which they were applied to its troubled sister. By February 1983, Voyager 2's platform was up and running smoothly again, albeit under new rules to restrict the frequency of its future movements and to limit its slewing rate. However, even after freeing the platform, the four-year cruise was by no means smooth sailing. Although it had been deemed possible to fly to Uranus with Voyager technology, neither spacecraft had been specifically designed or equipped to operate in such dark, distant parts of the Solar System.

Voyager 2 would be required to study and photograph a world where high noon is not as bright as dusk on Earth, together with a complex system of rings and moons several times darker than anything encountered before. More than one member of the imaging team has likened his job at Uranus to photographing a pile of charcoal briquettes at the foot of a Christmas tree, dimly lit by a single-watt bulb.

A JOURNEY OF UNCERTAINTY

The spacecraft itself exacerbated the problem. In the almost friction-free environment of space, its every motion – even turning its tape recorder on or off – caused a reaction, in the form of tiny nodding motions, which blurred its close-up pictures. At Jupiter and Saturn, such motions produced either negligible effects or could be handled by the gyroscopic reaction control system or the attitude-control thrusters, which fired short bursts to 'right' the spacecraft and keep their cameras as still as possible.

These helped to reduce their angular motions to a tenth of the speed of a clock's hour hand, but even that was too fast for the combination of high-resolution and longer (15 seconds or more) exposure times needed to reliably capture the Uranian system. Matters were worsened by the fact that the spacecraft would be hurtling non-stop past the giant planet at more than 64,000 km/h.

It was, however, possible to sidestep such obstacles by allowing Voyager 2 to 'settle' for longer after manoeuvres, scan platform movements or tape recorder activities and by halving the duration of its attitude-control firings to more gently correct the nodding motions. Shortening the length of thruster firings may have seemed the obvious solution, but in order to produce the same effect it required an increase in the number of individual firings, which led to real concern that this might wear out the thrusters before their time.

However, a series of more than 25,000 test firings by the thrusters' manufacturer throughout the winter of 1984 and into the spring of 1985 satisfied NASA that they could indeed hold up to an increased pace of usage. New command routines were also programmed into Voyager 2's CCS to make use of new Image Motion Compensation (IMC) techniques, first tried at Saturn's moon Rhea, as well as 'panning' the cameras under gyroscope control while tracking their targets, in order to minimise the blurring effects caused by the spacecraft's extreme flyby speeds.

The downside to improving Voyager 2's imaging capabilities was that, in addition to now being able to show up faint features at Uranus, it was also able to reveal inherent problems or damage in the cameras' optics. These typically took the form of annoying doughnut-shaped blobs caused by dust stuck on the cameras' lenses.

Most serious was a problem just a week before the spacecraft arrived at Uranus, when a computer error caused bright and dark streaks to crop up intermittently on several pictures. Fortunately, engineers were able to isolate and rectify the problem, which resulted from an incorrect memory location for new image data-compression instructions, and within three days the streaks vanished.

Of course, in order for pictures and data to be received by scientists, changes had to be made to the way Voyager 2 sent its information back to Earth. As it ventured deeper into space, its radio signal, received by the DSN, grew progressively fainter until, by the time it reached Uranus, it was several billion times weaker than the power of a watch battery. This could be compounded still further by bad weather, changes in Earth's ionosphere or increased solar wind activity, all of which could hamper the quality of the signal and the precious data it carried.

Further enhancements in CCS routines led to new measures whereby data was 'encoded' before transmission, thus helping engineers to identify and correct errors, and the transmission rate itself was reduced and data 'compressed' to ensure that the data that was sent could at least be received and understood with a measure of reliability.

DSN UPGRADES

Despite these ingenious solutions, probably the most vital development that maximised the scientific harvest from Uranus were changes to DSN data-capturing techniques, particularly 'antenna arraying'. In essence, this took the form of two or more ground-based antennas simultaneously collecting Voyager 2's weak signal, electronically synchronising and combining them and, in doing so, producing a stronger signal. During the course of the four-year cruise, the network received an extensive $100 million overhaul, in which all three 64-m stations at Goldstone, Madrid and Canberra were outfitted with adjacent 34-m antennas to facilitate arraying.

The extreme southerly latitude of Uranus in Earth's skies during the winter of

1985–1986 meant that Canberra was the main DSN station for the actual flyby, capable of tracking Voyager 2 for more than 12 hours per day, compared to only eight hours for Goldstone and Madrid. As a result, in addition to its main 64-m capability and dual 34-m subsidiary antennas, Canberra was beefed-up still further by a 400-km-long, NASA-funded microwave link to array it with the 64-m radio telescope at Parkes in central New South Wales.

This gave the vast complex a data-gathering area about equal to having a single, 100-m antenna, which in more practical terms translated to a 25 per cent increase over previous capabilities and allowed a faster bit-rate, which translated into an extra 50 pictures from Uranus to be received every day during the encounter. The benefits and limitations of using alternate DSN stations at different times meant that scientists' observation requirements had always to take into account the capabilities of the communications link with Voyager 2 at any one time.

With full antenna arraying in place, up to 8.4 kbps of data could be received by Madrid, compared to 14.4 kbps for Goldstone and 21.6 kbps for Canberra, the latter of which also had the longest tracking-pass time. However, if the network became overloaded with observation requests, conflicts arose between different scientists and painful compromises often had to be made about whether to reschedule them, cancel them or augment the DSN to accommodate them.

"Conflicts arose on a regular basis," Miner recalls. "For those that could not be accommodated by small alterations in the observation design which allowed both, or all, objectives to be met, the conflicts and alternative solutions were discussed by the entire Project Science Group. If consensus could not be achieved, the Project Scientist [Ed Stone] made the decision as to which alternative would be selected."

As with Jupiter and Saturn, this process became a difficult, uncomfortable juggling act, in which highly valuable observations lived or died, depending on how essential they were, how many resources they needed and how much they could contribute to the overall scientific yield from Uranus. During 1984, it became increasingly necessary for scientists to identify the observations that would reveal as many of the planet's mysteries as possible, and which instruments would be needed, in the short time span they had available. These were then set out in graphical timelines to ensure that the instruments were kept busy and the scientists happy.

THE 'BULL'S EYE' WORLD

Such planning was easier said than done, particularly as so little was known about Uranus before Voyager 2's visit. It had been known for some time, however, that the planet has one unusual feature that stands out from the rest: it has a rotational tilt of 98 degrees, which means that it is tipped on its side, like a celestial beer barrel, when orbiting the Sun. Exactly how this bizarre state of affairs came about has been a source of endless speculation, but one of the more plausible theories attributes it to a titanic collision with an Earth-sized object during Uranus' accretion period.

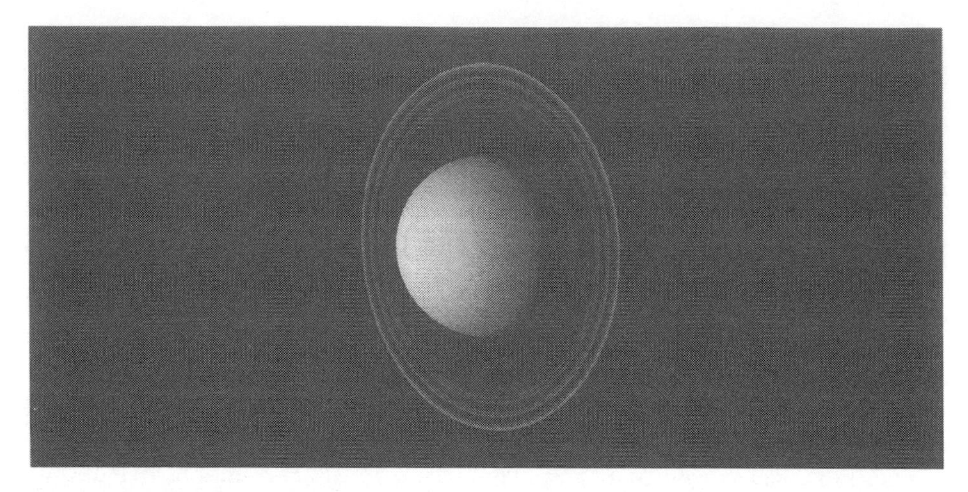

The unusual orientation of Uranus' rings and moons is primarily due to the 98-degree tilt of its rotational axis.

The fact that Uranus' five large moons orbit the planet within its equatorial plane implies that they formed much later from debris placed into orbit around the planet by this impact. Due to its peculiar orientation, the planet's poles receive considerably more sunlight than its equator. In fact, as Uranus rotates and progresses on its 84-year-long orbit of the Sun, one pole remains illuminated for 42 years, while the other is plunged into 42 years of frigid darkness. When Voyager 2 neared Uranus towards the end of 1985, only the southern hemisphere and those of the five large moons were in direct sunlight; their northern halves, shrouded in darkness, were in the dead of winter – a season that, in this tipped-over planet, lasts 21 years.

As all the moons orbited with Uranus' equatorial plane, and lay fairly close to the planet, the spacecraft would have less than six hours on 24 January 1986 to conduct the bulk of its close-range observations of the whole system. Time, sadly, was a luxury that scientists simply would not have at Uranus.

In this sense, it would be totally different from the work at Jupiter and Saturn, where scientists had several days to observe, and this limitation made the Pasadena conference in February 1984 all the more vital. It led to the creation of three working groups – essentially a Uranus task force – who worked through the spring and summer of that year to set out precise recommendations for a systematic survey of the planet and its entourage.

The groups published their report in late 1984. Its list of observations and questions that still needed answers forced the sheer paucity of knowledge about Uranus into stark relief. Even fundamental details such as the planet's chemical composition and its rotation rate were lacking, although scientists had made several informed guesses. Hydrogen and helium had long been suspected as the dominant chemical constituents in the atmosphere, together with traces of methane that gave Uranus its distinctive aquamarine hue, but exact ratios of one to another were needed to begin the complex process of determining the physical processes at work.

Observations from Earth in the early 1980s had inconclusively spotted what might have been clouds, together with bright, high-altitude hazes and even a dark, reddish-brown polar 'hood' in the upper atmosphere; the presence of all three were subsequently confirmed during the encounter. The polar hood in particular was similar to that previously seen on Titan. Voyager 2 revealed that it radiated large amounts of ultraviolet light in a phenomenon called 'day glow', which may be a thin haze triggered by chemical reactions between acetylene and methane in the upper atmosphere.

These observations, together with questions about Uranus' internal structure and whether or not it even possessed a core or magnetic field, were among the greatest unknowns facing Voyager 2. It was also hoped that its instruments would be able to determine if the planet had an internal heat source or if it simply absorbed and re-radiated solar energy.

Early studies suggested that, despite having both hemispheres subjected to long periods of intense sunlight and frozen darkness, its mean atmospheric temperature remained more or less the same, around 58 K. Understanding how Uranus heated itself had important implications for the existence of an as-yet-unknown atmospheric mechanism that percolates energy between the sunlit and dark poles when it is oriented pole-on to the Sun.

A SURPRISING DISCOVERY

Ironically, it was during a series of observations of the atmosphere a few years earlier to help to answer several of these questions that scientists found that Uranus is, like Saturn and Jupiter, a world bedecked with rings. On 10 March 1977, a team on board the Kuiper Airborne Observatory (KAO), a C-141 aircraft specially modified for astronomical research, were preparing to watch the planet pass in front of a star known as SAO 158687 in Libra. The 'SAO' indicates that it is listed in the Smithsonian Astrophysical Observatory's star catalogue.

The astronomers hoped that by measuring changes in its light as it passed through Uranus' atmosphere, it might be possible to derive clues about its temperature and chemical structure. Half an hour before the occultation was due to begin, the star mysteriously 'winked' five times on either side of Uranus. Another, ground-based team at Perth Observatory in Australia was watching the occultation and produced similar results.

It was a tell-tale sign that the star's light had been blotted-out by a series of five rings. The very fact that the star only winked and was not hidden from view for long quickly confirmed to the astonished teams that the rings must be extremely narrow. Only a few months later, Voyager 2 headed into space, and by the end of 1984 a detailed survey of the rings was close to the top of its scientific agenda for the Uranus encounter.

However, unlike Saturn's shining, frost-encrusted rings, which appeared to both Voyagers as a stunningly beautiful necklace with countless trillions of beads, those of Uranus bore a closer resemblance to clods of dirty charcoal dust, reflecting at best

CORNELL CHRONICLE

Vol. 8 No. 25 Thursday, March 31, 1977

Cornell Researchers Find Rings Around Uranus

Rings orbiting the planet Uranus — the first major structures in the solar system to be found since the discovery of the planet Pluto in 1930 — have been identified by Cornell University researchers flying aboard the NASA-Ames Research Center Kuiper Airborne Observatory.

Uranus is the seventh planet out from the Sun, one of the giants of the outer solar system. It is almost 1.8 billion kilometers (a billion miles) beyond Saturn, until now the only ringed planet, and is unique in "lying on its side" with its rotation axis almost in its orbit plane.

James Elliot, senior research associate at Cornell's Center for Radiophysics and Space Research, assisted by graduate student Edward Dunham and computer programmer Douglas Mink, made the discovery on March 10 while they were observing the temporary disappearance (occultation) of a faint star behind Uranus. The expedition was carried out by the Ames Center's Kuiper Observatory project team, headed by Carl Gillespie, expedition manager. The observations were made at 12,300 meters (41,000 feet) altitude, 2,000 km (1,200 miles) southwest of Australia over the southern Indian Ocean.

Elliot and his associates have inferred the presence of five rings orbiting Uranus, all of them in a narrow belt 7,000 km (4,400 miles) wide, lying 18,000 km (11,000 miles) out from the cloud tops of the planet. The five rings appear to consist of four thin inner rings, perhaps 10 km (6 miles) across, that follow nearly circular orbits around the planet, and one thick outer ring about 100 km (60 miles) wide, whose orbit may not be exactly circular.

Observations of the Uranus occultation made independently at Perth, Australia by Robert Millis of Lowell Observatory and by astronomers at Capetown, South Africa agree with the interpretation of at least five rings surrounding Uranus.

The rings are considerably smaller than those encircling Saturn. Elliot's data indicate that they are probably made up of fragments smaller than two kilometers (one mile) in diameter. They have never before been observed because the light reflected from the planet is sufficiently bright to obscure the lesser reflections from the rings under normal viewing conditions, Elliot said.

Elliot has named the rings for the first five letters of the Greek alphabet — alpha, beta, gamma, delta and epsilon.

Ames' Kuiper Observatory, from [illegible] astronomers. Named for pioneer planetary astronomer Gerard P. Kuiper, who discovered Uranus' fifth moon, it carries the world's largest airborne telescope, and has made various discoveries.

For the Uranus flight, because of its mobility, the C-141 provided the best solution. Flying far out over the southern ocean, it flew far enough south to be well within the shadow of Uranus and far enough into the Earth's night hemisphere to see occultations of the rings on both sides, as well as to be above any clouds. This combination of factors was impossible from any single ground-based observatory, and in fact several ground observatories were clouded out.

Viewing of the second ringed planet was done through the Kuiper Observatory's 91 cm. (36 inch) telescope, stabilized by gyroscopes and a tracking system that compensates for change in altitude of the plane during flight. The C-141 flew above 75 per cent of the Earth's atmosphere.

Data were displayed for the Cornell observers on a television screen, and variations in intensity of the light were recorded on magnetic tape and plotted automatically on graph paper.

The front page of the *Cornell Chronicle* for 31 March 1977, announcing the discovery of Uranus' rings.

only a few per cent of the weak sunlight that struck them. Very little could be done to study them before Voyager 2's visit; indeed, they could barely be seen at all in pictures taken from Earth.

PLANNING FOR THE ENCOUNTER

A similar problem faced scientists hoping to conduct the first comprehensive survey of Uranus' moons. Five were known to exist at the time, but in view of their extreme darkness, small size and immense distance from Earth, the simple fact that they existed was about all scientists could confidently say about them.

While the three working groups were busily completing their report, in June 1984 the Voyager Mission Planning Office set in motion a seven-month 'scoping product' to better estimate the resources needed to conduct the observations at Uranus. These resources included both those on board the spacecraft itself and those on Earth.

The scoping helped to ferret out and rectify any problems, as well as to fine-tune the sequence of events planned for the encounter. It was quickly realised that, as

Voyager 2 would have only a few hours to complete the most important close-range observations of Uranus for several decades to come, and in order to maximise its overall productivity, it should be given additional time to study the giant planet from afar.

NASA managers therefore decided to divide the encounter into four segments, spread over a 16-week period from November 1985 to February 1986. The first of these would be an Observatory phase, in which the spacecraft's instruments would be activated two months before closest approach for ongoing studies of Uranus, particularly in an effort to locate its elusive magnetic field. This would be followed by more detailed Far and Near Encounter phases, spanning the all-important fortnight from 10 to 26 January 1986 and requiring all 10 instruments to be running continuously.

The last segment was the Post Encounter phase. Ellis Miner has described this phase as "a little like the cleaning crew that does its work the morning after an all-night party". Yet, this last month in the Uranian neighbourhood, from 26 January to 25 February, would involve a considerable amount of critical work. Engineers needed to play back all the data stored on Voyager 2's tape recorder – and do so twice to ensure that it could be received and understood successfully – as well as turning off most of the instruments in readiness for the impending three-year journey to Neptune.

The decision to run a long-term series of observations before as well as during the encounter seemed to make absolute sense. When Voyager 2 was brought out of its semi-hibernation on 4 November 1985 to begin the Observatory phase, its pictures already greatly surpassed any taken from Earth in terms of resolution. Images acquired during July of that year, from a distance of 247 million km, had unveiled atmospheric features less than 4,600 km across.

Taking photographs of specific features on a planet, using a robotic spacecraft billions of kilometres from Earth, is an extremely complex process. "For all the planets, we had a prediction of where the spacecraft would be at each point in time," says Andy Ingersoll of Caltech. "We told the engineers what latitude and longitude we wanted to look at, and they told the camera to take a picture. These commands had to be worked out and radioed to the spacecraft weeks in advance. We had to predict the positions of interesting weather features weeks in advance."

Furthermore, unlike its observations at Jupiter and Saturn, which involved taking several narrow-angle shots during each rotation period, the imaging strategy at Uranus was somewhat different. Its south pole would always be facing Voyager 2 during the approach, and would change very little as the planet rotated.

In response, the spacecraft's cameras took a series of continuous, 38-hour 'movies' in order to better track cloud movements and determine wind speeds more accurately. At length, by early January 1986, the planet had grown so big in the narrow-angle lens that its wide-angle counterpart took over, although the former continued to produce high-resolution mosaics of scattered features, such as clouds and high-altitude hydrocarbon hazes.

During this period, Voyager 2's other instruments were being geared up for a fortnight of intensive activity. At intervals throughout the four-year cruise, they were calibrated to monitor their performance and sensitivity. These calibrations took a variety of forms. In the last few weeks before closest approach, for example, the PPS and UVS were trained on a number of stars to check their ability to pick out differences in brightness and chemical composition. Similar stellar observations planned for during and after closest approach would enable scientists to place the most critical, close-range observations reliably between two high-quality calibration points.

Other instruments, notably the MAG, were calibrated by turning the entire spacecraft through its pitch and yaw axes, while the cameras and the IRIS employed a target plate attached to the side of the main bus. This plate had precisely known reflective properties and helped to determine the instruments' ability to identify subtle brightness and colour differences. Still others, including the PRA, were fully calibrated before launch and needed no in-flight adjustment. In fact, according to Michael Kaiser of NASA's Goddard Space Flight Center in Greenbelt, Maryland, "turn 'er on and let 'er rip" has always been the operating strategy for such instruments!

A BLAND PLACE

As Voyager 2 marched through its final preparations, the excitement within the scientific community as January 1986 wore on reached fever pitch. Then, on the 24th, at 5:59 p.m. Greenwich Mean Time (GMT), Voyager 2 swept silently just 81,500 km over the Uranian cloud tops. Trajectory planners would later trumpet the impressive accuracy of the spacecraft's delivery to the planet: just 100 km off-target equated to sinking a golf putt from a distance of 3,630 km!

Two hours and 45 minutes later, scientists and engineers yelled with delight and broke into spontaneous applause as a stream of data, that had been transmitted across 2.9 billion km of space to Canberra and from there to JPL via NASA communications satellite, announced success in the form of stunning images on their monitors. The Near Encounter phase, which officially ran for four days but was in reality focused into those six, up-close-and-personal hours on the 24th, was in full swing.

Yet, the pictures themselves – at least the true-colour ones that were closest to how an astronaut's unaided eyes might see Uranus – revealed a misleadingly quiescent place, with very little apparently going on in its dense atmosphere. They were 'stunning', of course, because they revealed a world never before seen up close by human eyes, but they exuded little of the splendour of Jupiter or Saturn.

There were no violent storms, no zonal bands or swirling spots and eddies as on the larger gas giants; nor was there the same kaleidoscope of vivid colour as that seen on Jupiter. In fact, some members of the imaging team for the Uranus encounter were so disappointed that they wryly renamed themselves "The Imagining Team". False-colour pictures of the planet, on the other hand, had a somewhat different tale to tell. They picked out chemicals in the upper atmosphere and positively identified clouds beneath a high-altitude hydrocarbon haze, but the overall impression remained one of a bland, inactive world.

Nevertheless, Voyager 2's work revealed more about Uranus in a few feverish days than had previously been possible in more than 200 years of research. UVS observations of gases escaping from the upper atmosphere helped to gain a clearer insight into their chemical composition, while the IRIS and occultations performed by the RSS precisely measured their temperatures and pressures.

They revealed, as expected, a world made predominantly from hydrogen and helium (83 per cent and 15 per cent respectively by number), in true 'gas giant' fashion, but with a proportionately greater abundance of methane and heavier elements than either Jupiter or Saturn. The temperature was found to be relatively uniform throughout the atmosphere, around 57 K, although, curiously, the equator – which received less sunlight than the sunlit pole – was slightly hotter. It remains to be seen if this indicates the presence of an as-yet-unknown heat-transfer mechanism in Uranus' upper atmosphere.

It had been correctly predicted before Voyager 2's visit that the planet's plain aquamarine façade is in all likelihood caused by methane, which absorbs red light from the weak sunlight that strikes it and reflects back colours at the blue–green end of the spectrum. Indeed, the presence of methane within both Uranus and Neptune may be sufficient to produce some unusual atmospheric phenomena.

Laboratory experiments, performed at the University of California in Berkeley in 1999 'squeezed' samples of methane to several hundred thousand times Earth's atmospheric pressure and heated them to 3,000 K with a laser beam, hoping to replicate the kind of conditions found deep inside Uranus and Neptune. The experiments produced diamond dust and Robin Benedetti, a postdoctoral researcher and joint team leader, thinks the existence of a kind of diamond 'rain' on both planets is quite possible. "The diamonds would rain down through the water, methane and ammonia ocean like stones falling through honey," she says.

Benedetti and her team believe that these and other extremely complex chemical reactions could well take place inside Uranus and Neptune. Some scientists have seen the different structure of these giants from Jupiter and Saturn as evidence that they took much longer to form and evolve, and picked up different chemical constituents along the way. Current thinking is that shortly after the Sun had ignited, it blew lighter gases such as hydrogen and helium out of the Solar System.

Jupiter and Saturn had evolved early enough to protect their massive reservoirs of these gases but, according to Ingersoll, Uranus and Neptune formed more slowly, possibly because the primordial solar nebula was less dense at its outermost edge and took longer to pull itself into planet-sized objects. Others, including Bill Hubbard, have even gone so far as to regard Uranus and Neptune as a different breed of planetary giants altogether: 'ice giants', he thinks, rather than true 'gas giants'.

Indeed, Voyager 2's observations uncovered a much more complex picture of the internal structure of Uranus; a picture that has been clarified and modified in light of more recent research using HST. It seems likely, based on Voyager's findings and theoretical modelling, that Uranus' atmosphere extends to a depth of at least 3,000 km, to an icy, slushy or perhaps even liquid 'ocean' of water, ammonia, methane and other volatiles, which could in turn surround a small, rocky core about the size of Earth.

Many atmospheric specialists have speculated that the existence of such a super-pressurised ocean deep within Uranus, presumably heated by the planet's core, may be the source of its strange magnetic field, which Voyager 2's instruments slowly began to detect in the last week before closest approach.

It seems a pity that the spacecraft did not carry an infrared camera, as it may have seen Uranus very differently. Not until HST cast its high-resolution, near-infrared gaze on the planet in the mid-1990s did a totally different Uranus start to emerge from the gloom: a banded world very much like Jupiter and Saturn, with some of the brightest clouds, coldest temperatures and most violent storms ever seen in the Solar System.

DARK MOONS

Together with a large retinue of moons, all intriguing geological treasure troves in their own right, Uranus is a worthy candidate for future study. In fact, it is the planet's remarkable collection of moons that have, since 1999 at least, guaranteed it third place in the record books (behind Jupiter and Saturn) for having the most natural satellites in the Solar System: around two dozen so far. At the time of the

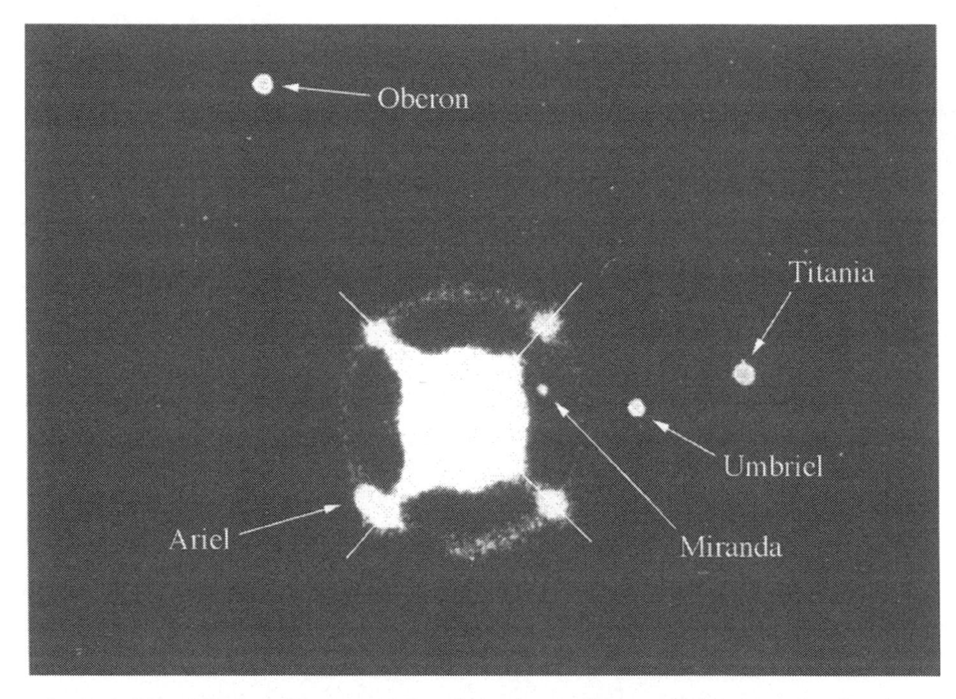

In 1948 Gerard Kuiper discovered Miranda in close to Uranus during a search using the 2-m reflector at the McDonald Observatory of the University of Texas. The planet's disk was overexposed in order to be able to seek faint moons. The symmetrical quartet are telescopic artefacts.

Voyager 2 encounter, however, only five were known. In order of their discovery, these were Titania, Oberon, Umbriel, Ariel and lastly, found in 1948 by Gerard Kuiper, tiny Miranda.

All five were named after characters in the works of William Shakespeare and Alexander Pope, a convention that owes itself to William Lassell, who discovered and named both Umbriel and Ariel in 1851. All five were barely detectable using even the best Earth-based instruments, but within days of beginning its Observatory phase, Voyager 2 swamped the scientists with reams of new information. Ten new moons were added to Uranus' tally by the end of January 1986 and, in view of their darkness and diminutive size, more were expected. Such expectations proved well founded, but had to await the 1990s for confirmation.

Voyager 2's instruments, particularly the UVS, which looked for the ultraviolet 'signatures' of gases escaping from the moons' surfaces in an effort to determine their chemical composition, revealed that they appeared to be made from about 50 per cent water-ice, 30 per cent rock and 20 per cent other elements, including, significantly, carbon and nitrogen.

An explanation for their compositional differences and extreme darkness, particularly in the case of charcoal-grey Umbriel, which reflects a mere fraction of the sunlight that hits it, could be the magnetic field of Uranus itself. Until 19 January 1986, just 7 million km away and five days before closest approach, Voyager 2's PRA and PWS instruments had still not detected a magnetic field around the planet. Such fields are thought to come from electrical currents produced deep inside the planets. In the case of Uranus, it was suspected to originate in the hypothetical, electrically conducting, super-pressurised ocean somewhere between its atmosphere and the core.

MAGNETIC FIELD

Although neither the PRA nor the PWS was able to find any conclusive evidence that such an ocean exists, the planet's peculiar orientation is consistent with theories that its magnetic field is generated at an intermediate depth in its atmosphere, where pressures are expected to be high enough to make water electrically conducting. At length, the two instruments picked up tell-tale radio signals produced by interactions between the solar wind and high-speed electrons moving in Uranus' magnetic field.

The discovery enabled MAG and PRA to calculate, with an accuracy of under a minute, the length of Uranus' 'day' at 17 hours and 14 minutes. "This is important for many reasons," explains Kaiser, "not least of which is to determine how fast atmospheric winds are blowing. You also have some measure of how fast the 'solid' planet is moving to compare against. Also, the people who measure energetic particles and magnetic fields need to know how fast the planet is rotating to properly analyse their data [and] for spacecraft operations, things are generally scheduled around the planet's length of day."

When Voyager 2 actually entered the magnetic field a few days later, it found a few surprises: it extended only about 600,000 km sunward, but wound downwind 10 million km into an extraordinary 'corkscrew' shape behind the planet. UVS

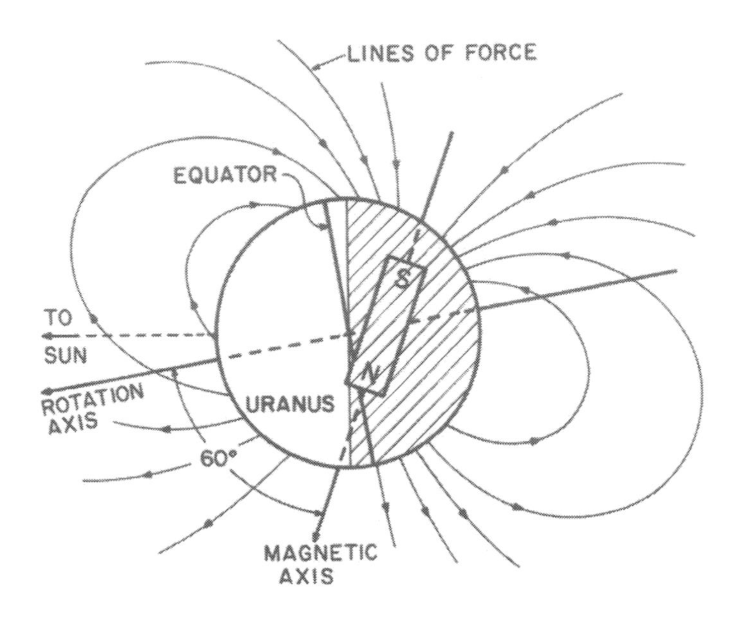

Diagram of the extraordinary tilt of Uranus' magnetic field. Note that the planet's axis of rotation lies almost in its orbital plane.

observations of auroral emissions from the poles also revealed that the field is 'skewed', with its magnetic axis tilted at a 59-degree angle to its rotational axis, a peculiarity that on Earth would be like having our north magnetic pole in Florida. Uranus' magnetotail should change shape depending on whether the planet (and its magnetosphere) is pole-on or broadside to the solar wind.

This, coupled with the 98-degree rotational tilt, could explain the bizarre corkscrew-like magnetic tail. One of the suggested reasons for this strange tilt was that the planet was in the middle of a periodic realignment of its magnetic axis; although some scientists expressed scepticism that such an event should have conveniently occurred just as Voyager 2 was hurtling through the system. This was undermined by the subsequent discovery that Neptune also has an inclined field; yet, in view of the fact that Earth's field reverses from time to time, it was not such a strange theory.

The field proved to have a powerful 'sting' in the form of intense, trapped radiation, which could have left its mark on the Uranian moons. The radiation belts themselves are of similar intensity to those of Saturn, although of different composition, and were actually found to be so intense that Uranian ions would very quickly break down and darken (within 100,000 years) any methane in the moon's icy surfaces, leaving a thick carbon surface 'dusting'. The fact that Umbriel is considerably darker than any of the other large moons could, therefore, indicate that it originally had a substantial methane content, or is bathed in the most intense radiation.

In spite of their kinship in terms of colour, the moons displayed a number of distinct differences from one another. However, it is dangerous to make sweeping generalisations here. Voyager 2, after all, flew through the Uranian system when all five moons had only their southern hemispheres in sunlight, so most of their surfaces – in some cases as much as 65 per cent – were totally shrouded in darkness.

This incomplete picture was hampered still further by the relatively poor resolution of the spacecraft's pictures: in the case of Oberon, for example, a strange little world with a host of mysteries unique to itself, even the smallest surface features detected by Voyager 2's cameras were about 20 km across.

MYSTERIOUS MIRANDA

However, one moon in particular was seen with greater clarity. Miranda – the last to be discovered in the pre-Voyager era, and at just 500 km in diameter the smallest of the five – proved to be one of the most remarkable worlds ever visited by an emissary from Earth. "Uranus' gravity increased the velocity of Voyager 2 by about 2 km/s," explains Miner. "The flyby distance was determined by the need to go on to Neptune and was close to the orbital distance of Miranda. It was for that reason that Voyager 2 was able to obtain high-resolution images of Miranda."

Before Voyager 2's arrival, scientists thought Miranda was the innermost Uranian moon and mission planners therefore executed a thruster firing in November 1984 to enable the spacecraft to fly within 29,000 km of it. They did not know if the tiny moon would be worth the effort, but luckily, they were not to be disappointed.

It would probably not be waxing lyrical to excess if one was to call Miranda a geological jigsaw puzzle or a mosaic of natural disasters, for these are just a couple of its most frequent alter-egos. The moon has one of the most bizarre surfaces ever seen by human eyes: one that surely testifies to a long history of tumultuous upheaval.

Merely glancing at Voyager 2's amazing images of Miranda shows clearly the traces of aeons of meteoroid bombardment so unrelentingly savage that the moon may have been literally torn to shreds, then forcibly hammered back together, as many as five times to produce its present appearance – a cataclysm that may also have occurred on Mimas. It may have been a result of these outside attacks that major disturbances took place within, for Voyager 2 spotted traces of internal melting and the occasional 'upwelling' of icy material from deep inside Miranda.

The evidence for such violent, remorseless activity, by whatever means it came, is seen in every one of Voyager 2's pictures. Vast, fault-like canyons up to 20 km deep; enormous oval, racetrack-like features, resembling scratches that wound their way across half of the moon's surface; and mysterious 'terraced' regions with a mix-'n'-match of old and young, dark and bright, heavily and lightly cratered surface types left scientists awestruck and baffled.

Certainly, several of the younger areas of lightly cratered terrain, dubbed 'coronae', have often been cited as evidence for the fire-and-brimstone theory of Miranda's evolution. One of these in particular, the Inverness Corona, was found to have a pale, chevron-shaped feature sandwiched between darker layers, suggesting

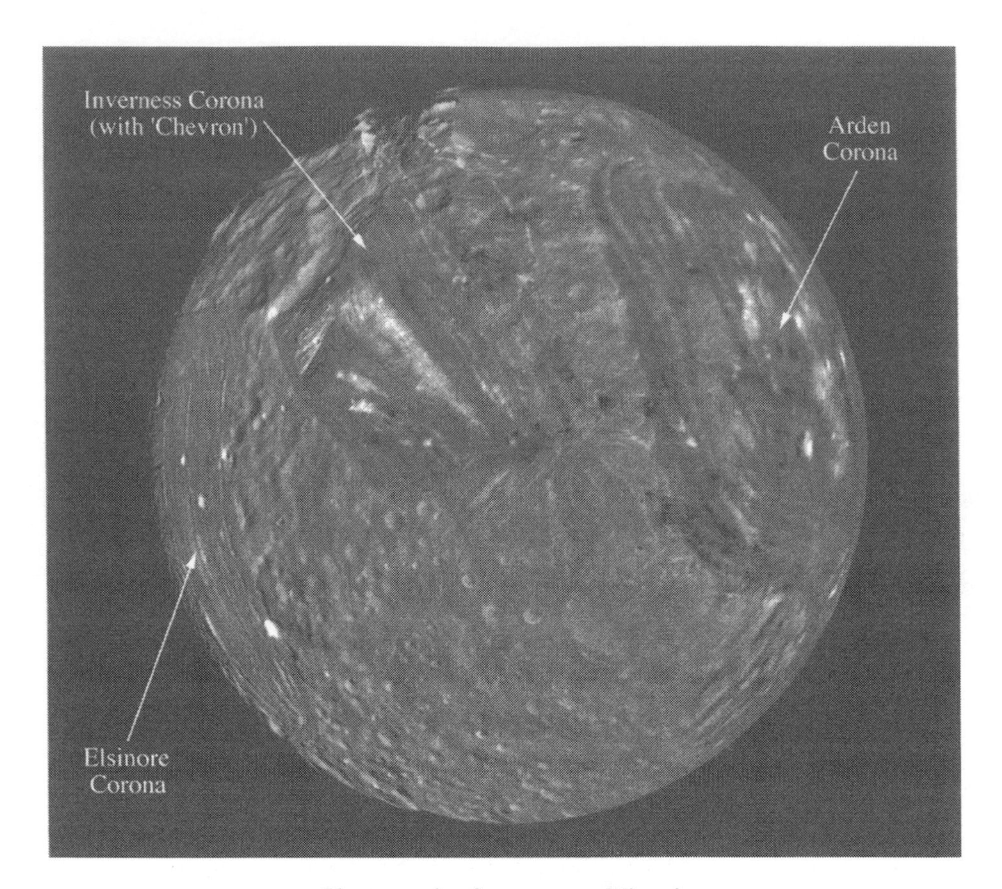

Photomosaic of coronae on Miranda.

that it may represent re-aggregated bits of the original moon exposed end-on, sticking out of the present surface.

Some scientists have seen in its striking appearance the result of the incomplete evolution of the moon itself. Perhaps, they argue, Miranda froze midway through a long geological process in which it almost literally turned itself inside out. Such processes are common in the early evolutionary histories of most terrestrial bodies. Proponents of this theory have pointed to the coronae as evidence that Miranda's growth was dramatically halted, like some Uranian Peter Pan, and eternally deep-frozen at some stage in its immature youth.

More recent interpretations have speculated that after an initial spate of heavy meteoroid bombardment to set them off, the coronae actually developed through either internal tectonic activity or some form of volcanism. This has led scientists such as Bob Pappalardo of the University of Colorado in Boulder and Bill McKinnon of Washington University of St Louis, to suggest that the ridged features could be icy volcanoes, implying that internal heating and 'rifting' of the crust could help to explain its bizarre morphology.

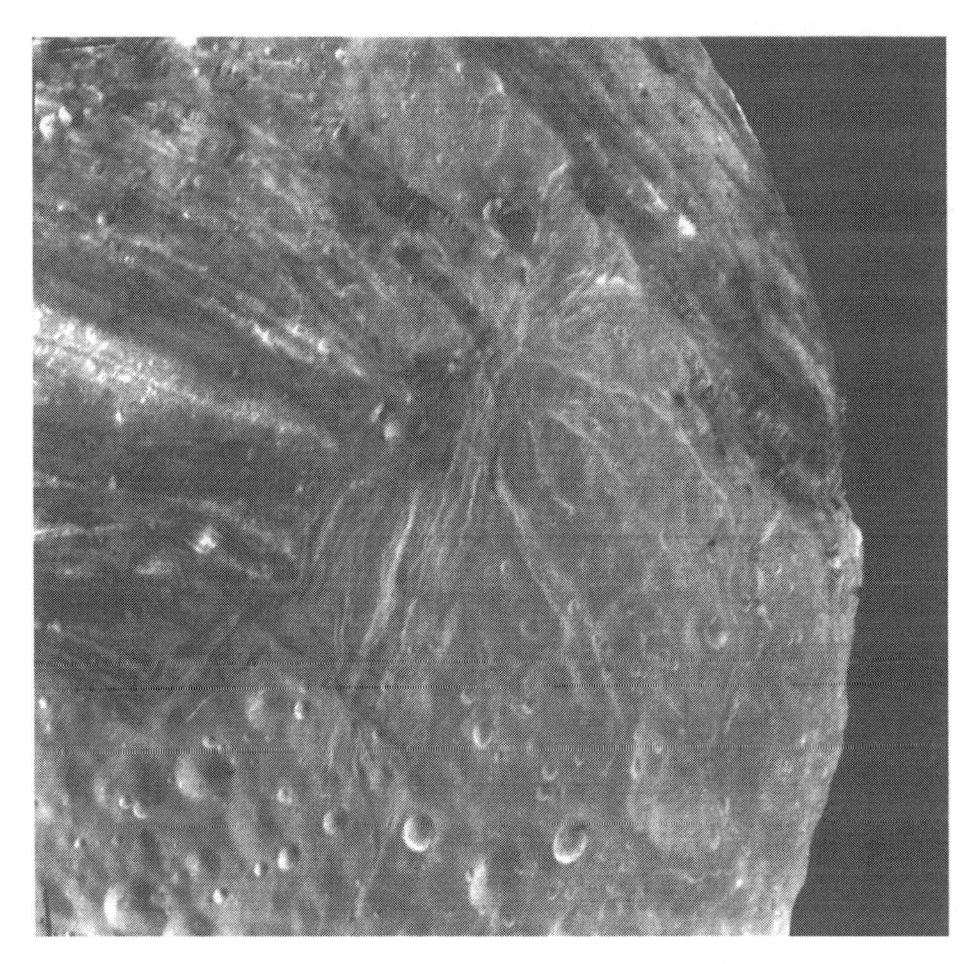

Rugged terrain of Inverness and Arden Coronae on Miranda.

Whichever of these theories one might wish to accept is compounded by a major problem: where did a moon as small and as cold as Miranda, at around 73 K, gain sufficient heat to sustain such intense internal activity? The only plausible answer, based on present knowledge, is Uranus itself. Perhaps, in the remote past, the tiny moon occupied a more unusual orbit around its giant host and was subjected to strong tidal heating in a manner not dissimilar to what both Voyagers saw so dramatically at Io.

Although Miranda was the scientists' best close-up look at a Uranian moon, all five came under intense scrutiny during the flyby. Voyager 2 managed to obtain high-resolution, full-colour pictures of them, although they all turned out to be predominantly dark grey, while the PPS measured their respective brightness under a variety of conditions and from different angles to determine how much sunlight they reflected. Umbriel, a character from Pope's *Rape of the Lock* that appropriately

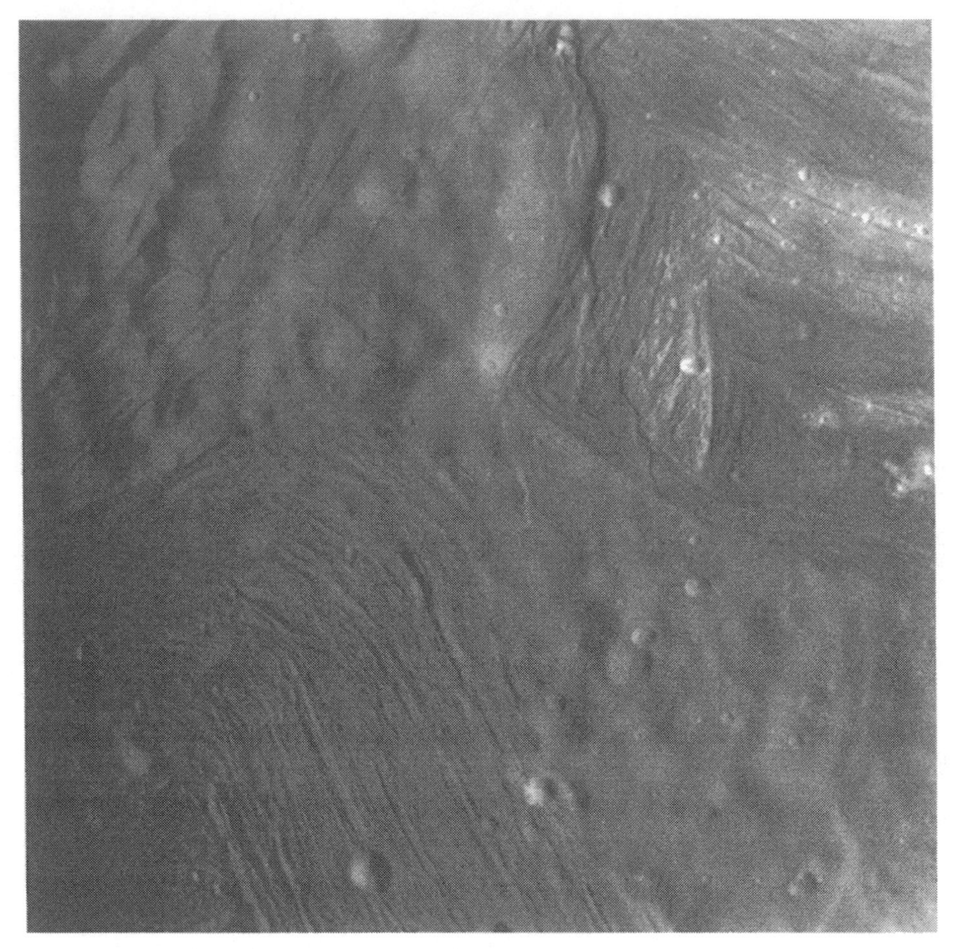

An astonishing juxtaposition of terrain types involving Inverness and Elsinor Coronae on Miranda.

means 'dark angel', turned out to be the darkest of them all, reflecting only a few percent of the sunlight that struck it.

DARK UMBRIEL

Certainly, Earth-based observations of the moons in the early 1980s singled out Umbriel as different from the others. Unlike Titania, Oberon and Ariel, whose spectroscopic signatures implied that they were almost completely covered with ice or frost, Umbriel revealed much less ice cover and a darker, rockier surface. Voyager 2's observations of it seemed to reinforce theories that Uranus' intense radiation belts are to blame, although the fact that Umbriel is uniformly dark throughout surprised many scientists.

Artist's concept of the tortured surface of Miranda, with Uranus dimly visible in the sky.

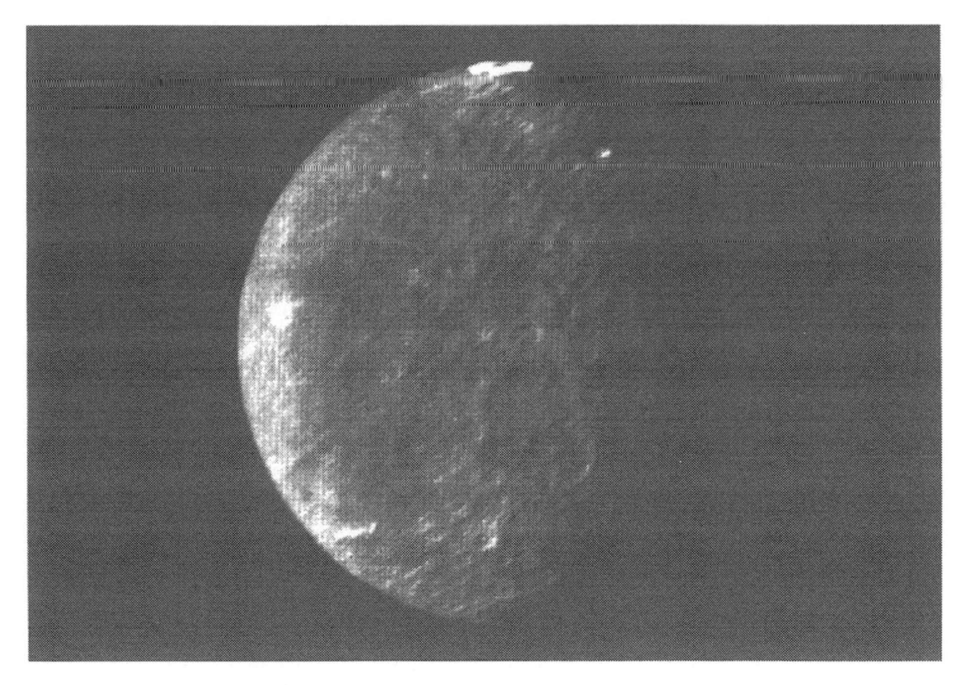

Charcoal-grey Umbriel, including the pure white 'Cheerio' near the top of the picture.

They expected that meteoroid impacts would have carved out cleaner, brighter subsurface material, or at least produced slightly paler ejecta through the fragmentation and pulverisation of rock. None of these expectations materialised. There is, however, a strange bright ring, nicknamed the 'fluorescent Cheerio', which may be the icy floor of a crater. However, the most pressing question about the Cheerio is not so much what it is, but why, surrounded by a vast sea of such uniform darkness, there is only one.

"We don't know the full answer to that question," Ellis Miner admits. "We suppose that it was the result of some sort of cryovolcanism which coated the floor of a relatively fresh crater, except for the central peak area, with relatively fresh, reflective water-ice. There are probably other places where such things have occurred, but they are either buried by subsequent deposits of meteoritic material, have evaporated away or were not large enough to show up in the [Voyager 2] images."

Umbriel's darkness has also led to suggestions that a coating of small debris, somehow created nearby and confined to the moon's orbit, is responsible for having 'dusted' the surface and hidden whatever markings were there previously. Where this dark material came from is unknown and will have to await future visitors to Uranus, although current theories suggest a global mantle of dust from recent impact craters or, perhaps, an explosive cryovolcanic eruption.

Umbriel, however, is a very ancient, 'dead' world, and so neither theory is fully consistent with what has actually been seen on its surface. It would appear that the most likely answer is that it simply contained more methane than the other four moons and, thanks to irradiation from Uranus, received more widespread surface darkening.

A VARIETY OF SURFACE FEATURES

The other three large moons displayed a diverse range of surface features and brightness. Titania, the largest at about 1,600 km in diameter, turned out to be marked by huge faults and vast, winding canyons 2–5 km deep and up to 1,500 km long, implying an era of intense tectonic activity at some stage in its history.

Like Miranda, this activity has often been linked to tidal heating from Uranus, rather than from an internal heat source. Titania's faults are probably quite young, appearing to cut through older crater fields and rarely having newer craters superimposed on them. A notable example is the enormous Messina Chasmata valley, which cuts through a 200-km-wide crater called Ursula.

Titania, despite being closer to Oberon in terms of size, actually shares closer physical ties with Ariel. The latter seems to have the youngest and brightest Uranian surface, and in general lacks craters much bigger than about 50 km across. This has led to suggestions that it may have been 'peppered' by low-speed debris, which helped to obliterate larger, very ancient craters.

Much like Titania and Miranda, it bears the scars of an intense geological history, in the form of huge fault valleys, together with traces of what might be flows of icy material. The nature of this material is still unknown, but may include traces of ammonia, methane or even carbon monoxide.

Since it flew much closer to Ariel than it did to Titania, Voyager 2 was able to observe its surface with greater clarity, and succeeded in distinguishing three main

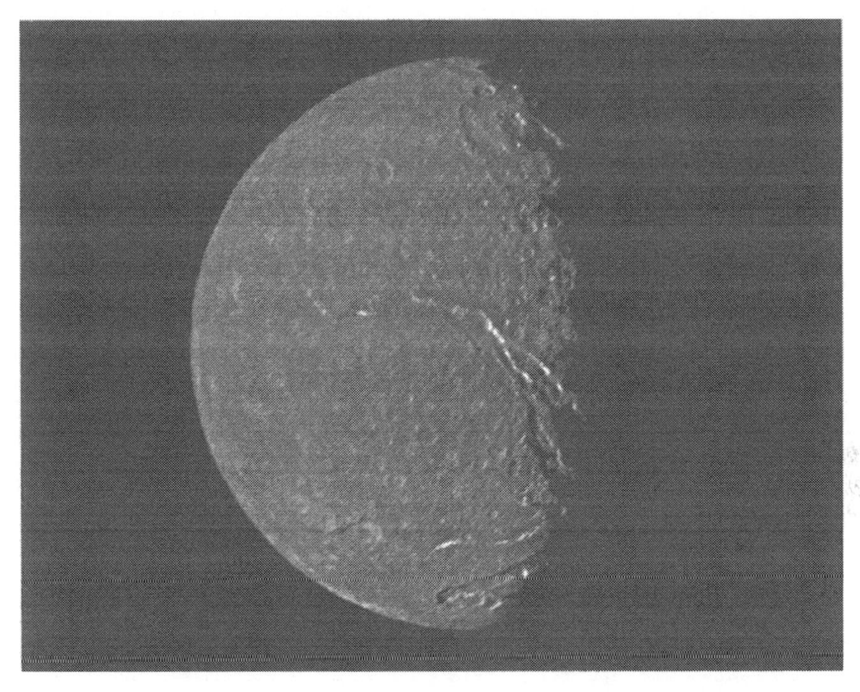

Titania.

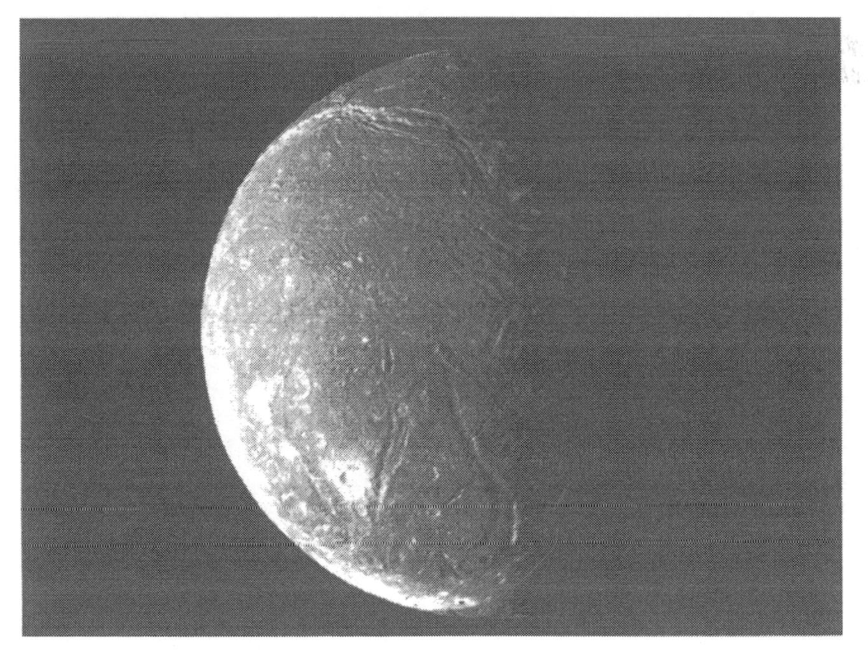

Ariel.

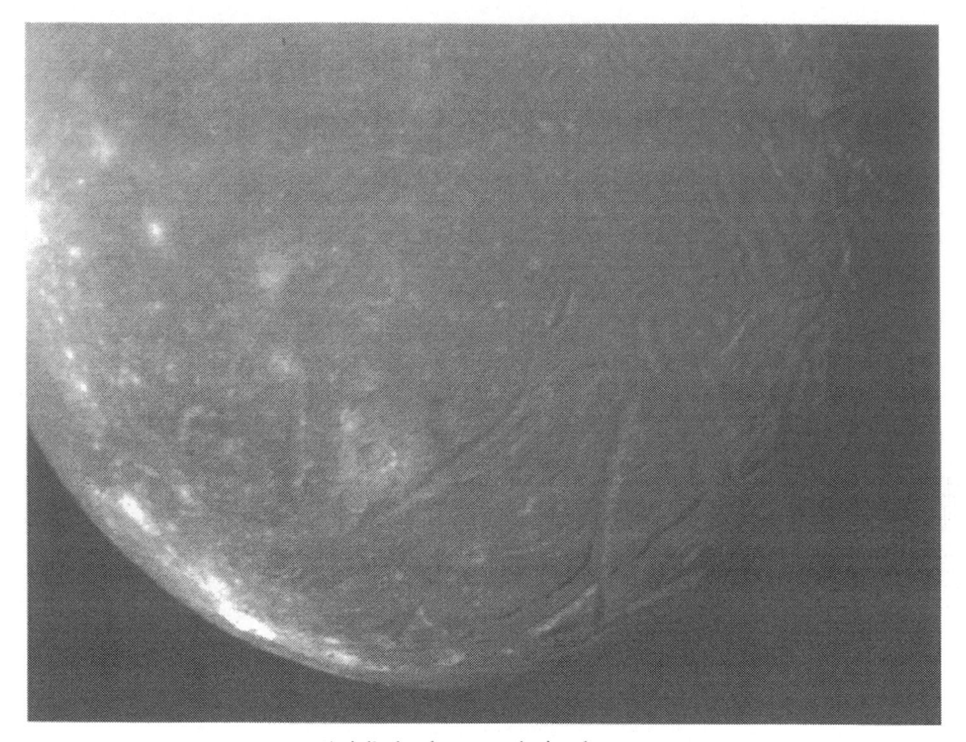

Ariel's broken terrain in close up.

types of terrain. The first is heavily cratered, pointing to its extreme age, although, curiously, its craters are actually much younger than those found on the other moons. It seems that older craters, possibly akin to those seen on Umbriel and Oberon, have been erased by a recent meteoroid bombardment.

The cratered regions are separated from one another by the second terrain type, which consists of parallel 'ridges', about 10–35 km apart, with shallow troughs between them. As the crater density here is about the same as on the cratered terrain, they are roughly about the same age. The last terrain type is a series of rolling plains, several of which have hinted at a history of volcanism over a long period of time.

Oberon, the last of the pre-Voyager Uranians, shows both light and dark surface material, which can be partially explained by impact craters formed during a long period of meteoroid bombardment. In many ways, its heavy cratering is reminiscent of our Moon's highlands, with little evidence of recent geological activity, apart from a strange dark material covering several crater floors. Its precise nature is unknown, but it may be an icy, carbon-rich deposit that volcanically extruded onto the crater floors.

In some ways, Oberon shares features with Umbriel. It has a predominantly dark surface, although pale ejecta blankets and bright 'rays' surround several of its younger craters. Perhaps these betray a dust-rich surface overlying an icy substrate or, as has already been speculated in Umbriel's case, irradiation of methane trapped in Oberon's surface by Uranus itself could be the culprit.

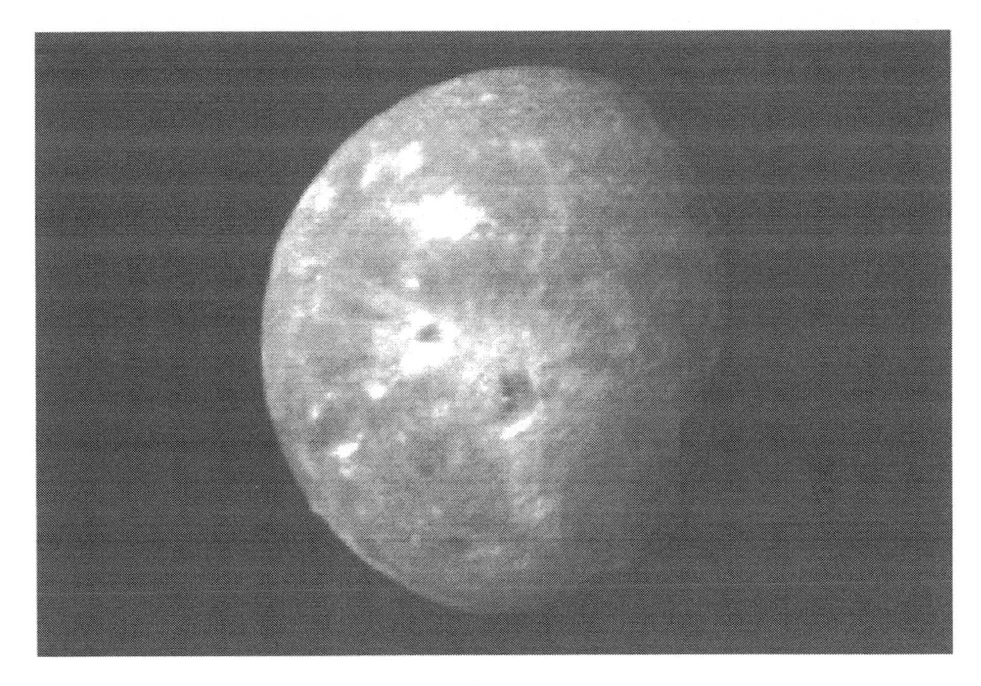

Oberon.

Theoretical models suggest that if intense irradiation is the cause of the extreme darkness, any methane in the upper 1.2 mm of either moon's surface could be chemically broken down and reduced to a charcoal-grey colour within about 10 million years. As well as helping to explain why Umbriel and Oberon are so dark, this also helps to reveal why no methane has been detected spectroscopically on their surfaces: it has long since been chemically separated into carbon and hydrogen.

NEW MOONS

Voyager 2 found 10 new moons during its flyby, the largest of which, unflatteringly named Puck, turned out to be only about 150 km across. More than a decade later, five more were discovered through a combination of ground-based observations, HST data and theoretical calculations. In September 1997, Philip Nicholson of Cornell University and his colleagues, working at the Hale telescope on Mount Palomar in California, discovered two irregular moons, estimated at between 80 and 160 km across. In accordance with the Shakespeare–Pope convention, they were named Caliban and Sycorax, which "precludes my preference", joked Nicholson, "for naming a moon after my cat, Squeaker".

Then, in May 1999, Erich Karkoschka of the University of Arizona in Tucson found another moon, about 40 km in diameter that had actually been photographed by Voyager 2 on 23 January 1986, only 19 hours before its closest approach to

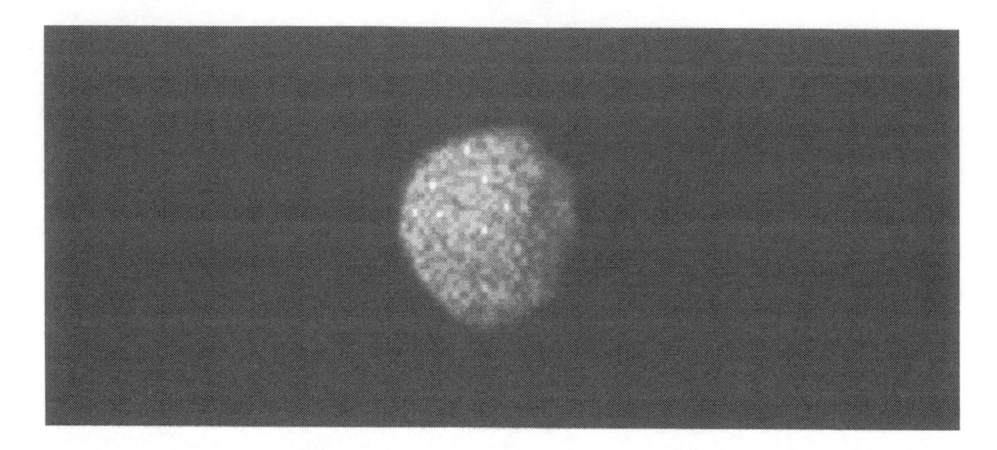

Puck.

Uranus. However, its small size and dark colour meant that it was not seen again for 13 years, until Karkoschka compared the Voyager pictures with more recent HST images. In fact, citing a lack of reliable data, in December 2001 the International Astronomical Union (IAU) took the unusual step of stripping the title of 'moon' from Karkoschka's find until more conclusive evidence could be acquired through HST imagery. Perhaps its most unusual characteristic is that it orbits 75,000 km from Uranus – in exactly the same orbit as Belinda, one of the 10 moons found by Voyager 2 in 1986. In fact, the two small worlds actually drift slowly past each other about once a month.

Of these new moons, very little is known about their composition, although Sycorax appears to be reddish in colour, perhaps indicating that its hydrocarbon surface has been bombarded by energetic particles from space. Most likely, the majority of them are of a similar darkness to the five pre-Voyager moons. If the irradiation-darkening theory is the cause of this, then in all likelihood it also left its mark on the planet's rings, which Voyager 2 revealed to be a similar colour and reflectivity.

URANUS' RINGS

Unlike Saturn's rings, which are covered with highly reflective ammonia and water-ice, those of Uranus are considerably darker and extremely difficult to photograph, even during the encounter. After the initial discovery of five rings in 1977, four more were found during Voyager 2's cruise to Uranus and another pair were picked up by the spacecraft's PPS instrument during its close-range observations. There may yet be more, but they must await the sensitive eyes of a future robotic explorer.

The nature and evolution of planetary rings, even after more than 300 years of research, is still poorly understood. They are thought to come from planetary material that formed with the planet itself, but for some reason was never

Uranus' narrow rings, as photographed by Voyager 2.

incorporated into it, or from ancient moons ripped apart either by violent tidal forces or meteoroid impacts. If the fire-and-brimstone theory of Miranda's evolution is anything to go by, it may have only narrowly avoided a similar fate.

Other possible explanations include cometary material passing close to the planet, as the violent impact of fragments from Shoemaker–Levy 9 into Jupiter's atmosphere in July 1994 dramatically demonstrated. At present, rings seem to be the exclusive preserve of the gas giants, although there is no reason to suspect that this will always be the case.

In addition to their darkness, the Uranian rings are very narrow, which has foiled many early Earth-based attempts to photograph them. They were, however, readily seen at infrared wavelengths because their dust re-radiates in the infrared the sunlight that it absorbs. Even Voyager 2's attempt to create a 'movie' of the rings failed to produce any useful data because they were much too dark. Many of the rings are no more than about 12 km across and only the outermost ('epsilon') ring is as much as 100 km wide.

Some scientists have seen this as evidence that they may be relatively young, at least compared to those of Saturn, and evolved much later than Uranus itself. One of the greatest surprises was their sharpness. In fact, the edge of the epsilon ring is so sharp that it can be no more than about 150 m thick and seems to lack particles smaller than about 20 cm in diameter. The bulk of the epsilon ring, according to PPS, UVS and RSS data, seems to be composed of large, icy boulders.

This conspicuous lack of very small dust particles in many of the rings, which has been noted by a number of scientists based on PPS results, could be due to Uranus

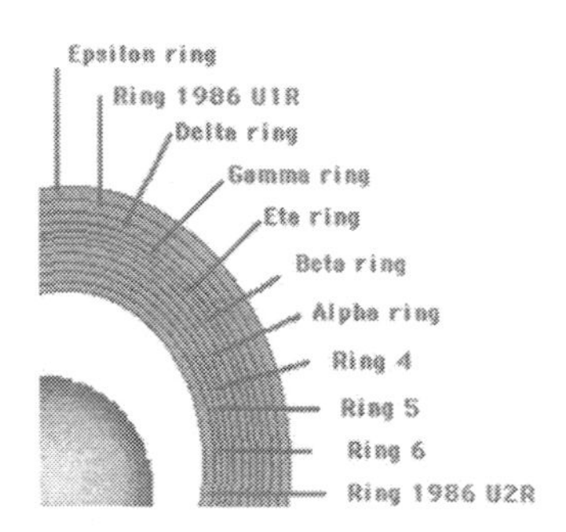

Diagram showing the layout of Uranus' rings.

itself pulling them down into its extended atmosphere. Certainly, UVS observations strongly suggested that in the case of the eta ring – the fourth outermost – the planet's atmosphere, coupled with neutral atom densities, were easily high enough to drag down a micrometre-sized particle every thousand years or so.

However, such minuscule particles were spotted between the two new rings (designated 1986 U1R and 1986 U2R) found by Voyager 2 during its flyby. The spacecraft also received numerous micrometre-sized particle impacts as it flew through the Uranian system, which received a maximum density of 20–30 hits per second.

SHEPHERD MOONS

It seems that the most likely explanation for the rings' sharpness is the presence of tiny 'shepherding' moons, in some cases only a little bigger than the ring material itself, whose gravitational influence prevents large particles from escaping into space. Yet, despite predictions of finding as many as 18 shepherds at Uranus, only two (Cordelia and Ophelia) were actually discovered. Neither was much larger than about 20 km across, although both straddled – and presumably 'bound' – the edge of the wide epsilon ring.

In fact, the two shepherds were so small that it was impossible to confirm their existence through Earth-based observations until March 2000, when scientists compared the HST images with Voyager 2's incomplete orbital models for them. Karkoschka rediscovered Ophelia by electronically 'stacking' HST images, while Richard French's team at Wellesley College in Massachusetts spotted Cordelia by examining and measuring faint ripples in the epsilon ring caused by the motions of the tiny moon.

Unperturbed by the apparent shortage of shepherds, scientists still think that they are responsible for keeping the rings in line, but that they are too small and faint to have been detected even by Voyager 2's instruments. There may be as many as 25 shepherds still to be found, say researchers from the University of London, although they would probably be no more than about 12 km across. It will almost certainly be left to future exploratory missions to locate them.

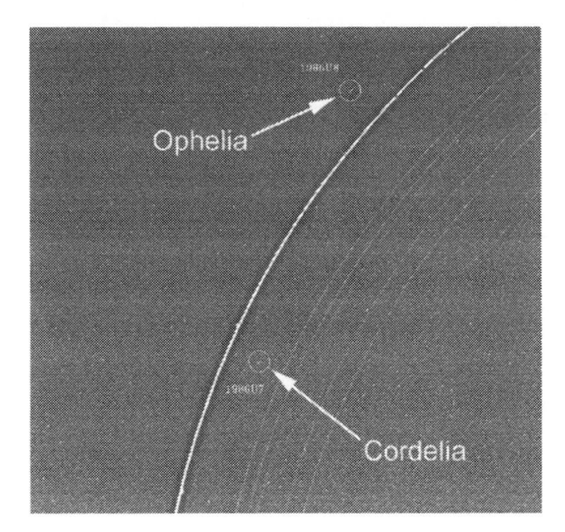

Apparently vindicating the idea that small moons exist to 'shepherd' ring material and keep it from escaping into space, Voyager 2 photographed diminutive Cordelia and Ophelia in January 1986.

Most of the ring observations utilised the RSS, whose S- and X-band radio signals viewed from Earth passed 'behind' them before and after the planetary occultation. This allowed the instrument to gain two high-resolution radial profiles of all the rings. Although the attempt to produce a ring 'movie' using Voyager 2's imaging system failed, a more recent attempt in March 1999 was more successful.

Using the HST data gathered between 1994 and 1998, Karkoschka and his team produced a time-lapse movie of the rings, which revealed for the first time that they seem to 'wobble' like an unbalanced wagon wheel. This has variously been attributed to either the shape of Uranus itself – a slightly flattened globe – or the gravitational pull of its five large moons or, more likely, a combination of both.

The research by Karkoschka and others over the last few years has been aided by HST's Near Infrared Camera and Multi-Object Spectrometer (NICMOS), fitted during the STS-82 Shuttle servicing mission in February 1997. However, from 1999 until March 2002, a lack of nitrogen ice coolant meant it had to be turned off. Then, the STS-109 crew brought it back to life and in May 2002, it returned its first rejuvenated images. This has led to increased optimism that studies of Uranus with this unique infrared camera can begin anew, perhaps leading to more ground-breaking discoveries in the next few years.

A DIFFERENT URANUS

Such optimism is not overly misplaced. As has already been seen, the Uranus of the late 1990s and today is very different from the bland place seen by Voyager scientists nearly two decades ago. The giant planet has now moved further along in its orbital

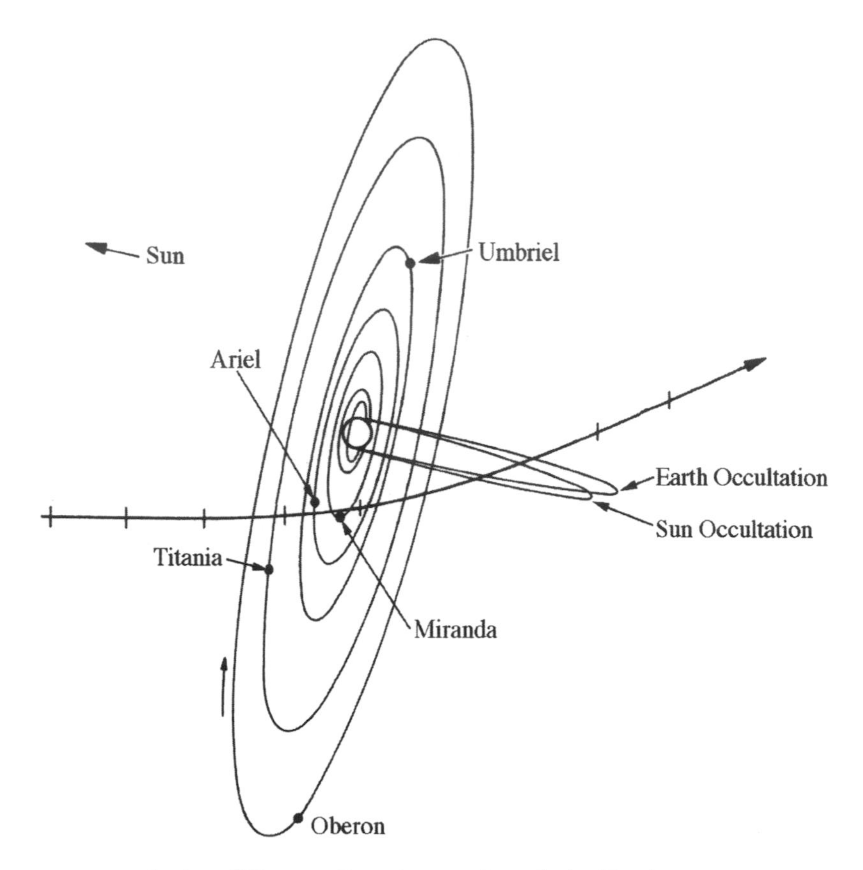

A plot of Voyager 2's trajectory through the Uranian system.

path around the Sun, and its northern hemisphere – shrouded in darkness and unseen by Voyager 2's instruments – is now stirring back to life after a decades-long winter.

Shortly before it was shut down in 1999, NICMOS revealed a faint striped pattern of clockwise-moving winds in Uranus' atmosphere, scooting along at over 500 km/h, together with enormous waves of storms the size of Europe and frigid temperatures of 88 K. Although they were nowhere near as pronounced as the striped weather on Jupiter and Saturn, they were nevertheless there and clearly marked the onset of northern spring on Uranus.

Latitudinal banding on the two larger gas giants comes from strong solar heating on their equators; Uranus, which spins on its side, currently receives more heat at its poles than its equator, explaining its less prominent atmospheric banding. The heating pattern changes as the planet pursues its orbit. It seems now that the differences between what Voyager 2 saw and what the HST has seen are purely seasonal ones. By 2007, the Sun will be shining directly over the planet's equator, so the most exciting discoveries may be still to come.

However, for the Voyager scientists and engineers in the last few days of January

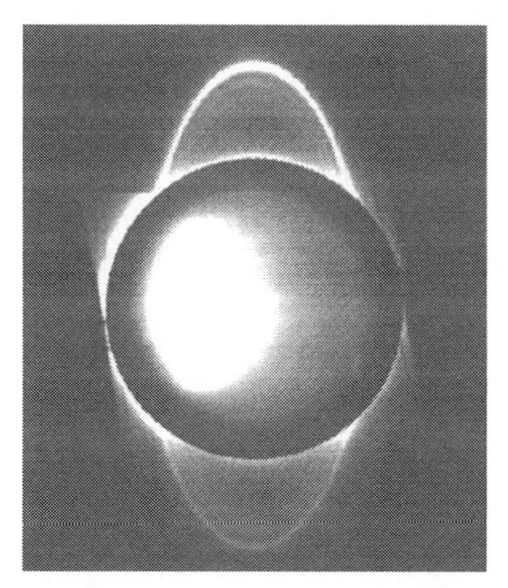

Infrared image at 2.1 microns of Uranus, taken by the Keck-II telescope on Mauna Kea in June 2000, showing the methane haze layer on the planet's south polar cap. Courtesy of Imke de Pater/UC Berkeley.

Uranus appears as a narrow sliver of a crescent in this parting shot from Voyager 2 on 25 January 1986. The spacecraft was about 600,000 km beyond the giant planet at this point.

1986, all that was in the future. For them, the immediate concern was managing and working through the vast amount of data from Uranus and beginning the long process of better understanding the giant planet and its entourage. Then, on the bitterly cold Floridian morning of 28 January, only four days after Voyager 2 flew past Uranus, Challenger and her crew began their fateful journey, "touched the face of God", and fell victim to one of the most public disasters ever witnessed in the human space programme.

Triumph and tragedy, within a week, had been perversely juxtaposed. The success at Uranus was quickly overshadowed by a very public and damaging presidential inquiry into the cause of the disaster; and the Shuttle fleet was grounded for more than two years as the Rogers Commission's recommendations were put into effect. So intense was public feeling against NASA that hardly anyone noticed when, towards the end of February 1986, Voyager 2 turned off its cameras and most of its instruments, bade farewell to Uranus and started the long, lonely, 42-month journey to Neptune.

6

The "boiling cauldron"

INAUSPICIOUS DISCOVERY

The strange circumstances that led to the discovery of Neptune are laced with irony and lost national pride. If the Italian scientist Galileo Galilei had not been hampered by cloudy skies during a few nights watching Jupiter in 1612–1613, he might have discovered the eighth planet in our Solar System, more than a century and a half before William Herschel found the seventh. Then, in 1845–1846, if Astronomer Royal George Airy and James Challis of the Cambridge Observatory had listened to a young mathematician named John Adams, the credit for finding Neptune should have belonged solely to England.

It was not to be. Galileo, however, can be forgiven for missing the faint, eighth-magnitude object that he thought was a star. Airy, on the other hand, foolishly ignored Adams' calculations; for reasons that remain unclear, Airy simply didn't believe that it was possible to 'reduce' observations to predict the location and mass of a perturbing body. Instead, he sent the young man to Challis, who, despite actually spotting Neptune four times in the summer of 1846, failed to realise that it was the new planet.

At about the same time, French mathematician Urbain Leverrier made similar, independent predictions to those of Adams and, after growing frustrated at the lack of interest in the Parisian astronomers in looking for the object, he passed the co-ordinates to Johann Galle at the Berlin Observatory, with whom he had previously corresponded. Galle found Neptune within hours of initiating his search on 23 September 1846.

Adams' and Leverrier's mathematical calculations not only proved that Neptune existed, but succeeded in plotting its position in the heavens, based on the way its gravity tugged at Uranus. It was the first time in the history of astronomy that a planet had been 'found' theoretically on paper, before being visually confirmed. Although Adams' work was later given the credit it deserved and he was rightly named as co-discoverer of the planet, the fiasco sent shockwaves through the

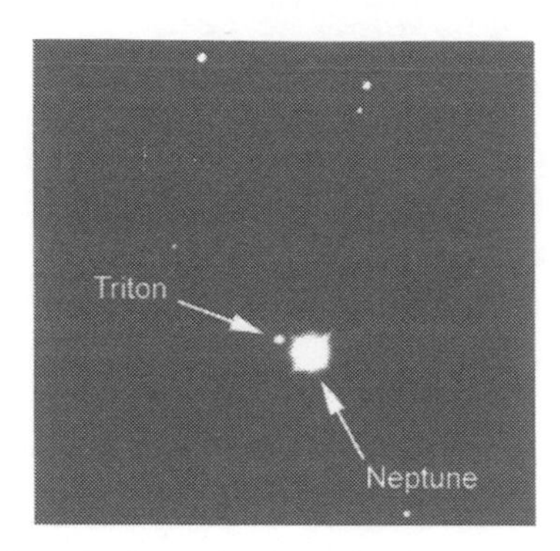

Ground-based telescopic view of Triton and Neptune.

scientific community as image-conscious Victorian England realised Neptune had been snatched from them by the French and Germans. The story of Neptune was very much the story of England's failure to find it.

The full account of the discovery and the scandal it caused, together with the inexplicable disappearance of letters between Airy and Adams – which prompted mutterings of a cover-up to protect the reputations of those involved – is told elsewhere. Nevertheless, the events surrounding the discovery are a useful introduction to the level of knowledge about Neptune before Voyager 2 arrived in August 1989. Apart from the fact that it had a moon called Triton, discovered a few weeks after the planet by William Lassell, very little else was known until the mid-twentieth century.

In the autumn of 1981, when NASA decided to proceed with the Grand Tour by directing Voyager 2 to Uranus and Neptune, knowledge of both worlds remained sketchy. With this in mind, the conference at Pasadena in February 1984 (discussed in the previous chapter) gave scientists an opportunity to look at what was known about Neptune and identify a comprehensive series of observations to tackle the most important unanswered questions about its atmosphere, moons, magnetosphere and ring system. Working groups for both planetary encounters were established and the Neptune team published its report in July 1987.

THE QUESTION OF RINGS

One of the key questions raised by this report was whether Neptune had rings. By the late 1970s, it was known that rings were a feature of the other gas giants. Saturn's magnificent system had been discovered 300 years earlier; Uranus' set of narrow rings was identified following a stellar occultation shortly before the Voyagers were launched; and Voyager 1 revealed Jupiter to possess a fine, dusty ring when it flew by the planet in 1979. Most scientists confidently expected to find rings around Neptune and it seemed that stellar occultations offered the best chance of seeing them from the ground. It was here that the search hit its first obstacle.

Neptune lies 1.6 billion km further from Earth than Uranus. This reduces its apparent size in the sky and, as a result, it only blots out a few stars each year. Nevertheless, stellar occultations have still been used for some time to divulge useful information about the planet. Astronomers had watched one occultation in April 1968, although their intention was to measure Neptune's diameter more precisely, rather than look for rings. After observing another occultation in May 1981, their combined data revealed a dip in the subject star's brightness just before it passed behind Neptune.

Part of the difficulty with Neptune was that, in contrast to Uranus, whose rings formed a bull's eye in the sky, Neptune's rings were viewed at a much shallower perspective, so there was rather less chances of a star passing behind them. Might the dimming have been due to the presence of rings? Unfortunately, this and other data gathered in the early 1980s returned inconclusive results and were unable to answer this question, although ring-hunter James Elliot of MIT optimistically envisaged a thin, dark ring system, akin to that found around Jupiter.

Then, in July 1984, came a breakthrough, when one occultation in particular produced intriguing results. Two teams of astronomers at separate locations in Chile watched as Neptune passed in front of the star SAO 186001. The first team noticed a momentary dip in the star's brightness. Not only that, but although it flickered just before Neptune passed in front of it, there was no subsequent flicker as the planet moved away. Whatever had blocked the star's light apparently did not run all the way around Neptune.

The astronomers initially thought it might have been caused by a tiny moon, no more than 15 km across, orbiting 75,000 km from Neptune's centre. Alternatively, it could have been a result of the star's light being blotted out by a semi-transparent

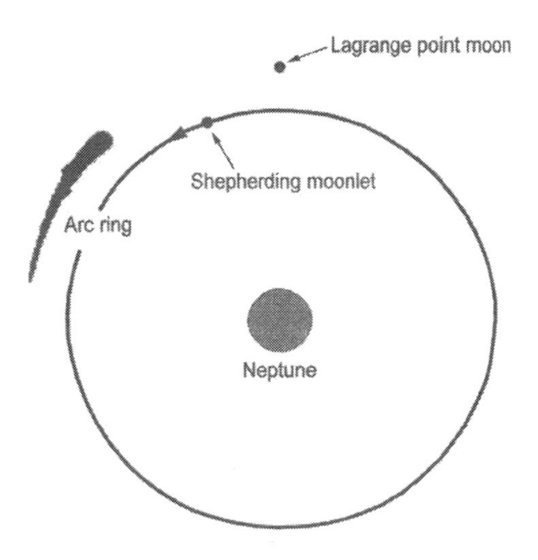

A possible explanation for the unusual 'arc rings' seen by ground-based astronomers during the 1980s. This diagram proposes that two tiny moons – a shepherd and a Lagrangian, neither of which would be directly visible from Earth – may be sufficient to keep ring arcs in place.

'arc' of ring material extending only partway around the planet. This last theory was supported by the observations of the second team, based 100 km further south. They suggested that from their perspective the dimming was more probably due to a partial ring arc than to a hitherto-unknown moon.

This notion of incomplete ring arcs was not new. Indeed, two arguments at the time postulated that (a) such rings were fully circular, but varied in thickness, which made them look incomplete when viewed from Earth, and (b) if ring arcs did exist, they would need something else to constrain them. Astronomer Jack Lissauer of NASA's Ames Research Center thought small 'shepherd' moons close to the arcs could fill such a need. With nothing else to hold them in place, Lissauer reasoned, it was unlikely they could maintain stable orbits and would quickly drift off into space or drop into Neptune's atmosphere.

It was even speculated that the second team might have seen a new ring being born, which at the time of the occultation took the form of an incomplete arc, but would be sheared into a fully circular shape by Neptune's gravity within a couple of years. Then, when two tiny shepherd moons – Cordelia and Ophelia – were found associated with the rings of Uranus in 1986, it seemed more plausible that a similar mechanism could be operating at Neptune. Mathematical estimates predicted that shepherds, only 80–200 km across, should be sufficient to keep errant ring material in place.

At this stage, of course, not even the rings themselves had been positively identified (nor would they be, until the summer of 1989), let alone hypothetical shepherd moons. This forced many scientists to grudgingly admit that if Neptune had either of the two, it would be virtually impossible to see them from Earth. As a result, it was with considerable excitement that they awaited Voyager 2's date with the distant world.

PREPARATIONS FOR NEPTUNE

Planning for this fourth and final planetary visit, as with the previous three, carried its own set of unique challenges. As Voyager 2 headed deeper into space and further from the Sun, it became harder to communicate with the spacecraft and increasingly necessary to improve its capabilities to cope with dramatic reductions in light levels. By August 1989, Neptune had for 10 years been considered to be the Solar System's outermost planet, a title it would retain for another decade, until Pluto, pursuing its eccentric orbit, moved out to reclaim this status.

When Voyager 2 arrived, the planet lay 4.5 billion km from the Sun and received less than half as much sunlight as even gloomy Uranus. Many of the improvements made to the spacecraft's imaging software to compensate for these lighting issues at Uranus were also brought into service at Neptune, but needed to be adjusted still further.

In particular, the software was extensively reprogrammed to take exposures up to 96 seconds long and, to avoid jolting Voyager 2 and smearing these priceless pictures, each attitude-control thruster firing was shortened to less than four milliseconds. Experience from Uranus had already taught the imaging team that, in such a dark environment, every movement by the spacecraft – even its tape recorder clicking on or off – could nudge the cameras off target and smear the close-up shots. In addition to the steps used at Uranus, a new technique called Nodding Image Motion Compensation (NIMC) was introduced.

This procedure was basically intended to hold Voyager 2 as steady as possible and keep its movements to the bare minimum when photographing Neptune. During

The expansion of one of the Goldstone DSN antennas to a diameter of 70 m.

each exposure, while the camera shutter was open, the whole spacecraft would be turned extremely slowly by short thruster bursts to track the motion of individual targets. It would then close the shutter and turn its high-gain antenna towards Earth, so that the image could be sent straight home without employing the tape recorder. The spacecraft would then 'nod' back to its target, reopen its shutter and take its next picture.

For all the benefits they provided, these improvements would have been of limited use without substantial changes to the DSN and help from two other tracking stations in Japan and New Mexico. The main Voyager complexes, at Goldstone in California, Madrid in Spain and Canberra in New South Wales, had already had their receiving dishes increased in diameter to 64 m before reaching Uranus, but were now expanded still further to 70 m to support Neptune operations.

The Australian station was especially vital, because Voyager 2 would make its closest approach to Neptune almost directly over Canberra. For this reason, NASA opted to retain the services of the 64-m radio telescope at Parkes in New South Wales, which had been electronically linked to Canberra before Uranus by a 400-km-long microwave communications system. Although this improved the chances of satisfactorily picking up Voyager 2's weak radio signal – estimated to be less than a billionth of a billionth of a watt – still more receiving clout was needed. Two other networks promptly offered to help.

One was the 64-m Usuda station on Honshu Island in Japan, run by that country's Institute of Space and Astronautical Science (ISAS), which was tasked with collecting the RSS occultation data as Voyager 2 passed behind Neptune and Triton as viewed from Earth and through their shadows. The other was the Very Large Array (VLA), near Socorro in New Mexico. During the Neptune encounter, this network – each of whose 27 dishes spanned 25 m – would be electronically linked to Goldstone on a daily basis to maximise its receiving power.

With these two additional facilities at their disposal, the Voyager team would quite literally be listening to their spacecraft with a gigantic radio 'ear' that stretched across the entire Pacific Ocean basin. That ear needed to be at its sharpest from 10:00 pm GMT onwards on 24 August 1989, when Voyager 2 was set to make its closest approach to Neptune and, indeed, its closest encounter with any celestial body since departing Earth 12 years before.

CLOSEST PLANETARY ENCOUNTER

Prior to the discovery of the ring material, it had been planned to fly very close over Neptune's polar region to set up a 10,000-km Triton flyby, but to preclude the chance of damage in passing through an unknown ring during the equatorial plane crossing, the polar range was increased and the Triton flyby opened to 40,000 km.

In fact, its path through the Neptunian system would be nothing less than a white-knuckle rollercoaster of a ride, skimming just 4,900 km over the planet's cloud-tops and undertaking a whistle-stop tour of Triton. In some ways, however, it would actually be the least risky of all four encounters, because Voyager 2 had no other planetary visits ahead of it. "It gave us the freedom to choose a flyby geometry that was best for the studies of Neptune and Triton," says Ellis Miner of JPL, "without having to worry about where the spacecraft would be going thereafter."

An extended trip to Pluto had already been ruled out because the tiny world no longer lay on either Voyager's flight path and too much fuel would be needed to get there. "Voyager 1's trajectory after the Saturn encounter remains well above the ecliptic plane in which essentially all the planets orbit," Miner recalls. "As we were approaching Saturn with Voyager 2, we could have gone directly to Pluto or we could engineer encounters of Uranus and Neptune. We didn't even know that Pluto had a moon at that time, but it wouldn't have made any difference. The combination of Uranus and Neptune were deemed far more important than a single flyby of Pluto. If we were making that choice today, I believe the choice would be the same. It was not possible to go from either Uranus or Neptune on to Pluto, which at that time was nearly one-quarter of the way further around the Solar System and actually closer to the Sun than was Neptune."

So Voyager 2 did not need to pass a specific distance from Neptune to pick up gravitational impetus and head off somewhere else. However, although the trajectory planners had a fairly free run, they were also keen to ensure the spacecraft's safety, because NASA intended to use both Voyagers for another 20–30 years in an extended search for the edge of the Solar System. This Voyager Interstellar Mission (VIM), as the post-Neptune effort had been dubbed, was due to begin later in 1989. In the meantime, the scientists were eager to make the most of their last planetary encounter.

Top of their list of priorities was getting as near as possible to Neptune and flying sufficiently close to Triton to acquire detailed, map-quality images of its frozen surface. Yet the question of whether the planet had rings remained unanswered. Triton would be on the way out, and so NASA did not want to collide with any ring material, no matter how small or diffuse, and there was also an alarming possibility that Neptune might have tenuous, but lethal extended atmospheric gases that could drag the little spacecraft to its doom.

Matters were compounded still more by the scientists' desire to collect occultation data for both Neptune and Triton. This would be done by transmitting the spacecraft's radio signal through their atmospheres as they passed in front of Earth and the Sun. Each occultation would provide invaluable information: the Earth pair would allow scientists to take temperature and pressure readings and the solar pair would enable the UVS to plot the ultraviolet 'signatures' of gases escaping from their atmospheres. This data was expected to help atmospheric specialists better comprehend the internal workings of both Neptune and Triton.

In order to do all four occultations and fly close enough to both worlds for detailed imaging, the trajectory planners went for a risky manoeuvre. After hurtling over the planet's north pole, Voyager 2 would plunge southward 'behind' Neptune, pass within 40,000 km of Triton a little over five hours later and ultimately leave the Solar System forever at southern mid-latitudes. It was an audacious plan that promised substantial rewards if it worked, balanced against the risk of losing Voyager 2 if it failed. In the event, the risk proved worth while.

Even after adopting this trajectory, there remained concern that a closer flyby of Triton was desirable if the spacecraft was to take high-resolution pictures. Moreover, it would also improve the chances of success for both Earth occultations, which could now be timed to occur directly over Canberra. NASA decided, therefore, to postpone Voyager 2's arrival at Neptune by five hours, until shortly before 3:00 am GMT on 25 August 1989, which would hopefully carry it somewhat closer to Triton. It would also make the encounter more easily visible from Goldstone.

On 14 February 1986, three weeks after leaving Uranus, the first of six planned trajectory correction manoeuvres was performed to set a course for Neptune. Only four of these were actually needed, thanks to the better-than-expected accuracy of earlier ones, as well as more precise observations from Earth which gave trajectory planners a clearer idea of where the planet's suspected ring material lay.

NEPTUNE'S MOONS AS NAVIGATIONAL AIDS

Navigating a path to Neptune proved to be much more difficult than it had been for the other three gas giants. For each of their planetary journeys, the Voyagers had used optical navigation to reach their targets. In other words, they took long-exposure images of each planet's moons in front of a known star field and this provided a determination of the exact positions of the planets, their moons and the incoming Voyagers. However, whereas Uranus was known to have five moons before the 1986 encounter, and Jupiter and Saturn clearly had more than a dozen each, Neptune, so it was thought, had only two. The smaller of these, Nereid, discovered in 1949, is a maverick in a highly eccentric orbit and takes 360 days to complete a full lap of Neptune. Triton, on the other hand, orbits its giant host in a more stable, relatively circular, but retrograde path every six days, making it more helpful to the Voyager team as a navigational aid.

Having only one reliable point of celestial reference to navigate the 1.6-billion-km gulf to Neptune seemed, at the very least, a risky endeavour. And yet, past experience had revealed each gas giant to have more moons in its personal entourage than all the terrestrial planets combined. Confident that more moons would be found as Voyager 2 drew closer to Neptune, the trajectory planners programmed optical navigation details into the spacecraft's computers, but omitted to specify their targets. These could then be updated at short notice when new moons were found and their orbital parameters accurately plotted.

Fortunately, the discovery of no fewer than six new moons came thick and fast in the summer of 1989, leading trajectory planners to execute a seven-minute thruster firing on 2 August to more precisely direct Voyager 2 towards Neptune. Ironically, one of the new moons, later named Larissa, had been seen from Earth during a stellar occultation in 1981, but could not be confirmed by more than one visual sighting. All six new moons are irregular in shape, resembling battered potatoes, and confined to Neptune's equatorial plane. Like the Uranian moons, they are as dark as soot.

By the time it reached Neptune, Voyager 2 had been studying the planet for some

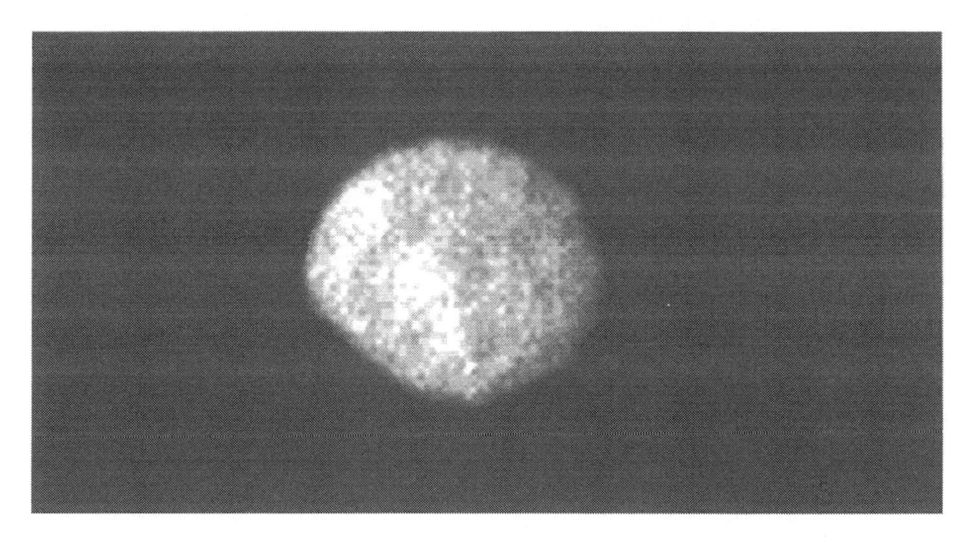

Larissa.

time. As early as May 1988, from a distance of 684 million km, its pictures were of better quality than those taken from Earth, revealing Triton for the first time as a pale reddish smudge. However, it was during the 12 weeks from June to September 1989 that it made its most spectacular observations of a world like no other: a beautiful, sky-blue planet with a misleadingly pacific façade, beneath which lurked violent storms, bizarre weather and some of the fastest winds in the Solar System.

A SURPRISINGLY ACTIVE WORLD

The nature of this weather, and what powers it, remain the greatest unsolved riddles from the Voyager 2 encounter and, despite recent HST insights, will probably be left to a future mission to be addressed in depth. School textbooks have long portrayed Uranus and Neptune as twins, virtually identical in size, colour and chemical composition. Yet, despite lying so much further from the Sun and receiving so little solar radiation to warm its atmosphere and drive its weather, Neptune's average temperature of 55 K is roughly the same as that on Uranus. Moreover, although it is slightly colder, Neptune is actually the more active of the pair.

This unusual activity first sprang into the headlines in early 1989, when NASA announced that Voyager 2 had spotted clouds in its atmosphere. This corroborated similar Earth-based observations a few years earlier, but much more was to come. By March, as the spacecraft drew closer, a large oval feature was discerned in Neptune's southern hemisphere.

Nicknamed the Great Dark Spot by analogy with the centuries-old storm on Jupiter, it was about the size of Earth and similar to its Jovian counterpart in several respects. Its diameter relative to the size of the planet it inhabited was the same, it too rotated counter-clockwise – implying that it was probably a high-pressure, fast-moving atmospheric vortex – and lay at approximately the same latitude (22°S) as the Great Red Spot.

As with the three earlier planetary encounters, Neptune operations ran like a military campaign, which kicked off on 5 June 1989 with a two-month Observatory period. By this time, in addition to the Great Dark Spot, several more cloud formations had been seen, including a vast, dark latitudinal band running across most of the southern hemisphere. Similar banding, albeit closer to the planet's equator, was also detected in the late 1990s by HST and the Keck-II telescope in Hawaii. The general consensus among scientists is that these represent active atmospheric regions, rather than single, long-lived storms.

During its two months of long-range observations, Voyager 2's cameras snapped five narrow-angle pictures during each rotation period. These were then combined to produce time-lapse 'movies' of Neptune's atmosphere in mesmerising action, which helped meteorologists to track the formation and dissipation of clouds and measure wind speeds at different latitudes.

On terrestrial planets, this kind of time-lapse imaging is frequently used to determine their rotation rates – the length of their 'days' – very precisely. Unfortunately, none of the four gaseous worlds explored by the Voyagers has a solid surface, and on Neptune in particular its high winds, which vary considerably at different latitudes, tend to hide the bulk motion of the planet. Furthermore, different regions of the atmosphere rotate at different speeds. As a result, Voyager 2 'watched' Neptune's core rotate by measuring interactions between charged particles trapped in its magnetic field and solar wind particles streaming from the Sun.

Pre-Voyager estimates for a Neptunian day ranged from eight to 24 hours, but had been refined to around 17 hours by astrophysicist Heidi Hammel's observations at Mauna Kea in 1986. Three years later, data from Voyager 2's MAG and PRA instruments pegged the planet's bulk rotation rate, with an accuracy of better than a minute, at 16 hours and 6.7 minutes. This extremely fast rotation is typical of the gas giants and causes Neptune's poles to appear flattened – a phenomenon called 'oblateness' – which in some regions makes its polar radii up to 400 km less than its equatorial radius.

By mid-August 1989, well into the Far Encounter phase and with Neptune's enormous bulk now more than filling the narrow-angle lens, Voyager 2's wide-angle pictures began to resolve greater detail. They revealed that the planet had not one dark spot, but two, as well as numerous other enigmatic cloud features that travelled rapidly through its atmosphere. The most noteworthy was the 'Scooter', a chevron-shaped, westward-moving structure that owed its name to its immense speed. It took less than 16 hours to circle the planet.

More than a decade later, its nature remains imperfectly understood, although most scientists think it was some kind of plume rising up from a deeper cloud deck. Whereas the source of the Scooter seems to have originated below Neptune's main layer of methane-rich cloud and haze, the Great Dark Spot apparently occupied a more lofty position. From close range, it seemed to be 10 per cent 'darker' than its sky-blue surroundings, implying that it lay as much as 50 km above the visible cloud deck.

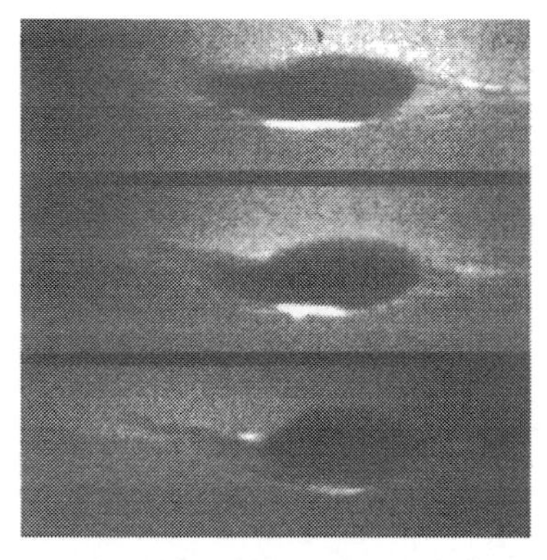
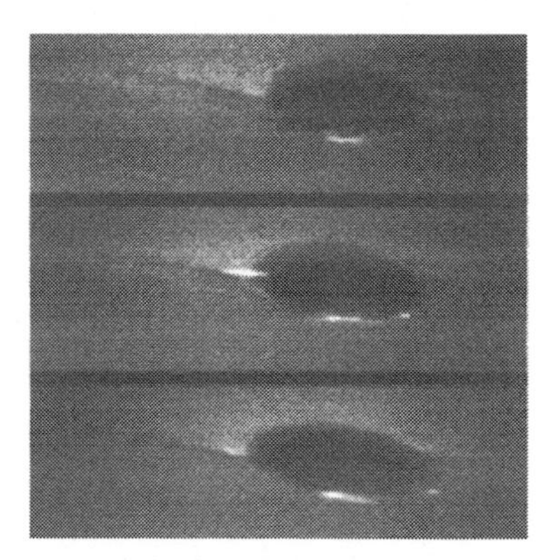

A series of Voyager 2 pictures, showing variations in the dimensions of the Great Dark Spot as they traverse Neptune's atmosphere.

Although the spot had been too small and indistinct to be seen with ground-based instruments, a bright white cirrus cloud, probably made of frozen methane, had been seen from Earth as early as January 1989. It seemed to 'hover' along the edges of the Great Dark Spot. In fact, the close proximity of this Bright Companion had made the spot invisible to ground-based observers, so its 'age' at the time of the Voyager 2 encounter cannot be determined with certainty. Voyager 2's cameras also noted that the companion seemed to change shape during each rotation period.

A FLOPPY, UNSTABLE STORM

Strangely, however, the Bright Companion always remained inextricably tied to the Great Dark Spot, often lingering around its southern rim. Yet the spot itself lay some 100 km 'below' the companion. This has led some atmospheric specialists to liken it to lenticular cloud formations often found in mountainous regions on Earth. Clouds of this type form as a result of rapid cooling when winds are pushed to higher altitudes by the presence of mountains. At Neptune, however, a world devoid of a solid surface, the only available 'mountain' seemed to be the rising column of the Great Dark Spot itself.

Other, less pronounced cirrus streaks, also of methane-ice, lay high in the atmosphere, forming and dissipating every few hours. At low northern latitudes (around 27°N), Voyager 2 watched them casting long shadows on the main sky-blue cloud deck 50–100 km below. This was the first time that cloud 'shadows' had been photographed on any of the gas giants. The streaks appeared to be 50–200 km across, casting 30- to 50-km-wide shadows. However, they did not form a continuous opaque layer, instead giving Voyager 2 a clear view as far down as the main cloud deck.

Watched with astonishment by thousands of mesmerised scientists, both the spot and its companion stretched and contorted restlessly as they journeyed across Neptune's disk. The storm was clearly nowhere near as stable as Jupiter's Great Red

Spot, which had endured for more than 300 years with relatively few physical changes. This fact would not go unnoticed as scientists continued their observations throughout the 1990s with HST and ground-based telescopes. "It was kind of floppy," says Hammel, now a senior research scientist with the Space Science Institute in Boulder, Colorado, "because it changed shape as atmospheric circulation carried it around the planet."

The material within the spot – including its bright central core – spun counter-clockwise every 18 hours and the whole storm, clipping along at 1,100 km/h, took a full eight days to travel around Neptune's atmosphere. In fact, it was in the vicinity of the spot and its companion that the fastest-known winds in the Solar System were measured: close to 2,000 km/h, surpassing even blustery Saturn, the previous record-holder.

Yet, despite its immense size, instability and the fact that it dominated the latitude it occupied, the spot did not appear to cause major disruption to its environs. Some atmospheric scientists, including Andy Ingersoll, saw in this apparent lack of turbulence, evidence of a 'flat', two-dimensional fluid mass with no shearing forces acting upon it. Atmospheric turbulence, says Ingersoll, is a three-dimensional phenomenon that usually produces shearing forces. Coupled with the age of these strange storms, which can endure for years or centuries, this reinforces the idea that small-scale turbulence is almost absent in the atmospheres of the gas giants.

Ingersoll points out that, on Earth, it is received sunlight that creates small-scale atmospheric turbulence, which helps to dissipate large storms. As this sunlight-driven turbulence is much weaker further from the Sun, winds on the distant planets are correspondingly stronger and storms last longer. Current thinking is that the spots are rapidly-rotating vortices, although those on Neptune remain puzzling because they are apparently powered by heat flow from the planet's interior. Neptune receives less than half as much energy from sunlight and its own heat reserves as Jupiter, although its internal energy is proportionately more significant. Observations throughout the 1990s, however, merely generated many more questions than answers.

As it rocked and rolled its way through Neptune's realm, the Great Dark Spot was also slowly migrating northwards in the direction of the equator, at a rate of about 15 degrees of latitude each year. According to predictions made at the time of the Voyager 2 encounter, it should have reached the equator – and the strong, westward-blowing jetstream known to reside there – by 1991 or 1992. However, HST pictures from early 1994 by Hammel's team produced unexpected results. The huge storm and its ever-present Bright Companion were nowhere to be seen.

A CHANGING ATMOSPHERE

At first, Hammel and her colleagues thought the familiar features might be just out of sight and about to emerge from around Neptune's limb. However, that seemed unlikely because HST had 'missed' only 20 degrees of longitude in its observations and the size of the Great Dark Spot was known to cover more than 40 degrees. Even

Close-up shot of the Small Dark Spot – D2 – from Voyager 2.

if it was not immediately visible, it should have taken a noticeable 'bite' out of the planet's limb. Other scenarios argued that it was temporarily overlaid by cloud. Towards the end of June 1994, however, further images confirmed that both spot and companion were indeed gone.

Not only that, but the second spot – dubbed 'D2' by the Voyager team, apparently in part to satisfy the legions of Star Wars fans – had also vanished. Though somewhat smaller than the Great Dark Spot, it was still larger than our Moon and shared some of the features of its big brother: a bright, rapidly-moving core and attendant clouds of methane-ice, carried along by 640-km/h winds. It would now appear that both spots and their companion clouds did not survive the perilous passage through Neptune's strong equatorial jetstream.

In November 1994, only a few months after the disappearance of the spots was reported, follow-up pictures from HST identified another storm brewing in the northern hemisphere. Like the original Great Dark Spot, this new arrival seemed to be its almost-exact mirror image north of the equator, also possessing a bright core and high-altitude cirrus clouds hanging around its rim. Atmospheric specialists were amazed at the endless list of surprises coming from the cold, remote world that had been expected to be bland, yet seemed to conjure up energy from nowhere to run an astonishingly active weather system.

"Hubble is showing us that Neptune has changed radically since 1989," Hammel says. "New features like this indicate that with Neptune's extraordinary dynamics, the planet can look completely different in just a few weeks."

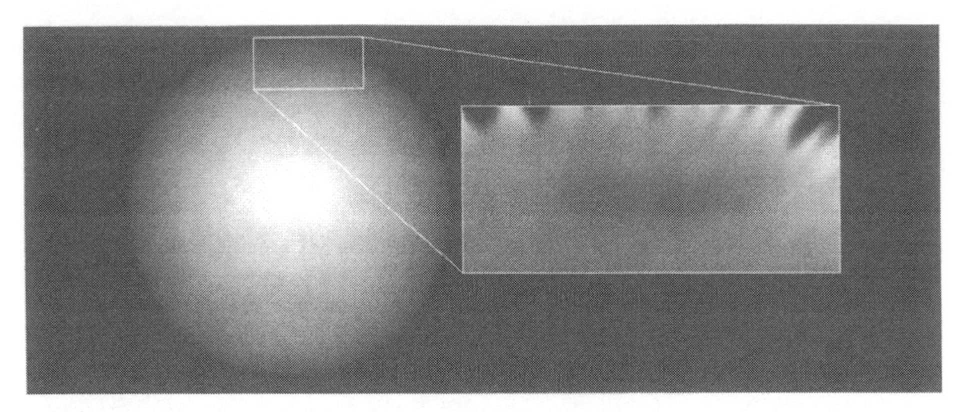

This HST image taken in November 1994 revealed the appearance of another dark spot in Neptune's northern hemisphere. Courtesy of Heidi Hammel at MIT.

In fact, many of these phenomena seem to endure for years. It is unknown how 'old' the first Great Dark Spot was when photographed in 1989 – the companion was first spotted in January that year, so it was presumably at least seven months old – but certainly its northern successor lasted for more than two years. HST images taken in mid-1996, combined with infrared observations from Mauna Kea, were used to produce a full-colour movie of Neptune's rotation. They revealed the second Great Dark Spot to be very much alive and well, as was the planet's powerful equatorial jetstream.

More recently, major storms the size of Europe, clearly capable of surviving for some months or even years, have been seen in the planet's southern hemisphere. In May 1999, the Keck-II telescope spotted such a storm and infrared images from the Hale observatory in California, taken three months later, confirmed that it was still going strong. Subsequent observations in June 2000 revealed the planet's atmosphere to be filled with spot-like vortices, waves and small-scale, narrowly-spaced bands of clouds akin to those on Jupiter.

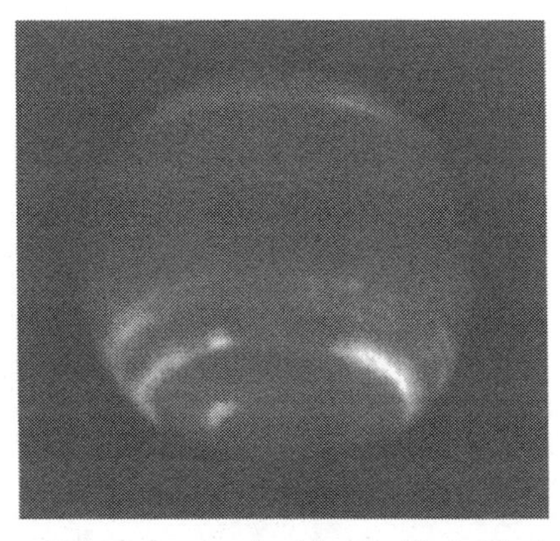

Adaptive optics image of Neptune, acquired by the Keck-II telescope on Mauna Kea in June 2000, showing an active atmosphere. Courtesy of Imke de Pater/UC Berkeley.

"We've never seen the detail we see now," says Imke de Pater, the team leader for the Keck-II observations. "This shows us how much structure there is in the planet's atmosphere [and] how dynamic it is – as dynamic as Jupiter."

In order to conduct their surveys of Neptune, both telescopes used the relatively new technique of 'adaptive optics', which has the ability to take HST-quality pictures from ground-based observatories. This technique works by using a primary mirror and light detector, accompanied by a secondary mirror whose position is precisely computer-adjusted several hundred times every second to compensate for the blurring effects of Earth's atmosphere.

INTERNAL HEAT

These findings clearly show that Neptune is dynamic. Atmospheric activity on the distant planet has by no means abated since Voyager 2's visit, but seems to have intensified, partly because it is now closer to the Sun than it was in 1989. More scientists are seeing closer ties between Neptune and Jupiter, although explanations are scarce. The only viable answer at present seems to lie with the planet's high 'energy balance'. Before the Voyager 2 encounter, observations from Earth and theoretical modelling had shown that, although Uranus is slightly bigger than Neptune (51,200 km, compared to 49,400 km), it is the latter which is more massive.

This strongly suggests that Neptune's interior is denser than that of Uranus, which in turn would be expected to influence its magnetic field and internal heat sources. Neptune emits 2.6 times as much heat than it receives from absorbed sunlight. This energy balance is more than twice that of Uranus because it is further from the Sun than Uranus and, in proportion to Neptune's size and mass, is larger than that of any other planet in the Solar System. One recent theory argued that temperature differences between the planet's internal heat source and its frigid clouds could trigger atmospheric instabilities that cause the large-scale weather patterns.

Even if this is the case, the nature of this internal heat source itself remains a mystery, and it is the search for answers that will make Neptune attractive to scientists when future missions are being planned. All that is conclusively known is that there must be significant differences in the internal workings of both worlds. The 'twins', it seems, are not quite so alike after all.

Nevertheless, a number of important advances in our understanding of what goes on inside Neptune were made by the IRIS and the RSS during Voyager 2's encounter. The IRIS started taking temperature measurements of Neptune's disk on 16 August 1989 and continued until early September, when it completed a series of observations of the planet's dark side. These measurements enabled scientists to observe how much received sunlight Neptune reflects back into space and allowed them to determine precisely its large energy balance.

Unfortunately, although HST has been able to trace the effect of Neptune's internal activity in the form of bright clouds and savage storms throughout the 1990s, our understanding of exactly how the planet works remains very much on a theoretical level. At present, most scientists think Neptune may have a rocky core,

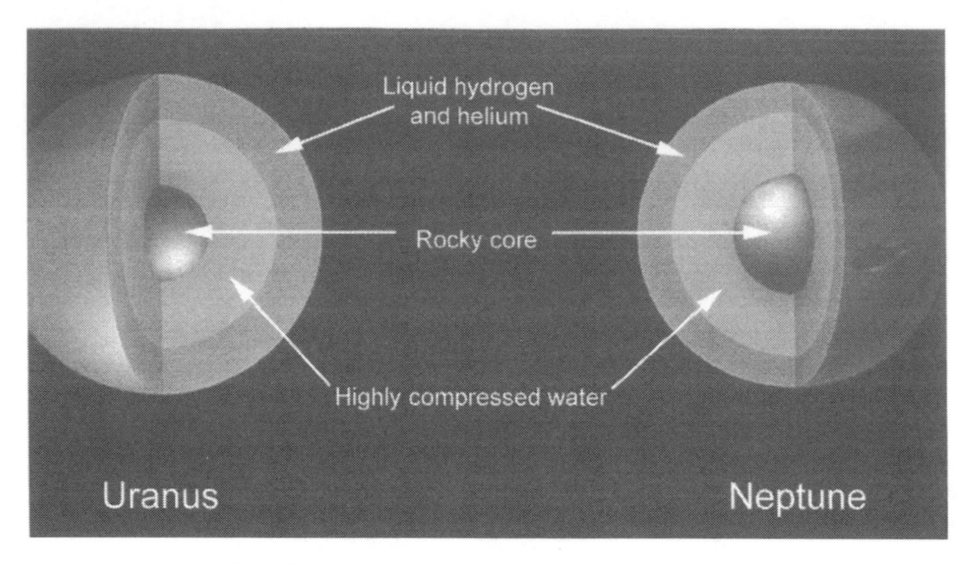

Possible internal structures of Uranus and Neptune.

although that is far from certain. If such a core does exist, it is possibly two or three times larger than Earth and constitutes only a few percent of the planet's overall mass.

Opinion as to the hypothetical core's temperature is just as divided, with estimates ranging from 6,000 to 10,000 K. It might then be surrounded by a thick, slushy 'ocean' of icy and rocky materials – comprising more than 90 per cent of Neptune's mass – and finally overlaid by a gaseous region, 3,000–6,000 km thick, of hydrogen and helium, together with smaller traces of water, ammonia and methane. Whether there is a sharp or gradual transition from the deep, hot layers to this outer gaseous 'skin' is unknown.

Other scientists, however, have argued persuasively that planets with extremely dense, rocky cores should have much shorter rotation rates than the 16 hours clocked by Voyager 2's instruments. This has led to increased speculation that, far from having a dense core, Neptune's icy and rocky components are mixed fairly uniformly in its interior. In fact, the only 'ice' in large enough quantities to be directly detectable during the RSS occultations was methane, although, like Uranus, water- and ammonia-ice were predicted at depths below those probed by Voyager 2.

A SKY-BLUE ATMOSPHERE

Further parallels with Uranus came with the discovery of hydrogen, helium and methane in Neptune's atmosphere, together with acetylene, hints of ethane and possibly haze-forming polyacetylenes. It will almost certainly be left to a future mission, presumably carrying a Galileo-type atmospheric-entry probe, to tell us more. Yet, several important differences between Uranus and Neptune were found by Voyager 2, the most obvious of which can be seen just by glancing at pictures of the two planets (see colur section).

As on Uranus, the absorption of sunlight by atmospheric methane is primarily responsible for Neptune's overwhelming blue colour. However, Neptune's hue is much deeper and richer than that of Uranus – "sky-blue", one might say, as opposed to "aquamarine" – indicating the presence of a slightly different colouring agent. Like so many things about Neptune, the nature of this agent is unknown, although Ellis Miner and Randii Wessen have speculated that a blue-tinted cloud layer in the planet's deep troposphere is most likely responsible for the deeper colour. They think the cloud could be composed of hydrogen sulphide.

Neptune also differs from Uranus in that its axial orientation is fairly normal, inclined at 29.6 degrees to the plane of its orbit. This means that, unlike Uranus, which rolls on its side like a celestial beer barrel, Neptune circles the Sun 'upright' like the other planets. As a result, it receives considerably more solar heating at its equator than at its poles, which scientists initially suspected would produce a significant temperature difference between them.

However, the actual difference is lower than predicted, implying that Neptune has an internal mechanism that transfers heat from its equator to its poles. It would seem that its convective interior is somehow connected to the parts of its atmosphere where sunlight is absorbed. This mechanism then blocks excess internal heat from escaping at the equator, pushing it instead to higher latitudes where it helps to warm the polar regions. Its precise nature is uncertain, but clearly it is the opposite state of affairs to that operating on Uranus, which directs heat from its sunward facing pole back to the equator and around to the dark pole.

RINGS AT LAST!

Neptune's long-awaited rings also turned out to be quite different from those of Uranus. On 11 August 1989, a fortnight before closest approach, Voyager 2's cameras spotted two ring arcs, apparently 10,000 and 50,000 km long, extending only partway around the planet. Already, late in July, NASA had announced the discovery of three new moons – Galatea, Larissa and Despina – that seemed to interact gravitationally with the rings. They ranged from 150 to 200 km in diameter:

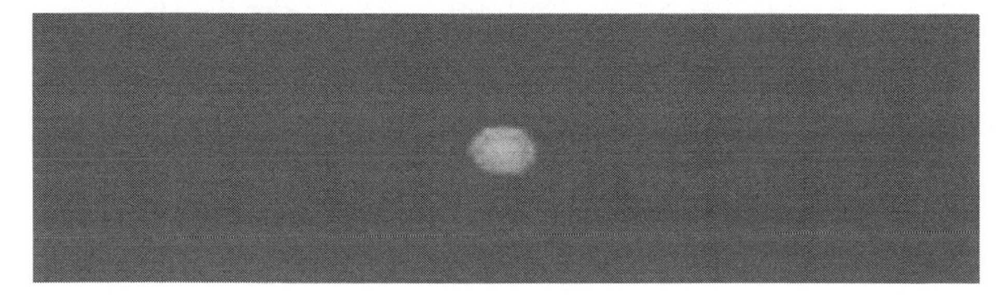

Despina.

Neptune's narrow rings, barely resolved by Voyager 2's camera, using a long exposure.

remarkably close to the sizes predicted by Lissauer and others for hypothetical shepherd moons a few years before.

As August wore on and Voyager 2's pictures became clearer, there were not two rings, but four, and they did in fact run fully around the planet. There were, however, peculiarities. All four seemed to be relatively uneven and 'clumpy' in terms of the distribution of material within them. Voyager 2's images were at last able to explain why so many ground-based observations had reported partial arcs: the outermost ring, subsequently named after John Adams, seemed to be accompanied by three to five 'clumps' of unknown material, each up to 50 km wide.

More recently, it has been suggested that debris thrown up from collisions with other objects – possibly recently-destroyed moons – could have caused or contributed to this unequal spread of ring material. This suggestion has been reinforced by Miner and Wessen, who speculate that two other tiny moons found by Voyager 2, now known as Thalassa and Naiad and each about 40 km across, could themselves be torn apart and incorporated into the rings at some stage in the future.

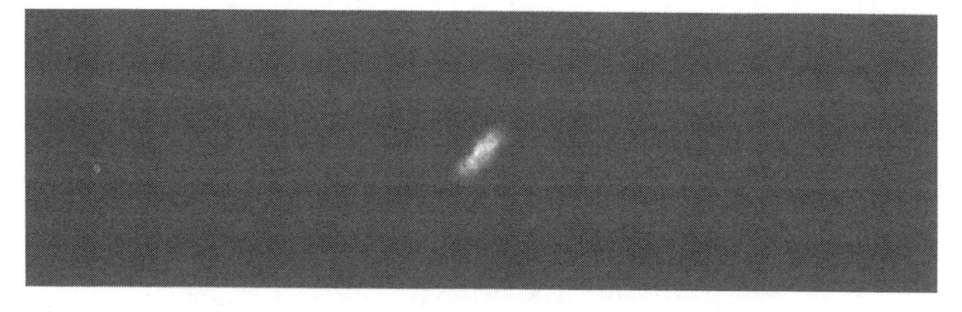

A somewhat smeared image of Thalassa from Voyager 2.

A somewhat smeared image of Naiad from Voyager 2.

Neptune's rings, seen by Voyager 2 on 25 August 1989.

Like the Uranian rings, those of Neptune are extremely dark, making it impossible to determine their colour and giving the imaging team a severe headache when trying to photograph them. In fact, they were so dark that prolonged 10-minute exposures were needed to resolve them adequately. Voyager 2's best images came around 4:00 am GMT on 25 August, as the spacecraft dipped behind Neptune, when the RSS took a high-resolution profile of their entire radial extent. Two time-lapse 'movies' of the rings were also recorded on 28 August as Voyager 2 headed out of the Neptunian system.

The rings are also very narrow, implying that they are probably quite young. In particular, when the PPS and UVS instruments watched the Adams ring pass in front of the star Nunki, they revealed its core is no more than 17 km thick. Miner and Wessen have also argued that the gravitational influence of the nearby shepherd

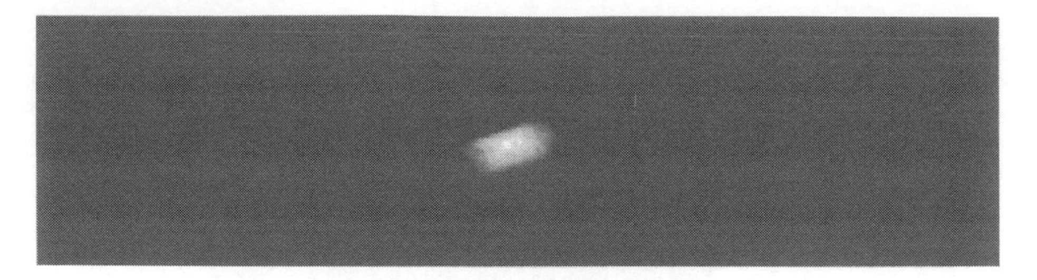

A somewhat smeared image of Galatea from Voyager 2.

Neptune's 'twisted' Adams ring.

moon Galatea would be sufficient to fully 'smooth out' the clumps of material into the Adams ring within a few years. Furthermore, Voyager 2's images revealed a broad, diffuse sheet of fine dust about the size of smoke particles around the planet.

Neptune's extended atmospheric gases, like those of Uranus, would be expected to drag down such tiny particles fairly quickly, which lends more credence to suggestions that they were only a few years old when Voyager 2 saw them. The first reports of ring arcs came around 1983, but have not been seen by either HST or ground-based telescopes, despite frequent observations throughout the 1990s. This suggests that the clumps of material have since been smoothed into the Adams ring and the diffuse dust has been pulled down into Neptune's atmosphere.

Voyager 2 flew closest to Neptune at 2:56 am GMT on 25 August 1989, when it passed over the planet's north pole, before turning south and crossing the ring plane at 26,000 km/h just over an hour later. As it did so, the PWS passively 'measured' the number of small particle impacts from ring material, which hit a peak of 300 per second during the quarter-hour on either side of the crossing. As each dust particle hit Voyager 2, it was instantaneously vaporised and ionised by the high-speed impact, and this plasma generated radio waves that were 'counted' by the PWS.

ORIGIN OF TRITON

After leaving the rings, Voyager 2 hurtled southward in the direction of Triton, which had been growing steadily bigger and brighter over the past 15 months until, by late August, it covered more than half of the narrow-angle camera's field of view. As the spacecraft drew closer, it became clear that Triton was much smaller than previously thought: only 2,700 km in diameter, which made it three-quarters the size of our Moon and the seventh largest satellite in the Solar System. Pre-Voyager predictions of its mass, based on the amount of sunlight it reflected, had yielded erroneous results.

"Triton has been shrinking as we approached," joked imaging team leader Brad Smith at the time, "until we feared that by the time we arrived, it might be gone!" The incorrect estimates had come to light following infrared observations by Dale Cruikshank, Bob Brown and others in the early 1980s, which revealed the tell-tale spectroscopic signature of nitrogen gas and ice on Triton's frozen surface. These observations started some scientists wondering if its diameter and mass were significantly smaller than first thought. The measurements were still being refined as Voyager 2 closed in on Triton.

Clearly, little was known about the amazing shrinking moon until 25 August 1989. Triton was first seen by William Lassell in October 1846, only three weeks after Neptune itself had been discovered, although curiously it remained nameless until the early years of the twentieth century. One of its peculiarities, first noted by Lassell, was its highly inclined, retrograde orbit: it circles Neptune 'backwards' or, more correctly, in a clockwise direction when viewed from 'above' (north). This has led to some interesting suggestions about its origins and evolution.

It has long been speculated that distant Pluto, which HST and ground-based telescopes have shown to be compositionally similar to Triton and roughly the same size, is an ancient moon of Neptune that escaped its gravitational clutches millions of years ago. When astronomers first began to ponder Triton's peculiar retrograde motion, they suspected that some ancient cataclysm had sent Pluto careering off into space and disturbed the orbit of Triton around its giant host. The later discovery that Pluto has a large moon complicated this proposal.

However, since the discovery of the Kuiper Belt – a swarm of debris beyond Neptune, which could harbour leftover chunks of material from the Solar System's formation – most scientists now think that Triton, and perhaps Pluto, originated in this region. It is also noteworthy that the six new moons found by Voyager 2 circle Neptune in well-behaved orbits close to its equator. Unless they formed significantly later than the hypothetical ancient cataclysm, their orbits would be expected to have been similarly disturbed, making a Kuiper Belt origin of Triton more likely.

A CAPTURED KUIPER BELT OBJECT?

Several scientists, including Bill McKinnon and Peter Goldreich, support the notion that Triton is a captured Kuiper Belt Object. The only other known retrograde-orbiting moons in the Solar System are four small objects at Jupiter, Phoebe and a few recently-discovered moonlets at Saturn and Caliban and Sycorax at Uranus, all of which are thought to be captured asteroids. On the other hand, unlike those seven tiny retrograde moons, Triton's orbit around Neptune is near-circular, its surface is covered with highly-reflective ice and its chemical composition is extremely diverse. Moreover, both Triton and Pluto are several times larger than the largest-known Kuiper Belt Object.

Goldreich argues that when Triton first encountered Neptune, it smashed violently into one of the planet's former family of moons. Rather than being destroyed, however, the intruder was slowed by the collision and captured in a retrograde orbit by the giant planet's gravity; then, over the next billion years or so, Neptune gradually reduced Triton's orbit from a lazy, looping ellipse into a near-perfect circle. The arrival of this celestial gatecrasher may then, says Goldreich, have flung some moons out of the Neptunian system or put them into eccentric orbits, like Nereid, whose orbit ranges millions of kilometres from the planet.

WHERE ARE THE 'MID-SIZED' MOONS?

The sizes of the eight moons found so far – six tiny equatorial ones, eccentric Nereid and giant Triton – offer persuasive evidence that some catastrophic series of events similar to that envisaged by Goldreich and McKinnon did occur in Neptune's distant past. Unlike the other three gas giants, the planet does not seem to have a system of what might conveniently be called 'mid-sized' moons, with diameters ranging from 500 to 1,600 km. Triton lies just above the uppermost limit for this breed of moons, while Nereid and the others fall slightly short of its minimum qualifying size.

Did Neptune once have a system of 'mid-sized' moons? If so, what might have happened to them? McKinnon believes that such a system did indeed exist, possibly similar in size and appearance to the five large moons of Uranus, but that the arrival of Triton and its after-effects either destroyed them, scattered them or ejected them from the Neptunian system altogether. Even the innermost 'original' moons, says McKinnon, would not have escaped the carnage: at the very least their orbits would have been disrupted and at worst they may even have been torn asunder.

To support his theory, McKinnon notes that Triton orbits its giant host at a distance of approximately 15 Neptune radii – some 370,000 km – which happens to be squarely in the domain occupied by the mid-sized moons of Jupiter, Saturn and Uranus. It is not unreasonable, therefore, to suppose that Triton's arrival dispatched any mid-sized moons Neptune may originally have had. Whatever debris remained from this catastrophe was unable to coalesce into a 'second generation' of moons until Neptune had circularised Triton's orbit into a better-behaved path.

PROTEUS AND NEREID

There is some evidence to suggest that Proteus – one of the six new moons found by Voyager 2 in mid-June 1989 – could be a member of this second-generation family. In fact, it measures 420 km across, which places it close to the minimum threshold for a mid-sized moon. It is roughly the same size as Mimas or Miranda and, indeed, is bigger than Nereid, but had not been seen from Earth because it lies close to Neptune and is lost in its glare. Yet, unlike Mimas or Miranda, Proteus is a 'primitive' body: dark in colour, heavily cratered and irregularly shaped.

Some dynamicists have argued that potato-like Proteus is about as large as an irregular moon can get before its own gravity starts to pull it into a sphere. Its shape is often regarded as a tell-tale indicator that it has not undergone thermal processing through much of its history, with little geological upheaval, although it has apparently suffered cratering on a grand scale from repeated meteoroid bombardments. Voyager 2 did not fly particularly close to it, and even its best pictures revealed little in the way of surface features: just a few blurred craters, trough-like depressions and tall, forbidding ridges.

If the 'gatecrasher' theory of Triton's arrival is accurate, where does eccentric Nereid fit into this evolutionary picture? Its precise origins remain uncertain, but some scientists think that if it was not thrown into its eccentric orbit by Triton, then it is probably a captured asteroid, despite the fact that its orbit around Neptune is not retrograde. Voyager 2 saw it only as a faint, faraway blob – the spacecraft flew no closer than 4.7 million km to the moon – and was unable to resolve any detail. Indeed, even after staring intently at it for 12 days, it was impossible to even determine its rotation rate, so indistinct was its surface.

Certainly, Nereid's shape and highly eccentric path around Neptune is consistent

Proteus from Voyager 2.

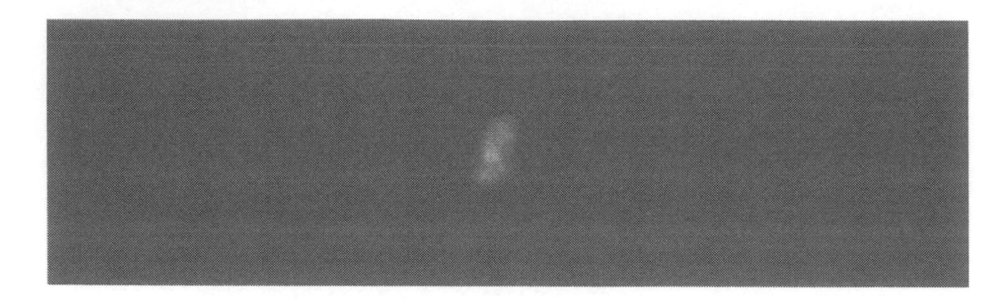

Nereid from Voyager 2.

with a captured body: it orbits its parent planet as close as 1.3 million km and as far as 9.6 million km. Voyager 2 was unable to determine its exact shape, but it is not spherical – a further indication of an asteroidal origin. However, recent spectroscopic observations revealed dirty water-ice on its surface, hinting that it may yet be an 'original' pre-Triton Neptunian moon.

CLOSE ENCOUNTER WITH TRITON

The questions of when and how Proteus and Nereid arrived and what subsequently happened to them will have to await a future mission. For Voyager 2, the mysteries of these two intriguing moons were eclipsed by its flyby of Triton, which came just a few hours after passing the planet itself. As early as 4:30 am GMT on 25 August

Artist's concept of Triton, with Neptune visible in the background.

1989, with closest approach to the moon still three hours away, the spacecraft's cameras began the meticulous process of taking high-resolution pictures of Triton's surface.

It was during this time that the image motion compensation procedures first tried at Uranus and modified for Neptune proved their worth, holding Voyager 2 sufficiently steady to shutter some of the most amazing images of the entire mission. At 8:10 am GMT, moving at 63,730 km/h, the spacecraft passed within 39,800 km of Triton's centre, revealing a world of astonishing complexity and a fitting finale to a spectacular 12-year planetary tour. "All we can say is: 'Wow! What a way to leave the Solar System!" exulted US Geological Survey planetary scientist Laurence Soderblom at the post-encounter press conference.

A TENUOUS 'ATMOSPHERE'

Two minutes before Voyager 2's closest approach to Triton, the PPS and UVS instruments watched the moon pass in front of the star Gomeisa, in the constellation Canis Minor (the Little Dog), and then our Sun, which helped to reveal clues about the composition of Triton's thin atmosphere all the way down to its frozen surface. At 37 K, this surface turned out to be the coldest known in the Solar System. Nevertheless, at the top of its 800-km-deep atmosphere, temperatures climbed slightly higher to 93 K.

This incredibly tenuous gaseous mixture can only really be termed an 'atmosphere' in the loosest sense of the word; indeed, its low pressure – just one 70,000th of the surface pressure here on Earth – is very nearly a vacuum and only barely capable of supporting thin clouds and haze. These clouds, made from particles of nitrogen-ice, were seen to 'hang' 13 km above Triton's surface, while the haze, says Dale Cruikshank, is probably some kind of photochemical smog akin to that seen on Titan.

What is the composition of this smog? Cruikshank thinks it might be organic matter produced when solar ultraviolet radiation and charged particles from Neptune's magnetosphere break nitrogen and methane molecules into fragments. These then recombine into more complex organic compounds like hydrogen cyanide and ethane, which eventually precipitate onto Triton's surface. The fact that none of these compounds has yet been detected spectroscopically (coupled with a notable lack of craters) implies that the present surface is geologically quite young: recent estimates, based on Voyager 2 results and announced in late 1999, suggest it to be just a few hundred million years old.

The main source of Triton's atmospheric gases seems to be the slow evaporation of nitrogen, methane, carbon monoxide, carbon dioxide and water, which exist as ices on its frozen surface. Of these, nitrogen is by far the most abundant, constituting approximately 99 per cent of the atmosphere. Winds, presumably caused by the movement of these gases across Triton, are thought to carry dust particles through the atmosphere and deposit them on the surface up to 50 km away. Ground-based spectroscopic observations in 1980–1981 also hinted that atmospheric ice or frost occasionally shower the surface with frozen nitrogen 'snow'.

These observations revealed Triton to be the most spectroscopically diverse icy body in the Solar System. Its large quantities of surface ice cause it to reflect 85–95 per cent of the sunlight that strikes it; by comparison, our Moon reflects, on average, a mere 11 per cent. The variation in Triton's reflectivity is caused by its highly inclined orbit around Neptune, which causes its seasons to vary quite considerably in severity. In addition to its large size, this extreme brightness was one of the reasons why faraway Triton was found telescopically so soon after Neptune's discovery.

TRITON'S SURFACE

As Voyager 2 approached, Triton's mass was determined with an increasing degree of accuracy by working out the changes it caused to the spacecraft's velocity. When the mass data was combined with Triton's observed diameter, the results revealed its density to be twice that of water. This led scientists to assume that three-quarters of its content is rock and one-quarter is ice – a much higher ratio than in the Saturnian or Uranian moons. This substantial rocky content provides further evidence to support the argument that Triton is a captured Kuiper Belt Object, rather than an 'original' Neptunian moon.

Voyager 2's highest resolution pictures covered only a third of Triton's surface and revealed a curious, ubiquitous greenish terrain, which was nicknamed 'cantaloupe' by the Voyager team because of its textural similarity to the ridged, scaly skin of a cantaloupe melon. Visually, the terrain looked like a series of roughly circular, interlocking depressions, known as 'cavi', each about 25–30 km across.

The cantaloupe terrain is criss-crossed by long, interconnected ridges, which are thought to have been caused by one or more epochs of melting and near-complete resurfacing of Triton. In view of its immense distance from the Sun, what heat source might have been responsible for this melting and resurfacing? Radioactive material in the moon's core is one possibility, although most scientists think the tremendous heat associated with Triton's capture by Neptune, and the subsequent circularisation of its orbit, is the most likely culprit. The melting might have then been accelerated still further by cometary or meteoroid impacts.

It is believed that this melting might have rendered Triton's surface liquid for as much as a billion years after the moon was first captured by Neptune. This has led to suggestions that a form of microbial life might have evolved and thrived in the distant past, although it is extremely unlikely that it would have survived to the present day in its frozen state.

The immense forces associated with Triton's capture by Neptune are also thought to be responsible for substantial crustal fracturing noted by Voyager 2. These forces caused localised melting of the moon's surface, which left large, smooth regions or filled basins with lava-like flows, probably of water-ice. At a much later date, 'outgassing' either from volcanic vents or other subsurface processes led to the gradual formation of its thin atmosphere.

Some scientists think that Triton had a thick atmosphere during most of its formative period, but that eventually temperatures near its surface began to cool.

This cooling led to the gradual formation of surface ice, which reflected away the already-weak sunlight that struck it and caused temperatures to decline ever more sharply. More surface ice formed as a result, overlaying large sections of cantaloupe terrain. Eventually, temperatures cooled to a point where elements of the tenuous gaseous nitrogen atmosphere began to freeze onto the surface, forming extensive polar ice caps.

Voyager 2 revealed this southern cap to be a curious pinkish colour, possibly caused by the evaporation of methane- and nitrogen-ice. Beyond the edge of the cap was a bluish crustal region, probably water-ice. This was pockmarked with interlocking cellular features, similar to melted chainmail. Criss-crossing this wasteland were vast canyons and long, straight ridges with central furrows, which may represent subsurface material pushed up through crustal fractures. Most of the features seen by Voyager 2 are thought to be solid water-ice, which is frozen so hard that it has the consistency of rock at these low temperatures.

Throughout this late period in Triton's history, low-level bombardment, mainly from debris in orbit around Neptune, coupled with meteoroid impacts, are thought to have provided the light sprinkling of craters seen today. The biggest of these, seen by Voyager 2, is a monster called Mazomba, which measures 27 km across. In general, however, the number of craters is relatively few, leading most planetary scientists to assume that Triton's present surface is only a few hundred million years old.

AN ACTIVE SURFACE?

Recent HST and ground-based observations have led to suggestions that, as Neptune moved through perihelion (its closest point to the Sun), Triton's surface has warmed up somewhat since Voyager 2's departure. "Since 1989, Triton has been undergoing a period of global warming," says James Elliot of MIT. "Percentage-wise, it's a very large increase." Spectroscopic data acquired in 1998 revealed that this warming has turned part of the moon's frozen nitrogen surface into gas, which, in turn, has increased its atmospheric density and pushed up its temperature slightly to 39 K.

"With Triton, we can more easily study environmental changes because of its simple, thin atmosphere," says Elliot. The global warming is believed to be caused by seasonal changes to the moon's polar ice caps as it approached high summer in its southern hemisphere. Summer arrives on Triton every few centuries and is characterised by more direct sunlight and greater heating of the polar ice caps. Cruikshank believes that this causes nitrogen, carbon monoxide and methane surface ice to sublimate and release heat, before precipitating to the colder regions and warming them.

The water-ice bedrock and local patches of carbon dioxide ice, on the other hand, are left behind as they do not readily sublimate, even in direct sunlight, at Triton's immense distance from the Sun. As a result, the ices migrate from region to region, 'renewing' the surface and maintaining its brightness. In October 1999, after a re-evaluation of Voyager 2 data, scientists revealed to the Lunar and Planetary Science

Conference in Houston, Texas, that cryovolcanic resurfacing is still underway on Triton.

"This is a crazy idea," says Soderblom, "but it's the best we have." It was speculated that liquid might be coming up from deep within Triton and freezing on its surface, thus obliterating all but the very youngest craters. In light of speculation during the mid-to-late 1990s about the existence of life-bearing oceans on Jupiter's moon Europa, the extraterrestrial bandwagon turned its gaze to Triton. However, although it is quite possible that the remote Neptunian moon may have an ocean of liquid methane or ammonia, the existence of life within that hypothetical ocean is rather unlikely.

Volcanism had previously only been seen on our own world and, a decade before the Neptune encounter, on the photogenic Jovian moon, Io. Yet, in front of the astonished eyes of scientists in August 1989, icy geyser-like plumes and their after-effects were detected on the surface of Triton. Voyager 2's pictures revealed large quantities of dark material and a 'powder' of rock particles transported up to 50 km across the surface by prevailing winds. The spacecraft's images clearly showed dark 'smudges' caused by this material, which ran across the surface of the southern polar ice cap.

Artist's concept of possible ice geysers on Triton.

Current thinking is that, despite having such a cold surface, some solar energy manages to penetrate Triton's near-transparent water-ice cap and creates a moderate greenhouse effect in exotic ices called 'clathrates' within the ice layer. This heating is often only a couple of degrees Kelvin, but is nevertheless enough to cause the clathrates to 'de-gas' and build up pressures until the liquid or gaseous nitrogen finds a vent and escapes explosively into the thin atmosphere. One such geyser, photographed by Voyager 2, spewed dark carbonaceous material 8 km above the surface and blew it 150 km westward.

Detailed observations of two plumes in particular, dubbed Hili and Mahilani by the Voyager team, revealed that the latter changed over a 90-minute period, suggesting that winds close to it blew at about 54 km/h. The relative abundance of

A possible geyser site in a 'filled basin', on Triton, photographed by Voyager 2.

dark streaks across Triton's southern polar cap also implies that this process of ejecting subsurface material into the atmosphere and carrying it over considerable distances is fairly commonplace. Measurements of the streaks led to estimates of several dozen active geysers, which were likened to those at Yellowstone Park in the United States.

DEPARTURE FROM NEPTUNE

Although Voyager 2 performed its closest approach to both Neptune and Triton early on 25 August 1989, it actually remained in the planet's sphere of magnetic influence for several hours. Magnetic fields are thought to come from a planet's rotation, together with the action of a convective internal 'liquid' layer. Earth's field is thought to come from interactions at the boundary of its molten iron core, and those of Jupiter and Saturn from their massive internal reservoirs of liquid metallic hydrogen. The Uranian and Neptunian fields may come from mixtures of ionised water, ammonia and methane, although exactly how they work remains unknown.

Long-range studies of Neptune's magnetic field started during the two-month Observatory phase, when the PRA and PWS instruments began to search for telltale radio signals generated by its interaction with solar wind particles. Although they had been detecting intermittent radiowave bursts since 26 July 1989, the first

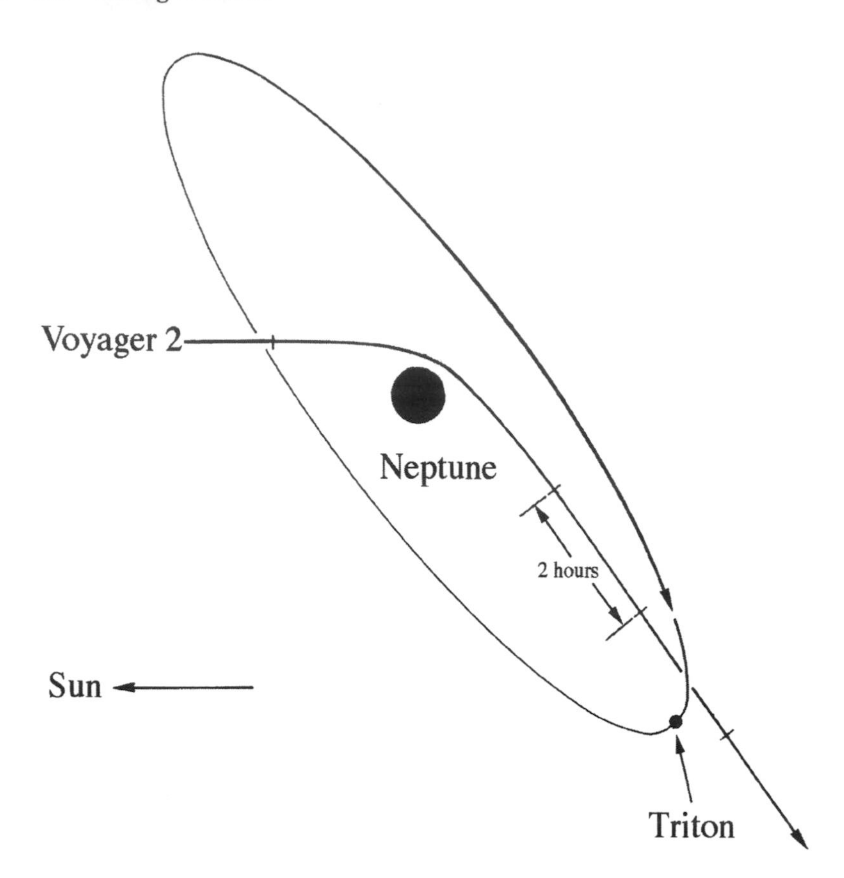

Schematic of Voyager 2's unusual flyby geometry at Neptune.

conclusive evidence for the existence of the field came on 17 August, when Voyager 2 was 11.6 million km from the planet. At 1:38 pm GMT on 24 August 1989, while still 879,000 km away, the spacecraft crossed Neptune's bow shock and entered the field.

During this time, the 'reference star' Canopus in the constellation of Carinae was used to align the PLS and LECP instruments for observations of aurorae over Neptune's north pole. Additional auroral studies were made by the UVS, which took profiles of the planet's limb and twice observed solar occultations. These auroral-emission studies helped determine the location of Neptune's magnetic poles and characterise the interaction between the atmosphere and magnetic field. After crossing the bow shock, Voyager 2 remained in the magnetosphere for 38 hours, before experiencing multiple bow shock crossings and re-emerging into the solar wind.

During the last few days of August and into early September, the PRA continued to measure the rotation of the magnetic field until its radiowaves became too weak to be detectable. One feature of Neptune's magnetic field is that it is not aligned with the rotational axis; in fact, it is highly tilted at an angle of 47 degrees, which on Earth would be like having our magnetic north pole in New York. Furthermore, the field is offset by about 13,500 km from the planet's physical centre.

Similar peculiarities were also found at Uranus, leading some scientists to think – however unlikely – that Voyager 2 might have just arrived there during a periodic magnetic field reversal. However, when it became clear that a similar situation was apparently afoot at Neptune, the coincidence that this could have been happening at both planets seemed too great. It seems more likely that the strange fields have more to do with the interiors of both planets: perhaps they are generated at a relatively high level in their atmospheres, close to the outermost boundary of their slushy liquid layer.

This argument suggests that circulating electrical currents within the conducting interior do not necessarily need to be based near to the planet's physical centre. In general, Neptune's field proved to be variable in terms of strength, but was considerably weaker than any of the other gas giants and, indeed, of Earth's own field.

Scientists initially thought Voyager 2 was on an ideal trajectory to detect aurorae around Neptune's north pole, but the discovery of its extremely tilted rotational axis altered the situation dramatically. Aurorae were actually found over wide regions of the planet, but all were very weak and diffuse. They averaged about 50 million watts, compared to 100 billion watts for Earth-based aurorae. The PWS detected them in the form of charged particle discharges, although none was confirmed visually.

Although its time within Neptune's magnetic field lasted less than two days, Voyager 2 continued periodic observations of the planet until 2 October 1989. Seven weeks later, its attention was turned to a new endeavour. Neptune's gravity bent its trajectory inexorably southward. If the Solar System is spherical – as many scientists believe – the spacecraft is heading south of the ecliptic plane at mid-latitudes, in search of the 'heliopause', the outermost boundary of the Sun's magnetic influence. This Voyager Interstellar Mission continues to this day.

7

"... this bottle into the cosmic ocean"

FAMILY PORTRAIT

On St Valentine's Day in 1990, as lovers the world over exchanged cards and tokens of affection, a truly remarkable and unprecedented event was unfolding nearly 9.6 billion km away. While countless bouquets of roses and boxes of chocolates were delivered to countless addresses across the globe, a new and in some ways even more poignant message was being sent to the whole of humanity. For nearly a full decade, since its spectacular November 1980 rendezvous with Saturn, the cameras on board Voyager 1 had remained dormant. That was about to change.

Operating through wide- and narrow-angle lenses, the spacecraft took 64 shots and, in doing so, captured a unique 'portrait' of six of the nine planets in our Solar System. It was the first time that most of the Sun's family had been seen from beyond the orbit of Pluto. Voyager 1 picked out the Sun, Venus, Jupiter, Saturn, Uranus and Neptune. Regrettably, distant Mercury was lost in the Sun's glare, Mars was a mere crescent and could not be positively identified and Pluto was too small to be resolved. The portrait also included a tiny, far-off pale blue speck that we call Earth.

The multi-frame mosaic was finally released by NASA in mid-June 1990, after intensive image-processing, and when displayed in its full glory would easily cover a wall measuring 30 by 45 m. It yielded little in the way of scientific value, but, as JPL Director Ed Stone commented at the time, "it's more of a historical event [that] will remain unique for decades to come".

Astronomer Carl Sagan summed up the portrait best, by pointing at Earth and stating simply: "This is where we live – on a small dot." Despite having covered such phenomenal distances and taken such an incredible family shot, neither Voyager was anywhere near the edge of the Solar System. Contrary to popular belief, it does not end with Pluto. If it did, then Voyager 1 would have officially left the Sun's realm in May 1988, when it passed quietly 'above' the orbit of the tiny, frozen world, and its sister ship would have followed suit two years later in August 1990.

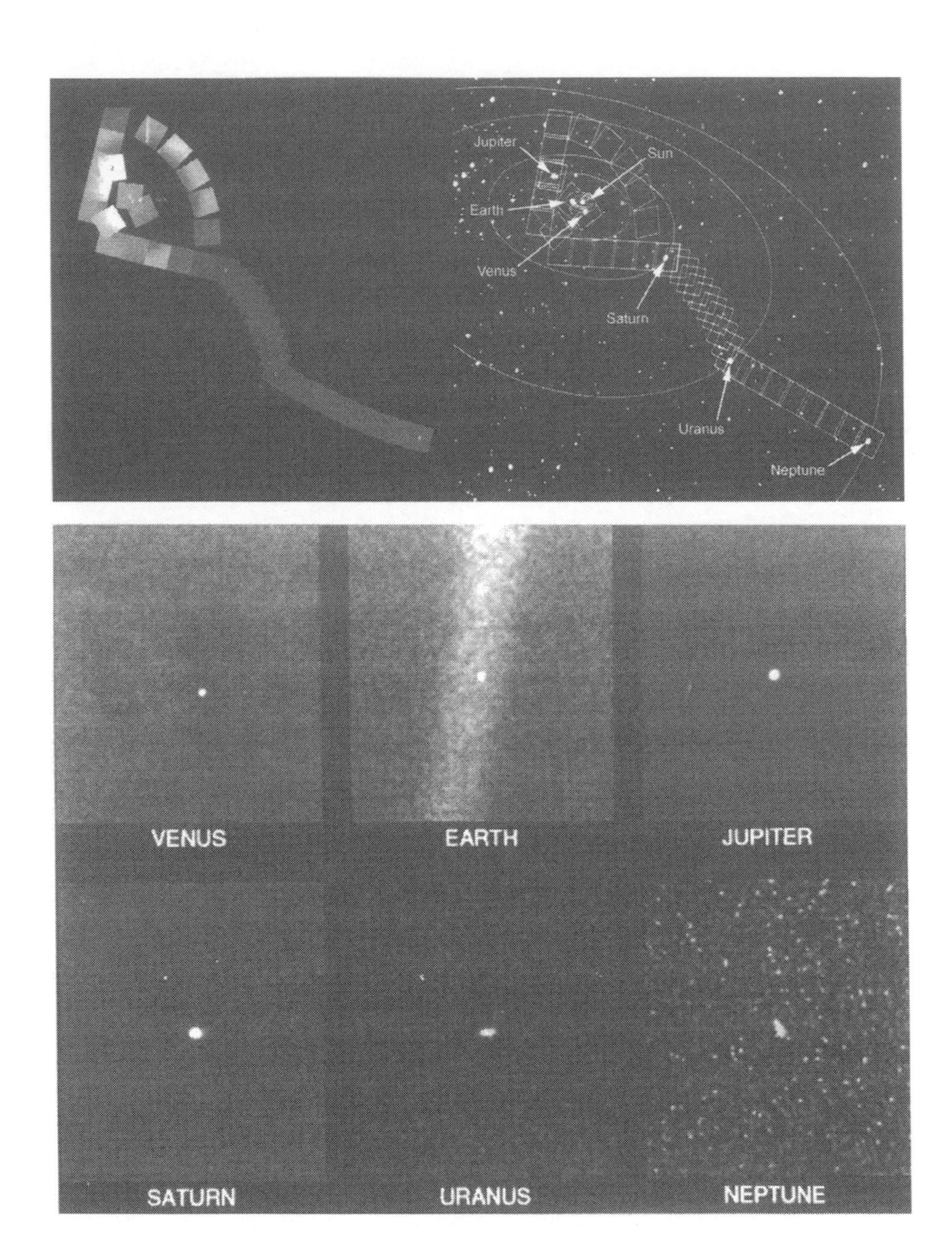

Composite Voyager 1 image of the Sun and six members of its planetary family.

THE 'INTERSTELLAR MEDIUM'

Nor does the Solar System end with the 'heliopause' – a poorly-understood 'outer atmosphere' that is thought to demarcate the edge of the Sun's sphere of magnetic influence – but in all likelihood extends more than a thousand times further into the cold, empty void where an enormous swarm of cometary debris, called Oort's Cloud, is feebly held in orbit by our parent star's gravity. Today, both Voyagers are fast closing in on the heliopause, which they may encounter anytime soon: perhaps even this year or next. No one is certain.

Even moving at speeds of 56,000 km/h, it will take both spacecraft 20,000 years to reach the mid-point of the shell-like cloud and twice as long to emerge through the *other side*! By this time, they will have journeyed for two light-years – some 19 trillion km – and crossed the equivalent of half the distance to Proxima Centauri, the closest star to our Sun (although they are not actually travelling in that direction). It is difficult for us to comprehend such immense distances. Unperturbed, in late 1989 the twins took their first faltering steps on the road to cross it, by starting a magnificent new venture: the Voyager Interstellar Mission (VIM).

Broadly, the goals of this new effort – which will continue until the RTGs on board both spacecraft can no longer supply enough electricity to run the scientific instruments – are to examine the outer Solar System, search for and locate the

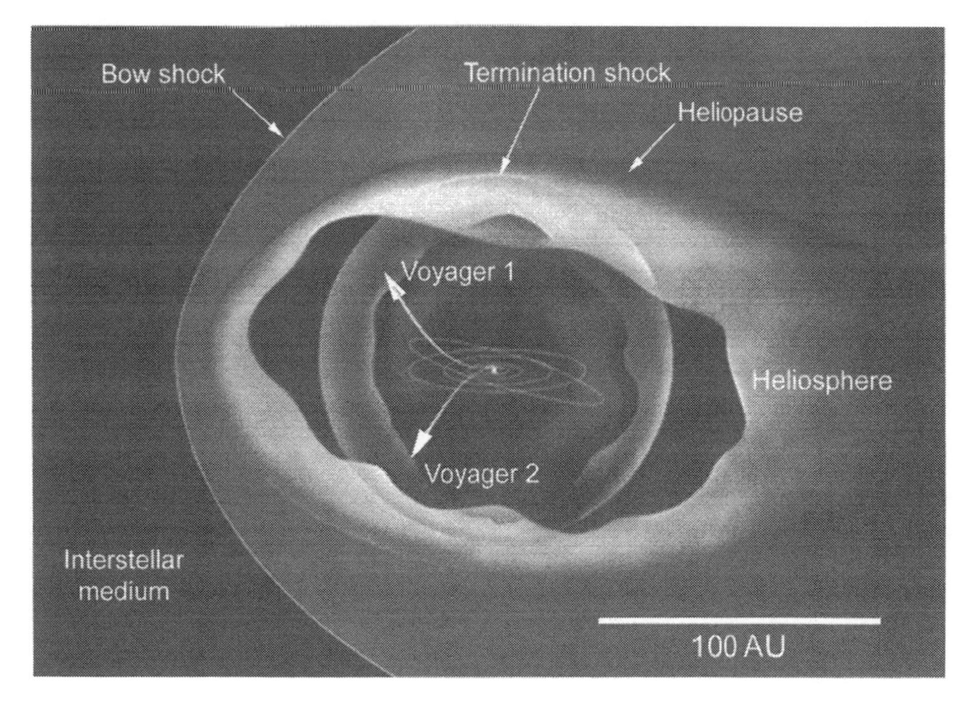

Hypothetical concept of the dimensions and positions of the termination shock, the windsock-shaped heliopause and the 'bow shock' churned up in the wake of our Solar System. The trajectories of both Voyagers are included for reference.

heliopause and, when this is finally passed, measure fields and particles of the 'interstellar medium', an umbrella term for the region of space beyond our Solar System unaffected by the solar wind.

All this operates on the long-held theory that the entire Solar System – Sun, planets, moons, asteroids, comets and assorted debris – is cocooned inside a huge protective bubble created and sustained by the solar wind. The exact dimensions of this bubble (called the 'heliosphere') are unknown, but Voyager data has provided several intriguing clues. One key discovery has been the ever-changing size and shape of the heliosphere. This is caused by the action of the solar wind, which pushes incessantly against the plasma and magnetic fields pervading the interstellar medium.

Although the properties and strength of the solar wind at various distances from the Sun are reasonably well understood, it is difficult to model the extent and workings of the heliosphere accurately without conclusive data from 'outside'. According to present estimates, both Voyagers have sufficient power-producing capability in their RTGs to operate until around 2020 or perhaps slightly longer. It may, therefore, be left to them to make the first *in-situ* measurements of what conditions are like in the interstellar medium.

Voyager 1 seems to be the favourite to make the historic first breakthrough into interstellar space because it is taking the most direct route out of the Solar System, heading 'north' from the ecliptic plane at a 35-degree angle. Its sister ship, on the other hand, was hurled south by Neptune's gravity and is now on its way 'south' at a 48-degree angle. Besides the Voyagers, only two other spacecraft are currently heading out of the Solar System at 'escape velocities': Pioneers 10 and 11. Tracking of the latter ended in January 1995 and NASA's last successful effort to contact the former took place towards the end of February 2003.

Pioneer 10 is leaving the Solar System in a direction 'opposite' the Sun's motion through its interstellar neighbourhood. The distance to the heliopause 'upstream' – in the direction that the Solar System is moving – is believed to be shorter than the

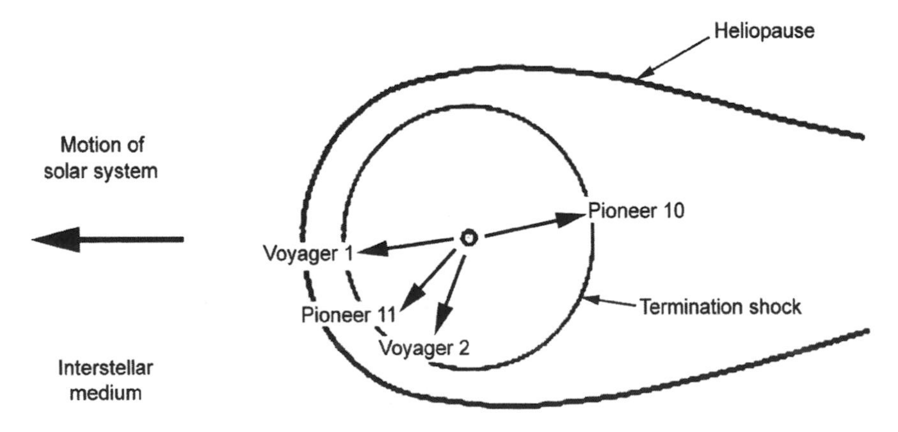

Diagram showing the directions of travel, viewed from over the Sun's north pole, of Voyagers 1 and 2 and Pioneers 10 and 11 as they depart the Solar System.

downstream run. The main reason for this is that the speed of the Sun in the upstream direction causes the pressure of the interstellar medium to be greater, pushing the heliopause inward. As a result, Pioneer 10, heading downwind, will probably take longer than its three comrades to reach interstellar space.

It is, of course, possible that the Voyagers, like the Pioneers, could well expire before they reach the heliopause. If this happens, another long-range mission is on NASA's drawing boards. This mission, at present rather unassumingly dubbed 'Interstellar Probe', will primarily demonstrate advanced nuclear electric or solar sail propulsion technologies to reduce deep-space journey times. It is expected that these will enable it to reach a distance of 200 AU – nearly 30 billion km – within a decade and a half of launch. Its scientific tasks include a detailed survey of interstellar magnetic fields.

The strength and direction of these fields may determine the size and shape of the heliosphere itself, and another mysterious feature known as a 'termination shock', where the solar wind slows dramatically to 400,000 km/h, just a quarter of its previous speed. This shock is thought to lie quite close to the heliopause and hopefully within reach of the Voyagers before they expire.

Information gathered by Pioneers 10 and 11 and both Voyagers suggests that the shock may be situated 80–100 AU (12–15 billion km) from the Sun and the heliopause itself 20–50 AU (3–7.5 billion km) beyond that. Computer simulations predict that, depending on their strength, magnetic fields in the interstellar medium may push or stretch the shock into a spherical or elongated, tadpole-like shape. Particularly strong external fields pushing against the solar wind could even churn

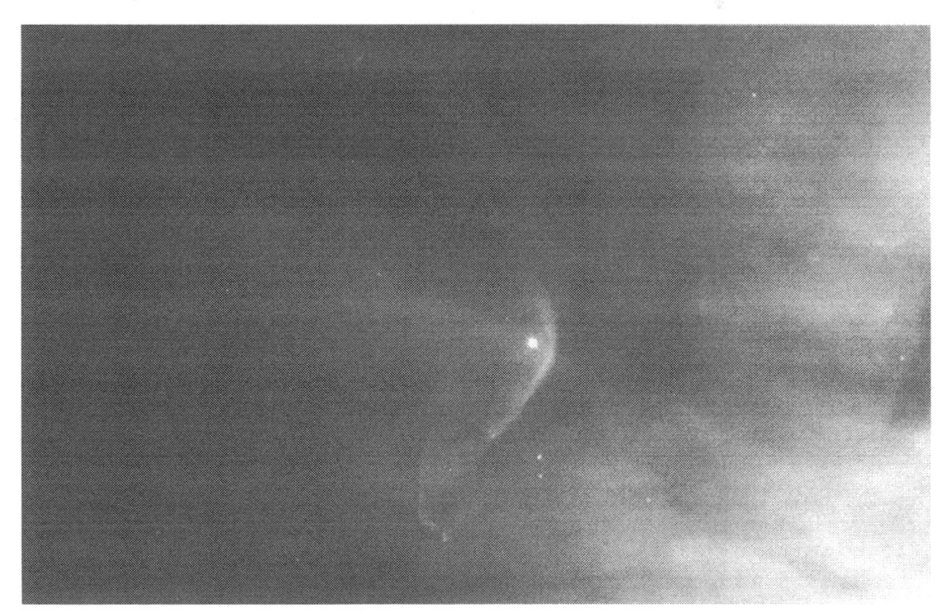

In February 1995, HST found this arcuate bow shock created as the vigorous wind emitted by the star LL Orionis interacts with the tenuous gas of the nearby nebula.

up a 'bow shock' along the leading edge of the heliosphere as it ploughs through its interstellar neighbourhood.

Although the Voyagers are now technically blind – their cameras having long since been switched off – their other senses are helping to keep track of conditions around them, and the signs seem to hint that the termination shock and heliopause are drawing inexorably closer as each year passes. Identification and exploration of the shock is the first major objective of the VIM and, in support of this end, the Voyagers' fields and particles instruments are being kept alive for as long as possible because they will be able to detect the extreme deceleration of the solar wind.

At present, both continue to plough through an environment still controlled by the Sun's magnetic field. In the near future, however, they may pick up changes in plasma-flow directions and the orientation of our parent star's magnetic field. More than a quarter of a century after they were sent aloft, the Voyagers are on the brink of making their most exciting and far-reaching discoveries of all.

When they pass through the termination shock, the twins will begin the 'heliosheath exploration' phase of the VIM. This is a region of space still dominated by the Sun's magnetic field and the solar wind, albeit considerably weaker by now. The size of this heliosheath is unknown, but the Voyagers will undoubtedly need to traverse several tens of astronomical units, perhaps requiring a decade or more after passing the shock, in order to cross it. On the other hand, it may come sooner, perhaps as early as 2008.

Finally, at the furthest edge of the heliosheath is the heliopause – the outermost limit of the Sun's domain. Exactly where this resides is a conundrum that might be solved by the Voyagers. After crossing this region, the two spacecraft will technically be operating in interstellar space, the ultimate goal of the VIM. How long this will require is anyone's guess, although PLS Co-Investigator Ralph McNutt of the Johns Hopkins University Applied Physics Laboratory in Baltimore, Maryland, thinks the heliopause could reside up to 120 AU (18 billion km) from the Sun.

However, his prediction could change dramatically as new data comes back from the two spacecraft. Both have about another decade and a half before their data-collecting capabilities are significantly impaired by reduced power levels and lack of attitude-control propellant. By that time, Voyager 1 will be almost 20 billion km, and its sister ship around 17 billion km, from the Sun. They will then be abandoned to float, dead, perhaps for all eternity.

When the VIM started towards the end of 1989, Voyager 1 was 40 AU (close to 6 billion km) and Voyager 2 was 31 AU (4.6 billion km) from the Sun. By mid-September 2001, Voyager 1 had reached a distance of 82.3 AU and its sister ship was 65 AU from our parent star. Already, as early as February 1998, Voyager 1 had become the most distant man-made object in the heavens, by cruising further from Earth than the previous record-holder: the slower-moving, 1972-launched Pioneer 10 spacecraft.

THE POWER MANAGEMENT LOTTERY

Both Voyagers are today moving at more than 56,000 km/h through space, covering about 3 AU each year. As they continue to cruise gracefully in the solar wind, their

MAG, LECP, CRS, PRA and PWS instruments are constantly studying conditions around them. Additionally, the PLS instrument is still returning data from Voyager 2, but an identical device on its sister ship has long-since failed. Broadly, they are gathering and evaluating data on the strength and orientation of the Sun's magnetic field, the composition, direction and energy-spectra of solar wind particles and the hydrogen distribution in the outer heliosphere.

Another instrument, the UVS, was also used, although, like the PRA, no science team was directly assigned to it. The UVS instruments were used to give astronomers a unique vantage point to look at ultraviolet-emitting celestial objects. Unfortunately, as part of power-conservation procedures, Voyager 2's UVS was turned off in November 1998 and that of its sister ship in 2000. By the turn of the millennium, the IRIS and PPS instruments had also been turned off. Over the next few years, supplemental heaters will be switched off, but at length the electrical power management lottery will become more difficult as fewer non-critical operational components are left.

In order to maximise on what they have left, and keep their crucial fields and particles instruments – our key indicator of where the termination shock and heliopause may lie – running for as long as possible, the first spacecraft 'loads' to be switched off are non-critical ones. Next, in 2010 (for Voyager 1) and 2012 (for Voyager 2), when the DSN becomes unable to continue collecting data at the high 115.2 kbps rate, the digital tape recorders will be turned off. This will be followed by the deactivation of the gyros needed to counter perturbations caused by the tape drives' movements a year or so later.

Without gyro support, the spacecraft will find it impossible to properly calibrate the MAG instruments, among other things. Normally, every six months, each Voyager is rolled 360 degrees to allow engineers to determine its intrinsic magnetic field. This is then subtracted from the solar magnetic field to give an accurate measurement for the latter. Such exercises are vitally important, because each spacecraft's own field overwhelms that of the Sun's field being measured. An end to gyro operations also means that there will be no way of ensuring that the high-gain antennas are being pointed accurately towards Earth.

Circumstances will get worse as power levels continue to drop. Sometime between 2016 and 2018, the electricity-generating capability of the RTGs will fall so low that whatever scientific instruments are left must be alternately turned on and off to 'share' the remaining power. A couple of years later, according to present estimates, neither Voyager will be able to run a single instrument. After this stage has been reached, with no useful data being transmitted from either spacecraft, there will be no point in maintaining communications with them. The DSN's list of higher-priority deep-space missions requiring its attention is already overflowing.

FIRST HINTS OF THE HELIOPAUSE?

This is a pity, because those instruments that are still running have in recent years produced some of the most exciting data from the entire mission. One ongoing activity of note has been the detection of radio emissions in the outer heliosphere, possibly coming from the heliopause, and this brought the Voyagers into the headlines again in the summer of 1992.

In May and June of that year, the Sun experienced a particularly intense period of activity, which emitted a cloud of rapidly moving charged particles. When this cloud of plasma reached the heliopause, it interacted violently with the interstellar plasma and generated intense, low-frequency radio emissions that were picked up by the Voyagers' PWS instruments. Might the emissions pinpoint the whereabouts of the heliopause?

No one could be sure for some time, not even the scientists monitoring the incoming data. "We've seen the frequency of these radio emissions rise over time," PWS Principal Investigator Don Gurnett of the University of Iowa finally told a press conference in May 1993. "Our assumption that this is the heliopause is based on the fact that there is no other known structure out there that could be causing these signals. Earth-bound scientists would not know this phenomenon was occurring if it weren't for the Voyager spacecraft."

Interpreting the emissions as being caused by solar wind particles mixing with their counterparts in the interstellar medium, Gurnett revealed they were probably the most powerful radio sources in the Solar System. He estimated their total output at more than 10 trillion watts, but revealed that their frequencies are so low – no more than a couple of kilohertz – that they cannot be detected using Earth-based instruments. The Voyager radio data from 1992–1993, combined with other measurements made by the spacecraft since then, has led Ralph McNutt at least to suggest that the heliopause lies no further than 90–120 AU from the Sun.

Another useful instrument during VIM has been LECP, which helped immeasurably in efforts to better understand anomalous cosmic rays, which were particularly prevalent in the outer Solar System during the mid-to late 1990s. These are thought to intrude into the heliosphere as our Sun's realm ploughs through its local interstellar neighbourhood. Some of them become ionised, are picked up by the solar wind and are accelerated at the termination shock. Others are redirected towards the inner heliosphere, where the Sun subjects them to dramatic changes. The LECP, therefore, has revealed a great deal about the heliosphere, termination shock and our parent star.

SEMI-AUTONOMOUS OPERATIONS

Reaching the termination shock and heliopause are both major milestones, as no human-made vehicle has yet reached them and the Voyagers will hopefully gather the first direct evidence of what they are like. As the two spacecraft head further from the Sun, numerous upgrades have been made to them to ensure that every drop of data can be squeezed from their instruments. One of the most important of these has been changes to the operating sequences of their computers. In view of the fact that the twins are 1970s technology, the available on-board computer memory is limited by modern standards.

Indeed, less than 1,700 words are available between both CCS memories for storing instrument-sequencing instructions and pointing commands for the high-gain antenna. Fortunately, many operations carried out during the VIM are repetitive, so observations or instrument calibrations can be repeated cyclically essentially for the remainder of the mission. Moreover, this can be done at pretty much the same memory-word cost as a single observation or calibration. It has also been possible to store high-gain antenna-pointing instructions up to 2020 in the CCS, so even if ground controllers lose uplink contact with the Voyagers, the spacecraft can still send data home.

On the other hand, as long as the ground can successfully transmit commands, new sequences can be stored on board or existing ones can be adjusted. If command capability is lost, the Voyagers must continue operating with the sequence on board at the time. In recognition of this, a 'baseline' series of instructions are always stored in their CCS memories to enable them to carry out basic 'housekeeping' functions and other operations critical to their survival. One of these critical functions is ensuring that their high-gain antennas remain pointed towards Earth until at least December 2020 for Voyager 1 and August 2017 for Voyager 2.

If communications are lost and ground stations are unable to transmit instructions to the twins, the Command Loss Timer automatically starts counting down from six weeks. When it reaches zero, if it has not yet received a command from Earth, its on-board computers assume that a communications failure has occurred and start a Backup Mission Load (BML) program. This essentially lets the spacecraft run itself without interaction from the ground, using the command sequences already stored on board. Ordinarily, however, each time the spacecraft receives a command from Earth it resets its timer back to six weeks and starts counting down again.

Additional 'critical' functions include gathering and transmitting a minimum amount of heliospheric data using whatever scientific gear is operating at the time and self-calibrating and resetting its instruments and high-gain antenna periodically throughout each year. If, in a worst-case scenario, a permanent loss of commanding capability occurred, the Voyagers' on-board sequences could conceivably run the spacecraft for the remainder of its operational lifetime, gathering and regularly transmitting data to Earth. This could even include shutting down instruments and other components, one by one, when RTG power levels fall too low to continue operating them.

Looking on the brighter side, if commanding capabilities are not lost, or are re-established after having been lost, then non-repetitive sequences can be set up. Known as 'overlay' sequences, these were first introduced in 1990, early in the VIM campaign, and are typically put in place for six months. In the past, overlays were used to provide additional UVS observations, change the number of MAG calibration roll exercises each year, make additional playbacks of the on-board digital tape recorders or PWS instruments and improve or adjust the baseline sequences.

In this sense, overlay sequences focus on 'single' requirements and are not regularly scheduled, but done on an *ad-hoc* basis. When the Voyagers encounter the termination shock, for example, it is anticipated that ground controllers will put an overlay sequence in place to increase the frequency of PWS measurements from weekly to once every nine hours or so. Another overlay sequence might also be added to make repeated playbacks of this critical plasma-wave data from the shock region.

At the time of writing, both spacecraft continue operating, with some loss in the 'redundancy' of their subsystems, but still able to return data from an almost-full set of VIM instruments. Due to the natural radioactive decay of the plutonium pellets inside each RTG, their electrical power declines rapidly: they produced 470 watts at launch, but this had fallen to around 312 watts for both Voyagers by mid-September 2001. Their supply of hydrazine propellant had also fallen from 100 kg at launch to around 32 kg by March 2001. At present, the RTGs are losing about 7 watts per year and the quantity of propellant in the on-board hydrazine tanks drops by between 6 and 8 g each week. This consumption increases if MAG roll exercises are performed. The latter involve turning the entire spacecraft through ten 360-degree rolls, six times every year, to calibrate the magnetometers. Fortunately, the thrusters are not required to maintain the Voyagers' velocity: the gravitational boosts they received from Jupiter in 1979 gave them more than enough energy to leave the Solar System.

FUTURE ENCOUNTERS

Eventually, in the far future, and by then long-dead, they should pass other stars. Assuming that nothing stops or gets in the way of its trajectory, in 40,000 years' time Voyager 1 will drift within 15 trillion km – nearly two light-years – of the obscure star AC + 79 3888 in the faint constellation Camelopardalis (the Giraffe). Meanwhile, Voyager 2 is heading south of the ecliptic plane towards the constellations Sagittarius (the Archer) and Pavo (the Peacock). Forty thousand years from now, it will pass within 1.7 light-years of the dim red star Ross 248 in the constellation Andromeda. Interestingly, although Voyager 2 is heading in a southerly direction, Ross 248 is actually at present in Earth's *northern* skies! It is not only the Voyagers and our Solar System that are in motion; so too are the stars.

As Voyager 2 creeps outwards at 56,000 km/h, so Ross 248 will also be heading towards our Solar System at a speed more than five times faster. Shortly after the spacecraft emerges from its long, lonely trek through the Oort Cloud, it will encounter Ross 248 as the star passes just 3.25 light-years from our Sun, at which time it will no longer be in Andromeda as seen from Earth.

At this time, Voyager 1 will pass AC + 79 3888 – a star with no name, just a catalogue number. It presently resides close to Polaris, the North or 'Pole' Star, and is a little bigger and brighter than Ross 248. Yet they are similar: both are small, cool, red stars, less than a quarter as massive as our Sun and neither can be seen with the naked eye. The situation may change by the time the Voyagers encounter them: both stars will pass within four light-years of the Solar System, which should make them detectable to Earth-based astronomers at some time in the 420th century AD.

It is also possible that, a quarter of a million years from now, Voyager 2 may fly within 40 trillion km of Sirius, currently the brightest star in our sky. Meanwhile, Pioneer 10 will head for the constellation Taurus (the Bull), possibly passing one of its stars in two million years' time, while Pioneer 11 might traverse the constellation Aquila (the Eagle) some 40,000 centuries from now. By that time, doubtless other emissaries from Earth will also have left the Solar System and the extent of our knowledge of our celestial backyard will have advanced beyond our wildest dreams.

Perhaps one of these stars, or one of their neighbours, has a planetary system. Perhaps on one of these planets is a space-faring civilisation that will find the Voyagers. Perhaps at this moment, that civilisation is still in its infancy: with its own version of the big-brained, bipedal, stone-tool-wielding hominids that strode the plains and rift valleys of Africa a few million years ago. By the time the long-dead Voyagers reach them, that civilisation may have evolved into something similar to our civilisation now, sending its own Voyagers into space and perhaps finding ours. It is even conceivable that our star-faring descendants will retrieve them.

It does, admittedly, seem unlikely that either AC + 79 3888 or Ross 248 will have the right environment for life to evolve and thrive. Not only are they considerably smaller than our Sun, but they emit so little heat that a life-bearing planet would need to be very precisely located at a specific position in a perfectly circular orbit. Furthermore, this 'precise location' would bring the hypothetical planet so close to its parent star that it might be tidally locked to it. One side might fry and the other freeze under such conditions.

Consequently, most scientists do not expect planets around low-mass red dwarfs like AC + 79 3888 or Ross 248 to harbour anything more advanced than microbial life. Additionally, even if by some stroke of luck there was an intelligent species on a planet around one of the two small stars, it seems unlikely that they would detect a tiny, silent spacecraft drifting beyond the outermost fringes of their own Oort's Cloud. On the other hand, the Voyagers could be picked up before or after they reach these stars.

MURMURS OF EARTH

What then? During the development and construction of the Voyagers, it was always realised that, like the Pioneers before them, they would continue to fly through the heavens after they had ceased to function. The 1979 Jovian gravity assists would see to that. At the same time, it was also felt there was a slight possibility that an advanced alien civilisation would find them. As a result, like the Pioneers, each Voyager carries a gold-plated copper disk, containing sounds and pictures to tell the story of our world and the iridescence of life here.

This amazing creation, which lasts under two hours and was the subject of Carl Sagan's book *Murmurs of Earth*, contains 122 images and a range of natural sounds – from surf, wind and thunder to birds, whales and other animals – to illustrate the sheer diversity of life on Earth. To this, a committee chaired by Sagan added 90 minutes of music from numerous cultures and eras, together with greetings in 55 languages, spoken by people over a 6,000-year time span of recorded human history, and printed messages from then-US President Jimmy Carter and then-UN Secretary-General Kurt Waldheim.

It was designed for NASA by Carl Sagan, Frank Drake, Ann Druyan, Timothy Ferris, Jon Lomberg and Linda Salzman Sagan. Its images were encoded as 'vibrations' on the disk and the remainder of the contents in audio form. The whole record can be played back at a speed of 16 rpm and, made from copper and gold, should endure for a billion years. Each was encased in its own protective aluminium jacket, together with stylus and cartridge and instructions in symbolic form were included to explain to the recipients how to play them and from whence they came.

If either Voyager is ever found by alien recipients, they will probably not have too much difficulty locating the record as it is mounted in plain view on the spacecraft's main bus, together with a spider mounting for the stylus and cartridge. Our aliens might also find it slow, requiring a full 12 seconds to play back a single colour picture. By comparison, a modern CD, spinning nearly 30 times faster, can play back 360 video frames in a similar length of time. The record is composed of two copper 'mother' disks bonded back-to-back.

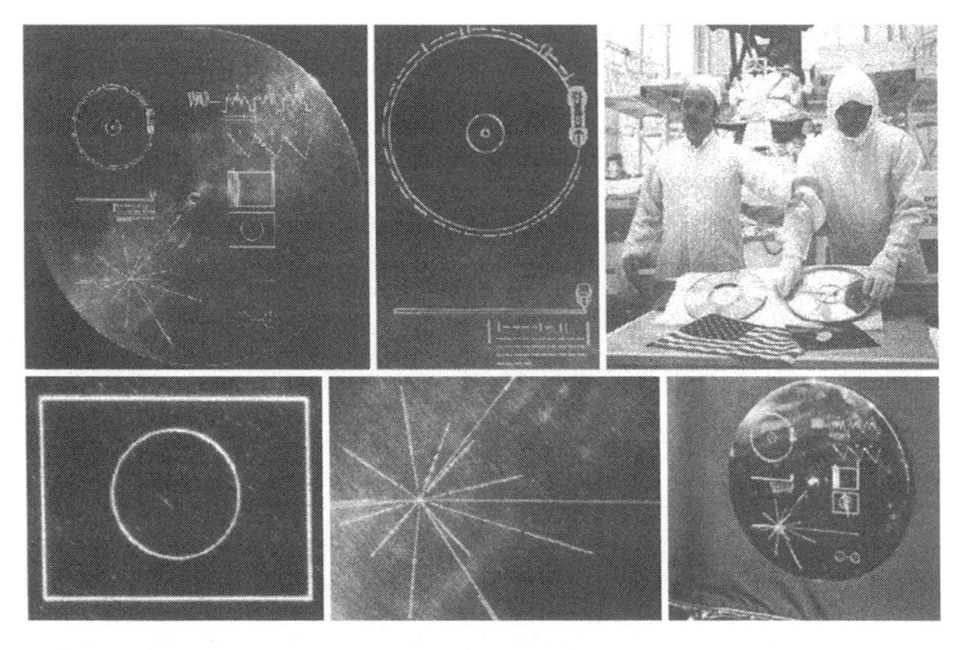

(*Top left*) The Voyager record. (*Top centre*) Diagram on the Voyager record, showing the stylus and how the record is to be played. (*Top right*) The Voyager record is prepared for attachment to the spacecraft. (*Bottom left*) The first picture on the Voyager record is this 'calibration circle', which will help any alien recipients to ensure that they are playing the record correctly. (*Bottom centre*) This pulsar map on the Voyager record helps to identify where the two spacecraft came from. (*Bottom right*) The golden record mounted on the side of the Voyager spacecraft's main bus.

The side that faces 'inward', against the spacecraft's main bus, contains all of the images stored on the record, as well as the sounds from Earth, the greetings – ranging from Akkadian, spoken in ancient Mesopotamia 60 centuries ago, to a modern Chinese dialect known as Wu – and a third of the music collection. The outer, space-facing side consists of the remainder of the music. On the lower right-hand of the record cover is a key to help the alien recipients to interpret the engraved illustrations.

In the upper-left corner is a drawing of the record itself, together with the stylus, which is shown in exactly the correct position needed to play it from the beginning. Written around the illustration, in binary arithmetic, is the time for a single rotation of the record: 3.6 seconds, expressed in time-units of 0.70 billionths of a second (a period also associated with a fundamental transition between the two lowest states of hydrogen, which it was judged was something that a space-faring civilisation would be sure to know). It also shows that the record should be played from the outside inwards and provides a binary number equivalent to the time needed to play one side.

Information on the upper-right portion shows how the pictures can be reassembled from the recorded vibrations. The top drawing shows the typical signal that occurs at the start of a picture. The picture is made from this signal, which 'traces' the image as a series of vertical lines, much like TV. Picture lines 1, 2 and 3 are provided in binary numbers and the length of one of the 'picture lines' – about 8 milliseconds – is noted. The illustration directly below shows how these lines are to be drawn vertically, with staggered 'interlace' to give the correct picture rendition.

Beneath this is a drawing of an entire picture raster, revealing that 512 vertical lines comprise a complete picture, and underneath that is a replica of the first image recorded onto the disk. Assuming that the alien recipients are able to follow the instructions this far, this should at least allow them to convince themselves that they are on the right track. A circle is also included to ensure that the recipients are using the correct horizontal-to-vertical height ratio when reconstructing the pictures.

Next, in the lower-left corner is a 'pulsar map', also added to similar records flown on board Pioneers 10 and 11. It displays the location of our Solar System with respect to 14 pulsars, whose periods are precisely given. Each pulsar has its own distinct and rapid pulsing frequency that very slowly changes with near-absolute linearity. Civilisations at least as advanced as ours should be sufficiently familiar with pulsars and their rates of change to appreciate this map for what it is.

A drawing containing two circles in the lower right-hand corner shows the hydrogen atom in its two lowest states, with a connecting line and the digit '1' to show that the time interval associated with the transition from one state to the other is to be used as a time scale for the time given on the cover and for the decoded pictures.

Another ingenious touch was a tiny, ultra-pure source of uranium-238, with a half-life of 4.51 billion years, electroplated onto the record's jacket. Its steady decay over time into its 'daughter' isotopes should provide a radioactive 'clock'. Thus, assuming the Voyagers are recovered by a civilisation at least as advanced as ours, by examining this 2-cm-diameter area and measuring the amount of daughter elements to the remaining uranium-238, our aliens could determine how much time had elapsed since the source was placed on board the spacecraft. Coupled with the pulsar map, this provides a check on the epoch of launch.

Assuming that our aliens have succeeded in finding a path through these pictorial instructions, they will be rewarded by indecipherable greetings in Sumerian, Akkadian, Hittite, Hebrew, Aramaic, English, Portuguese, Cantonese, Russian, Thai, Telugu, Arabic, Romanian, French, Burmese, Spanish, Indonesian, Kechua, Dutch, German, Bengali, Oriya, Urdu, Hindi, Vietnamese, Sinhalese, Greek, Latin, Japanese, Punjabi, Turkish, Welsh, Hungarian, Italian, Nguni, Sotho, Wu, Korean, Armenian, Polish, Nepali, Mandarin, Gujarati, Czech, Ila, Nyania, Swedish, Ukrainian, Persian, Serbian, Luganda, Amoy, Marathi, Kannada and Rajasthani.

Next, the aliens can decode the pictures on the record. These are too numerous to discuss in great detail, but it is remarkable that they capture the essence of Earth in *only* 122 images. Firstly, there is a circle provided by the record's co-creator Jon Lomberg for calibration purposes. The remainder includes pictures of the planets – including, of course, our own – together with chemical definitions, the structure of DNA, anatomical diagrams of men and women, human sex organs, numerous other species, various geographical locations on Earth and aspects of everyday life throughout the world.

Finally, spread across both sides of the record, is the music: ninety minutes of it, set aside to cover different types of music and song from traditional and classical to modern. Of particular note are Aboriginal songs, an initiation verse sung by pygmy girls from Zaire (now the Democratic Republic of the Congo), Chuck Berry's *Johnny B. Goode*, numerous classical pieces (from Bach and Beethoven to Mozart and Stravinsky), Scottish and Azerbaijani bagpipes, Peruvian and Solomon Islands panpipes and drums, Native American chants and many others from several different cultures.

SILENT WANDERERS

At present, however, the records are silent. They have not been played for nearly three decades. The two spacecraft that carry them continue to sail gracefully on the solar wind, taking scientific measurements, searching for our Solar System's termination shock and heliopause, elusive at present, but drawing inexorably closer as each year passes. And thousands – maybe millions – of years hence, or perhaps for all eternity, they will continue, like unquiet spirits from a past era, their lonely, restless wanderings across the sky.

The slow passage of time will gradually lose its meaning for them. Back on Earth, countless generations of humanity will come and go. The scientists and engineers who made the two spacecraft will be long-dead, as will the author of this book, and the Voyagers themselves will likewise have long since ceased to operate. Yet they will drift onward. What is in store for them, what they might encounter – or, perhaps more intriguingly, what will encounter *them* – is impossible for even our space-age, alien-obsessed culture to fathom.

As impossible, maybe, as it would have been for Galileo, Huygens, Cassini, Herschel, Galle and d'Arrest to have foreseen that machines built by their descendants would one day be exploring the worlds they first saw as tiny, far-off points of light in the sky – or, in the case of Leverrier and Adams, as mathematical plots on sheets of paper. The Voyagers' most remarkable discoveries – the edge of the Solar System, what conditions are like beyond it and whether the Universe is lifeless or teeming with alien civilisations – might be just around the corner.

And if some intelligent species does happen upon them in the distant future, the only means either spacecraft will have to describe the world from which it came will be a small disk holding a handful of pictures and a few strange sounds in unknown tongues. Undoubtedly, by the time the Voyagers have spent hundreds of centuries crossing interstellar space, the planet and Solar System about which they will speak will have changed beyond recognition. It has *already* changed beyond recognition, even in less than three decades, but humanity might have changed even more fundamentally.

Simply perusing a textbook of hominid evolution allows us to see the significant physiological and anatomical differences between our ancestors living a couple of million years ago and ourselves today. Several million years into our future, our own descendants may look quite different again. And yet, none of that really matters. Whether the Sounds of Earth are understood or not, whether or not the two spacecraft are someday found by a small alien boy at the shore of a great cosmic ocean, the long journey of the Voyagers will not be in vain.

One has only to look at tens of thousands of pictures and rooms full of data from Jupiter, Saturn, Uranus, Neptune and their multitude of moons – most of which is still being analysed – to realise that fact. Any planetary scientist will explain that our knowledge of the gas giants and their realms has increased beyond all comprehension, thanks to the Voyagers. And at a grand total of $865 million, spread across 30 years and two generations of the United States' 280 million taxpayers, it has cost as much per person as a lunch in McDonald's. Financially and scientifically, Voyager has more than paid for itself.

As the late Carl Sagan once remarked, the launching of "this bottle into the cosmic ocean" – not just the two spacecraft themselves, but also the positive, far-sighted message they seek to convey – says something very hopeful about *us* as a species.

8

Future Voyagers

RETURN TO SATURN AND TITAN

It is the first day of July 2004. An enormous spacecraft, larger than any that came before, has just arrived at Saturn and eased itself into orbit around the giant ringworld. Equipped with a set of scientific instruments to rival a small laboratory, the spacecraft – named Cassini – weighs 5,650 kg and is roughly the size of a 30-passenger bus. Its size and complexity has been dictated almost entirely by one goal: the need to undertake a comprehensive, four-year survey of the Saturnian system and to do so *in-situ*, more than 1.6 billion km from Earth.

At first glance, getting a spacecraft of Cassini's size across such a vast gulf would be expected to make both factors mutually exclusive. In fact, mission designers never seriously planned to deliver their spacecraft directly to Saturn, because the energy required would have been phenomenal. Instead, they opted for using the gravity of the giant planet Jupiter to provide the necessary impetus to get there. The trip has spanned nearly seven years since Cassini was launched in October 1997.

The mission actually consists of *two* independent spacecraft: the main Cassini vehicle, which will remain in orbit around Saturn, and a small craft called Huygens, which will plunge into the atmosphere of its mysterious moon Titan and attempt to land there. This monumental effort is an international one between three space agencies, covering 17 nations and including a team of more than 250 scientists to manage and analyse the data. Cassini itself was developed by NASA and Huygens by the European Space Agency (ESA). Additionally, the Italian Space Agency (ASI) built Cassini's 4-m high-gain communications antenna and associated data-relay system for the probe.

Turning firstly to Cassini itself, the orbiter is named after the seventeenth-century Franco-Italian astronomer Giovanni Cassini, and measures 6.8 m high. It carries 12 scientific instruments, although many of these will have diverse tasks at Saturn. Communications will be carried out primarily through the Italian-built high-gain antenna, although there are also two low-gain antennas. These provide a lower-rate

The Cassini spacecraft undergoes final preparations for launch. Note the Huygens probe – the conical craft mounted on the side of Cassini – at the left of the picture.

link while the high-gain antenna is not aimed towards Earth, as happened when it was shielding the spacecraft from fierce solar heating in the early stages of its journey and as it will be when Cassini turns to make observations of Saturn. Electrical power to run the instruments and subsystems comes from a trio of RTG power plants, similar in design to those on board the Voyagers, but considerably more powerful.

Yet it is the instruments themselves that will really show what Cassini can do. On numerous occasions, the mission has been compared with the way our senses operate: with our eyes and ears represented by its remote-sensing equipment, our abilities to touch and taste equating to its direct-sensing gear and our noses covered by both. Cassini's scientific gear can be broadly placed into two categories: remote-sensing and fields and particles instruments. The first category includes cameras, spectrometers, radios and radar equipment, while the latter measure magnetic fields, electrical charges and dust and atomic-particle densities.

Turning to the remote-sensing gear first, these are subdivided into two categories: optical and microwave instruments. There are four optical instruments: the Composite Infrared Spectrometer (CIRS), Imaging Science Subsystem (ISS), Ultraviolet Imaging Spectrograph (UVIS) and Visual and Infrared Mapping Spectrometer (VIMS). In addition, there are two microwave instruments: the Cassini Radar and Radio Science Subsystem (RSS). Then there are six fields and particles instruments, known as the Cassini Plasma Spectrometer (CAPS), Cosmic Dust Analyser (CDA), Ion and Neutral Mass Spectrometer (INMS), Dual Technique Magnetometer (DTM), Magnetospheric Imaging Instrument (MIMI) and the Radio & Plasma Wave Science (RPWS) unit.

Cassini's first photograph of Saturn, taken in November 2002 from a distance of 285 million km with arrival at the giant planet still 20 months away. Notice Titan, upper left.

During its four years at Saturn, Cassini is expected to complete 74 orbits of the giant ringworld, although – if the Galileo mission is anything to go by – it may be extended, depending upon its success. Its instruments will collect detailed information on the planet and its moons, measure the size of its enormous magnetosphere, examine the intricate structure of the rings and study Saturn's physical composition and its atmosphere. Every time the spacecraft's on-board data recorder fills up, it will point its high-gain antenna towards Earth and transmit several gigabytes of data home on a daily basis.

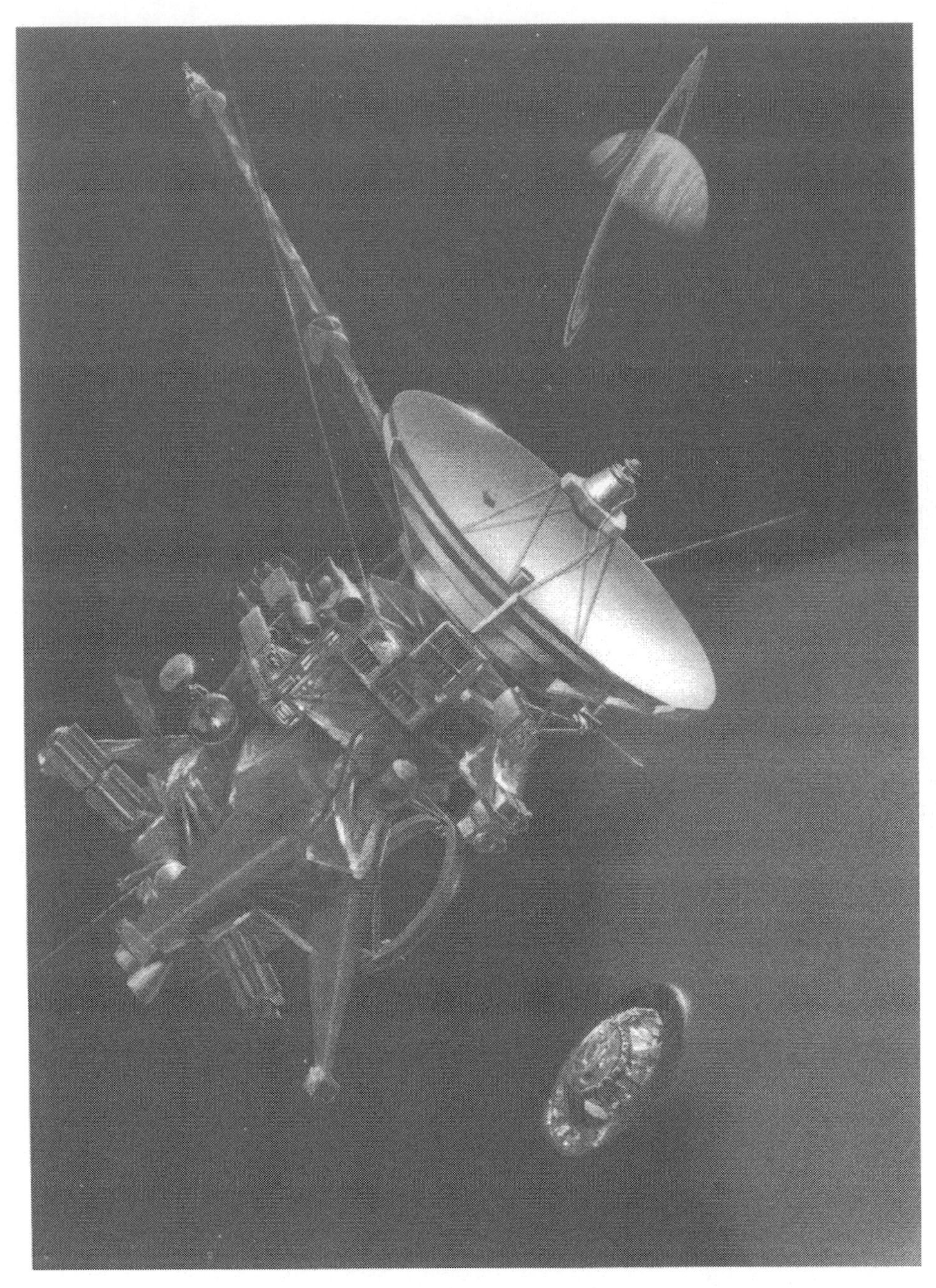

An artist's impression of Cassini releasing the Huygens probe. Note that a certain licence has been taken concerning the range from the Titan when this occurs.

FIRST LANDING ON TITAN

Perhaps even more intriguingly, Cassini's on-board radar device, which uses the HGA dish, will enable it to produce the first detailed maps of Titan's surface, at present hidden from our eyes by a dense canopy of gases and particulate haze. Neither Voyager spacecraft was able to see what Titan looked like beneath this thick atmosphere – they both revealed it as little more than a fuzzy orange tennis ball – although more recent HST and ground-based infrared observations have revealed some 'landforms', whose precise nature is unclear.

Although Cassini is expected to offer us our first global maps of Titan *beneath* its atmosphere, which should reveal a surface in at least as much detail as Magellan achieved when it surveyed Venus with synthetic-aperture radar in 1990–1994, it will be the Huygens landing vehicle that will hopefully provide some of the best data. A few days before Christmas 2004, Huygens – a gold-coloured, cone-shaped craft presently bolted onto the side of Cassini – will be ejected from its mother-ship to begin a 22-day cruise to Titan. On 14 January 2005, it will slam into its atmosphere at 20,000 km/h.

Like a spacecraft entering Earth's atmosphere, the compression of gas in front of the vehicle will slow Huygens' descent to a stately 1,400 km/h. Then, when the

Cassini's on-board radar will provide the first high-resolution map of Titan's surface beneath its opaque atmosphere.

spacecraft reaches an altitude of about 160 km above Titan's surface, the first of three parachutes will automatically open to stabilise it for its first scientific observations. The subsequent deployment of two other parachutes will set Huygens up for a reasonably soft, 25-km/h touchdown on Titan, somewhere near the equator. The descent through the murky clouds is expected to take a few hours.

During this time, Huygens will transmit data constantly to Cassini over the Italian-supplied relay, passing overhead, which will record it and later replay it to the 70-m DSN tracking stations around the world. As has already been seen in Chapter 4, no one is quite sure of the nature of Titan's surface at present. It is almost certain that the moon does not have a global ocean, but that does not mean that substantial bodies of liquid hydrocarbon volatiles do not exist. The shape and structure of Huygens has, therefore, been designed to support an amphibious landing on a variety of different terrains.

Assuming that the lander survives the impact of touchdown (or even 'splash-down'), it may continue to return data to Cassini for up to half an hour, when either its on-board batteries will expire or the orbiter will pass out of communications range. Huygens is a small vehicle, measuring less than 2.7 m in diameter and weighing 318 kg, but is nevertheless packed with scientific equipment to capture as much data as possible from the strange moon. Fittingly, it is named after Christiaan Huygens, the Dutch astronomer who discovered Titan in 1655.

For the majority of its journey to Saturn, Huygens has remained dormant, although it has been subject to twice-annual health checks during the seven-year cruise. Even after Cassini releases Huygens on 23 December 2004, the small vehicle will remain passive. In fact, the only device running will be a timer that will, 45 minutes before it reaches Titan on 14 January 2005, awaken the vehicle from its long slumber. Half an hour later, the timer will power up the rest of Huygens' on-board systems and scientific instruments.

From this point onward, it has 153 minutes of battery life to complete the descent and, if possible, some surface science activities. Two and a half hours have been set aside for the descent and a minimum of three minutes to return data from the surface. Of course, if it takes less time to reach Titan's surface and Huygens survives the impact, there may be more time available to take measurements and perhaps panoramic photographs of the landing site. However, after three hours of listening to the probe's signals, Cassini will turn its high-gain antenna away from Titan and refocus it on Earth.

Despite its small size, Huygens is packed with equipment. One piece in particular that will sell the mission to the public (if the Voyager experience is anything to go by) is the camera, which is expected to take more than 1,100 pictures during the descent and landing phases. There are a total of six scientific devices to sniff Titan's atmosphere during descent: the Huygens Atmospheric Structure Instrument (HASI), the Doppler Wind Experiment (DWE), the Descent Imager/Spectral Radiometer (DISR), the Gas Chromatograph Mass Spectrometer (GCMS), the Aerosol Collector and Polariser (ACP) and a Surface Science Package (SSP).

These will be employed for a variety of tasks, including the determination of atmospheric densities and the existence of winds at various altitudes. During the early stages of the descent, high-altitude aerosols will be captured by the ACP, heated to vaporise volatiles and decompose the complex organic matter within them, and then passed to the GCMS for a detailed chemical analysis. The GCMS will also sample the atmosphere and its built-in mass spectrometer will model the molecular masses of the captured gases. The instrument will continue to run after landing as part of efforts to analyse Titan's surface composition.

In the meantime, in addition to shooting in excess of a thousand visible-light photographs, the DISR will undertake infrared observations as Huygens descends through Titan's murky atmosphere. On-board Sun sensors will also measure the intensity of light around our parent star as it peeks through the haze as part of efforts to better understand the scattering properties, distribution and size of smog particles hanging up to 400 km above the surface. As Huygens closes in on its landing site, the vehicle – slowly spinning beneath its parachutes – should be able to capture a series of panoramic view of the ground track as the probe drifts in the prevailing wind.

A SOLID OR LIQUID SURFACE?

At very low levels, scientists expect conditions to be quite dusky. Just before touchdown, a powerful floodlight will be switched on to augment the weak sunlight and aid the camera and spectrometer. After Huygens has landed and settled, the SSP sensors will carry out a brief survey of the physical characteristics of the landing site. The most obvious piece of data will be whether the surface is solid or liquid. An acoustic sounder will start running 100 m above the surface to work out the decreasing distance, rate of descent and nature of the terrain.

"Huygens will verify whether there is really liquid on the surface of Titan," says JPL's Ellis Miner, a science adviser for the Cassini mission. "We can only conjecture now, but there has to be a source for all that ethane and methane in the air, which would probably be lakes or oceans of it."

If the surface happens to be liquid in nature, the sounder will measure the speed of sound in the 'ocean' and, if possible, also its depth. The SSP also includes an accelerometer which will accurately determine the deceleration rate at impact, indicating the structure and 'hardness' of the surface. Tilt sensors will measure any pendulum motions during the descent and indicate Huygens' attitude after landing to reveal any motions caused by the action of waves. Meanwhile, other instruments will measure its density, temperature, light-reflecting properties, thermal conductivity, heat capacity and electrical permittivity.

The Huygens mission will last a fraction of Cassini's own odyssey of discovery and although its sojourn will last under three hours, the mother-ship will perform numerous surveys of the mysterious, planet-sized moon. In addition to the radar, there are 11 other instruments on board Cassini. The CAPS will measure the energy and electrical charge of particles it encounters. Specifically, it will measure molecules originating from Saturn's ionosphere and determine the configuration of the planet's peculiar magnetic field. Additionally, it will investigate plasma levels in the magnetosphere and the behaviour of the solar wind in these regions. The CDA is designed to measure the size, speed and direction of minute dust grains – many with the consistency of cigar smoke – in the vicinity of Saturn as part of efforts to better comprehend their nature and sources.

Although similar dust-analysis devices were also carried on board Galileo and the joint US/European Ulysses solar polar orbiter, launched in 1990, both could only ascertain that they were entering regions of cosmic dust and determine their approximate directions, masses or sizes. The much more advanced Cassini dust analyser, on the other hand, can also figure out the trajectories and speeds of the particles, which in turn enables scientists to identify their sources by measuring their chemical composition.

A COMPREHENSIVE SURVEY OF SATURN

Returning to the planet itself, the CIRS instrument will measure infrared emissions from the atmosphere and rings of Saturn, as well as from the surfaces of its many moons, to determine their temperatures, chemical compositions and thermal properties. It will map Saturn's atmosphere in three dimensions to work out temperatures and pressures at differing altitudes and in regions of different gas composition or distribution of aerosols and clouds. In many ways, this instrument is similar to the IRIS flown on board the Voyagers, consisting of far- and near-infrared interferometers, which share a single telescope and scanner.

However, the Cassini instrument is much more advanced than the IRIS. It has a spectral resolution 10 times higher than was available to either Voyager and covers a much broader wavelength range. In précis, the result is that much finer spectral detail can be extracted from the CIRS data. Another instrument that operates in the infrared – as well as the visible and ultraviolet – portion of the electromagnetic spectrum is the ISS, which includes wide- and narrow-angle cameras capable of returning hundreds of thousands of pictures from the Saturnian system.

Stunning images have already been returned by Cassini's cameras when the spacecraft flew within 10 million km of Jupiter in December 2000 to gain a gravitational push to Saturn. As the spacecraft approached Jupiter, it snapped a long series of images for a movie of that planet's dynamic atmosphere. Planetary scientist Carolyn Porco of the University of Arizona at Tucson, the leader of the imaging team, was astounded. "This spacecraft is steadier than any I have ever seen," she enthused. "It's so steady that the images are unexpectedly sharp and clear, even in the longest exposures taken in the most challenging spectral regions."

There was some frustration soon after leaving Jupiter, when the narrow-angle camera became slightly fogged, but this was eliminated by thermal conditioning during the long interplanetary coast towards Saturn. The spacecraft also took its first picture of Saturn on the first day of November 2002. Even from a distance of 285 million km, the picture was astonishingly sharp. "Seeing the picture makes our science planning work suddenly seem very real," said Cassini camera-team member Alfred McEwen of the University of Arizona at Tucson.

The quality of this spectacular shot of Jupiter and its moon Ganymede, taken by Cassini during its December 2000 encounter, has whetted the appetite of scientists awaiting the spacecraft's arrival at Saturn.

"Now we can see Saturn and we'll watch it get bigger as a visual cue that we're approaching fast. It's good to see the camera is working well." Cassini Project Scientist Dennis Matson of JPL exulted: "We have Saturn in our sights!" Both the wide- and narrow-angle cameras are fitted with spectral filters on a wheel that can be rotated to a specific setting. This enables them to view different bands of the electromagnetic spectrum, ranging from 0.2 to 1.1 microns.

Other multi-spectral imaging equipment includes the UVIS, which will measure ultraviolet light reflected by Saturn's clouds and rings to better determine their structure and composition. It is also expected to prove useful when probing the distribution, aerosol content and temperature of Saturn's and Titan's atmospheres. It is different from other types of spectrometers because it can take both spectral and spatial readings and, in fact, can take so many images that it can create 'movies' to show how material is moved around by other forces.

While the UVIS is sensitive to ultraviolet wavelengths, the VIMS is particularly suited for visible-light and infrared studies. This is expected to yield further clues to the composition of Saturn's and Titan's atmospheres, as well as the icy surfaces of many of the moons. It will also be used to probe the structure and distribution of material in the planet's rings, by observing sunlight or starlight peeking through them. The instrument will play an important role in searching for the minor building blocks of life – known as 'pre-biotic' materials – and it will measure the locations and conditions under which they are found.

Another device that will be used to investigate the structure of Saturn's rings is the RSS, which will use ground-based antennas to observe how Cassini's radio signals change as they are directed through them. The RSS will employ radio receivers and transmitters operating at three separate wavelengths – an X-band communications link, together with an S-band downlink and Ka-band uplink and downlink – which will enable it to observe compositions, pressures and temperatures of the Saturnian and Titanian atmospheres, gravity fields and gross masses of the giant planet and its moons.

Clearly, in addition to the three hours of 'direct' *in-situ* measurements of Titan's atmosphere and surface by the Huygens lander, the instruments of Cassini itself will contribute a great deal to our understanding of this strange world. The INMS instrument will measure charged and neutral particles close to Titan, as part of efforts to identify constituents of its thick canopy of gases. Meanwhile, the MIMI and RPWS instruments will monitor the interaction of the solar wind with Titan's atmosphere and the DTM will determine the effect of Saturn's magnetic field on the planet-sized world.

These last four instruments will be used for a wide range of tasks, including studies of the other moons, rings, Saturn's atmosphere and magnetic field and the influence of the solar wind on the whole system. For its part, the DTM will offer the closest possible way to investigate where Saturn's intrinsic magnetic field comes from. To the best of scientists' present knowledge, it is generated partly in the planet's core and it is hoped that DTM data will enable them to develop three-dimensional models of Saturn's magnetosphere and its effect on the magnetic states of the other moons.

The nature of the planet's magnetosphere and atmosphere are also a primary focus for the MIMI's activities, which will survey and investigate all known sources of energy in and around Saturn. Such a survey is expected to turn up crucial data that scientists require for an understanding of the dynamics of the ringworld. Among other things, the MIMI will investigate storm patterns in its atmosphere, such as the mysterious white spots that revealed themselves to the HST and ground-based observers in 1990 and 1994. Meanwhile, the RPWS will monitor radio waves given off by the interaction of the solar wind with Saturn.

It is not difficult to see from the names of many Cassini instruments that they owe their genesis to the Voyager missions, although the Saturn-bound spacecraft is considerably more advanced. For instance, Cassini uses computer chips for data storage, whereas the Voyagers and even Galileo employed magnetic tape. This means that Cassini can store a million times more information than the Voyagers. It is also better at navigating its journey to the ringworld. Whereas the Voyagers have a tenth-of-a-degree field of view that they use to find stars and their position relative to the Sun, Cassini's vision is 15 degrees wide.

It also has a 'star map' and can lock onto moving objects to keep them in view. This eliminates the need to steer 'blindly'. "When you don't have a person behind the lens," says Miner, "it is handy to have a camera that tracks your subjects automatically. We had to point the camera on each object for Voyager, and we don't have to do that [for Cassini]. It does it on its own." Ultimately, Cassini's data from Saturn should be in the range of 2.5 terabytes – a million, *million* bytes – compared to about 125 gigabytes for each of the Voyagers.

A MULTI-FACETED SPACECRAFT

As we have seen, Cassini's instruments have been designed to expect the unexpected; they are robust and can be applied to whatever conditions await them at Saturn. Of course, although new discoveries will undoubtedly be made serendipitously, a number of key scientific objectives have been outlined for the primary four-year tour. Tasks to be carried out at the giant planet itself include determining the temperature, cloud properties – including growth and dissipation over time – and composition of the atmosphere, measuring wind speeds, working out the relationship between the ionosphere and the magnetic field and investigating atmospheric phenomena such as lightning.

Saturn's rings will also be subjected to intensive scrutiny. Cassini will attempt to test numerous hypotheses of their formation by studying their composition and changes over time, investigating the role that shepherd moons play in constraining or increasing their extent, mapping the composition and size distribution of ring material – including dust particles – and examining their interaction with the planet's magnetosphere, ionosphere and atmosphere. This might lead to the development of new models about how rings form and precisely what impact a planet's Roche limit has on their evolution.

Scientific objectives at Titan primarily include a comprehensive survey of chemical constituents in its atmosphere as part of efforts to better understand and model its formation and evolution over time. Data in support of this task will be returned by both Huygens and Cassini. Additionally, ratios of one chemical to another will be determined and their locations in the Titanian atmosphere will be plotted, much like the layers of a cake. A search for complex organic molecules will be undertaken and possible energy sources for atmospheric chemistry – including the entry of ultraviolet sunlight – will be identified.

Both spacecraft, but primarily Huygens, will also work to produce the first accurate data on wind speeds, the formation, dissipation and behaviour of clouds, the existence of lightning and the physical nature, topography and composition of Titan's surface. Cassini's radar maps, and possibly even Huygens' observations, are also expected to determine whether the surface itself is composed of solid or liquid material, as well as working out ratios of one to another. Cassini may also shed further light on the radar-bright, Australia-sized 'continent' first seen by the HST and ground-based teams in 1994.

The bulk of scientists' attention in Saturn's moons seems to be focused on Titan.

This is not surprising. "Part of the excitement is that it's thought Titan in many ways may have started out like early Earth," says Matson. "But because it's so far from the Sun, things chilled early on and got frozen out. Perhaps Titan is an *in-situ* example of early Earth's development." Despite the tantalising possibility of finding long-dead microbial life or its building blocks, a concerted effort will also be made to perform detailed surveys of Saturn's other moons.

In fact, there are no fewer than eight planned flybys of six Saturnian moons: Phoebe, Rhea, Dione, Iapetus, Hyperion and Enceladus. As shown in Chapter 4, all six have their own mysteries and are worthy of detailed study. The survey of Iapetus – which displays mysterious 'dark' and 'bright' terrain – will be particularly exciting because, unlike Voyager 2, which flew no closer than 100,000 km, Cassini will swoop much closer to just 1,000 km! It will approach to within 500 km of the shining, frozen world Enceladus, more than a hundred times closer than Voyager's investigation.

JIMO: A MISSION TO JUPITER'S MOONS

Although Cassini is presently the only spacecraft *en route* to one of the gas giants visited by the Voyagers, numerous other ventures are planned in the near to middle future. One of these, unveiled by NASA in February 2003, is Project Prometheus, otherwise dubbed the Jupiter Icy Moons Orbiter (JIMO). Assuming that it goes ahead and is allocated the required funds by the US Congress, this mission will build on knowledge gained from the Voyagers and Galileo by conducting a detailed orbital

Artist's concept of JIMO at Jupiter, with Ganymede visible at far right.

survey of three Jovian moons – Callisto, Europa and giant Ganymede – thought to harbour vast oceans beneath their icy surfaces.

As well as offering an exciting opportunity to return to the realm of Jove, the JIMO venture is also an important milestone in demonstrating advanced propulsion technologies to reach the outer Solar System. It is perhaps significant that JIMO was announced by NASA only a few days before Galileo, the agency's Jupiter flagship, was officially retired in preparation for its kamikaze dive into the giant planet's atmosphere in the autumn of 2003.

JIMO is expected to employ technology known as 'nuclear electric propulsion', which allows the spacecraft to orbit three Jovian moons, one after the other. If successful, it could be used for follow-up missions to planets further from the Sun. A Neptune orbiter, possibly equipped with an atmospheric-entry probe, has for several years been deemed an important target for the second decade of the twenty-first century. JIMO is particularly significant because one of NASA's central aims is the search for extraterrestrial life, and Europa in particular may have the three ingredients considered essential for microbial life to evolve: water, organic compounds and adequate internal heat.

Certainly, Galileo suggested that melted water on Europa may have been in contact with the surface in geologically recent times and may still lie quite near it. This led the National Research Council to publish a report in 2002, which ranked a Europa Orbiter as a high priority for future 'flagship' missions, in view of the likelihood that it possesses an ocean and perhaps life. Although work on the Europa Orbiter has now ended, it would appear to have been resuscitated under a different name, and with orbital missions of two other intriguing Jovian moons thrown in for good measure.

JIMO's three main objectives are (1) to conduct detailed surveys of Europa, Callisto and Ganymede to determine whether conditions are sufficient for microbes to evolve, (2) to investigate their origins and interiors, including full geological, geophysical and geochemical analyses, to determine, among other things, if water oceans do exist and (3) to measure the radiation environment around them to see if it is sufficiently benign to sustain life. Possible instruments for the JIMO spacecraft have yet to be confirmed, but could conceivably include a radar for mapping the thickness of surface ice and a laser altimeter to measure surface elevations.

Other possible analytical tools might include cameras, infrared imaging equipment, a magnetometer and instruments to study charged particles, atoms and dust encountered by the spacecraft in the vicinity of each moon. In this sense, at least, the instruments would draw on their Voyager, Galileo and Cassini heritage. JIMO's nuclear electric propulsion device would provide a much higher power supply than has been possible with past and present RTG technology, which would assist with scientific data-collection and improve transmission rates to Earth.

In theory, a nuclear electric power source could allow JIMO to carry much more advanced and higher-powered instruments, perhaps including radars and imaging equipment that could map the moons' entire surfaces and resolve details as small as the size of a detached house. A kind of electric propulsion – 'ion propulsion' – was successfully tested by the Deep Space-1 mission, which used solar panels to generate electricity for an experimental ion engine. JIMO, on the other hand, would fly too far from the Sun for solar panels to be practical, so it would derive power for its ion thrusters from a nuclear fission reactor.

It is estimated that this system and the spacecraft itself would not be ready for launch until at least 2011. A heavy-lift launch vehicle would carry JIMO into high Earth orbit, after which the spacecraft's ion thrusters would spiral it away from our planet and onto a direct trajectory to Jupiter. After inserting itself into orbit around the giant planet, it would then undertake numerous exploratory ventures to Callisto, then Ganymede and finally Europa.

INTERSTELLAR PROBE

The development of advanced propulsion technologies to achieve short journey times to the outer planets and even the edge of our Solar System is not new. In fact, an obscure mission called 'Interstellar Probe' already exists on NASA's drawing boards, which calls for a spacecraft powered by either nuclear electric technology or solar sails. Although it is not expected to be launched anytime before 2007, it is hoped that such technologies may be able to deliver the spacecraft at phenomenal speeds across the Solar System. At top speed, it is expected to travel five times faster than any previous craft.

If the Interstellar Probe flies, it may carry out the first detailed survey of conditions outside our Solar System. A solar sail seems the most likely propulsion device for the spacecraft and would allow it to journey 200 AU – nearly 30 billion km – from the Sun within 15 years of launch, overtaking even the Voyagers. Other missions that might employ nuclear electric propulsion or solar sails could be Uranus or Neptune orbiters. Of these, Neptune and its enigmatic moon Triton are perhaps the most attractive of the two. Moreover, many questions still remain to be answered about Neptune's surprisingly active atmosphere.

ATMOSPHERIC BALLOONS ON TITAN?

Future missions are so exotic that it is an impossible task to investigate each one here. One interesting proposal, however, has been an effort to develop an all-weather inflatable balloon to 'float' in Titan's thick atmosphere around 2010–2011. Engineer Jack Jones of JPL and Ralph Lorenz of the University of Arizona's Lunar and Planetary Laboratory brought up the idea during the Space Technology & Applications International Forum, held at JPL in February 2002. "Titan has a density about four times that of Earth's surface atmosphere," said Jones. "That makes it ideal for floating balloons. Any balloon on Earth of a given size would carry four times the payload on Titan."

At present called the Titan Aerover, the balloon would fly at an altitude of about 10 km above the moon's surface, completing a full circuit through the atmosphere every one or two weeks. By travelling underneath Titan's opaque cloud deck, it could train its on-board cameras and multi-spectral imaging equipment on the surface features. It might also include a landing wheel to touch down periodically at interesting sites.

The Titan Aerover would closely resemble an airship and would be made from very strong fabric, along with a special glue and sealant capable of withstanding the low atmospheric and surface temperatures prevalent on the planet-sized moon. "This material would allow us to fly at all the altitudes that we're looking at," said Jones. He added that the balloon could be delivered to Titan in a manner not dissimilar to the 'aeroshell' that dropped NASA's Sojourner rover onto Mars in July 1997. Furthermore, unlike Richard Branson's ill-fated round-the-world trips, the Titan Aerover would encounter far more benign atmospheric dynamics.

THE FINAL FRONTIER: PLUTO AND NEW HORIZONS

At present, however, the main outer-planet mission on NASA's agenda is a flyby of distant Pluto. Such a venture has been under discussion for many years, not least because the planet's atmosphere is expected to freeze midway through the second

Artist's concept of the New Horizons spacecraft at Pluto. Courtesy of Johns Hopkins University Applied Physics Laboratory and the Southwest Research Institute.

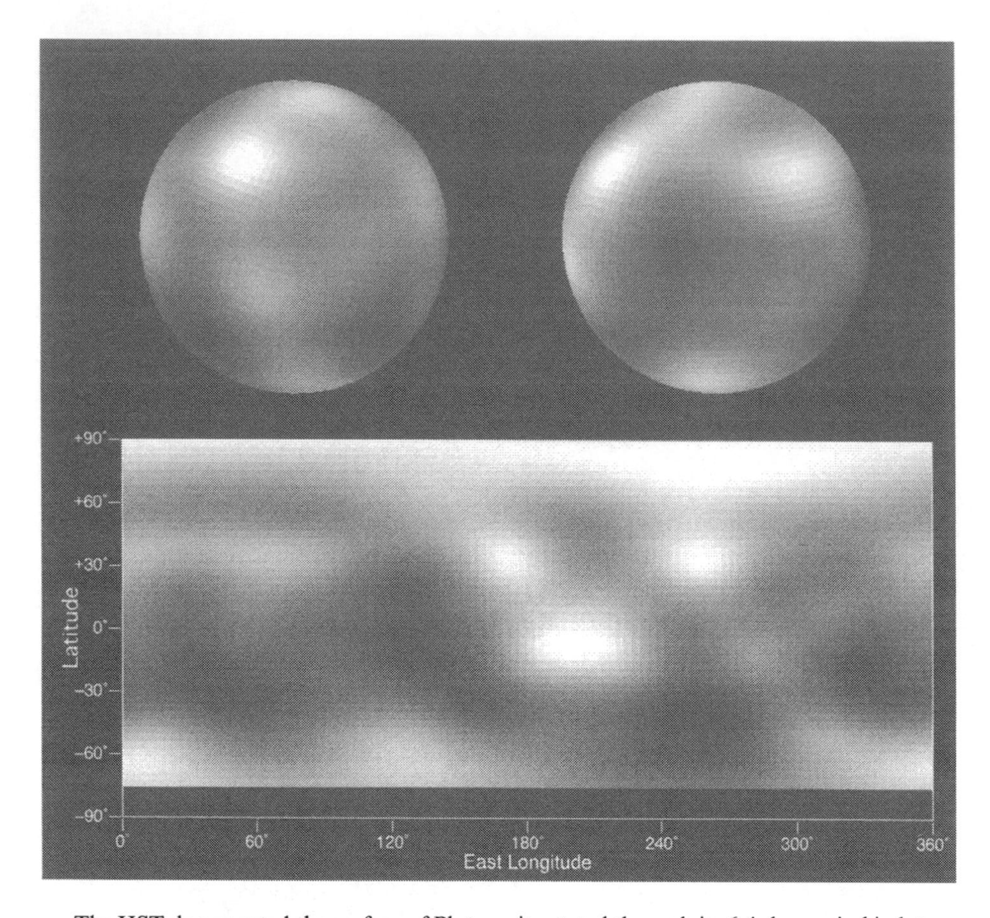

The HST documented the surface of Pluto as it rotated through its 6.4-day period in late June and early July 1994, and then a global map covering 85 per cent of the surface was constructed by image processing (two hemispheres of which are shown here) revealing bright and dark surface features. Courtesy of Alan Stern of the Southwest Research Institute and Marc Buie of the Lowell Observatory, and NASA/ESA.

decade of this century. In January 2006, the New Horizons spacecraft will be launched on a nine-year journey to Pluto and its moon, Charon. After hurtling past both worlds – which are actually more likely to be a 'double planet' than a planet and its moon – the spacecraft will head outwards to encounter one or more Kuiper Belt Objects (KBOs) around 2026.

The Pluto mission has come about primarily because of the planet's unusual elliptical orbit around the Sun, which brings it as close as 30 AU (4.5 billion km) and as far as 50 AU (7.5 billion km) from our parent star. This broad difference in solar distances causes profound changes to the appearance of Pluto and its thin atmosphere. As it moves further from the Sun, the atmosphere actually sublimates as frost onto its surface. Getting a spacecraft to Pluto before this occurs is crucial.

This will mark the first attempt by an emissary from Earth to visit Pluto and scientists are still unsure of what to expect. Even the best-resolution pictures, taken by HST, revealed little more than a few bright and dark blotches on its surface, although it may be somewhat similar to Neptune's moon Triton in nature. Yet if these bright and dark features are visible from such immense distances (Pluto currently lies 5.9 billion km from the Sun), they must be significant and scientists expect the small world to reveal evidence of a very complex geological history.

Pluto's thin nitrogen atmosphere was first identified in 1988. Its surface pressure is about 100,000 times smaller than the Earth's own, but even that is sufficient to provide weather, winds, particulate haze and perhaps an ionosphere. However, its gravity is so weak that whatever atmosphere is present is probably not very close to the planet. Charon, too, is difficult to assess. Pluto has an equatorial diameter of only about 2,370 km and Charon measures somewhere between half and two-thirds the size of its parent planet.

In many ways, therefore, Pluto and Charon can better be classed as 'companions', rather than a planet–moon system like Earth and our Moon. Although both evolved in the dark, cold outer reaches of the Solar System, they are quite different from one another, probably due to their differing masses. Charon consists primarily of frozen water, with a very thin (or even non-existent) atmosphere. It was discovered by James Christy of the US Naval Observatory in 1978, nearly half a century after Pluto itself was found by Clyde Tombaugh. Pluto, on the other hand, is covered with nitrogen frost and some methane and carbon monoxide ices.

Both have densities about twice that of water, implying they are roughly two-thirds rock and a third water-ice. Unlike every other planet except Uranus, Pluto rotates on its side, as does Charon in its orbit around its bigger host. The distance between planet and moon is also quite small – only 19,636 km, compared to the 384,400 km that separates Earth and our Moon – and Charon orbits at exactly the same rate as Pluto rotates. This is an extreme case of 'synchronous rotation': extreme because not only is the moon's rotation tidally locked, but so too is the rotation of Pluto. In other words, to an observer standing on Pluto's surface, Charon would perpetually hang in the sky, never rising or setting.

A HELPING HAND FROM JUPITER

The best route to get to Pluto, assuming New Horizons is sent aloft in January 2006, would involve picking up a gravitational assist from giant Jupiter a year after launch. As we have already seen, meeting this launch window could get the spacecraft to the Pluto–Charon system within nine years, which may be just soon enough before the atmosphere starts to freeze. Scientists are keen to arrive while there is still chance to see a 'substantial' (by Plutonian standards) atmosphere. The relatively quick arrival should allow the majority of the surfaces of Pluto and Charon to be mapped before they are covered with frost.

At some point in 2014, within a year of the Pluto–Charon rendezvous, the spacecraft's cameras would be switched on, probably resolving the 'double planet' as little more than a pair of bright dots at first, but eventually growing large enough for surface features to be discerned. Then, about three months before closest approach, from a distance of 105 million km, New Horizons' imaging gear will begin to produce its first detailed maps of the surfaces of both worlds.

A fortnight before the spacecraft reaches its nearest point to Pluto, cartographers will compile maps and take spectral measurements twice daily to track changes over the course of the planet's 'day'. Then, from 12 hours before until 12 hours after closest approach, it will look for ultraviolet emissions from Pluto's atmosphere and produce its best-yet global maps for both worlds, as well as undertaking multi-spectral surveys of their chemical and mineralogical compositions and surface temperatures. According to present estimates, the spacecraft will fly 9,600 km past Pluto and 27,000 km past Charon.

FIRST GLIMPSE OF A KUIPER BELT OBJECT?

During this period, it will also photograph surface features on Pluto as small as 60 m across. As it passes beyond the planet, it will look back at the darkened far sides of both worlds in an effort to reveal upper-atmospheric haze, look for rings and determine whether their surface textures are 'rough' or 'smooth'. Following the encounter, the spacecraft may be redirected to visit one or more KBOs – small lumps of primordial rock and ice left over from the Solar System's formation – and could encounter the first of these as early as 2026.

These objects are thought to be quite similar to whatever planetesimals formed the cores of Jupiter, Saturn, Uranus and Neptune, so a detailed survey of one or more of them might yield clues about the deep interiors of the gas giants. Although the first KBO was only discovered in 1992, their existence had been predicted several decades earlier by Kenneth Edgeworth and Gerard Kuiper. Their colours vary from red – possibly due to chemical composition or the effects of bombardment by solar radiation – to grey, which might be caused by impacts that extrude new, cleaner frost from beneath their surfaces.

Edgeworth was first to speculate that a vast reservoir of comparatively small objects might reside beyond the orbit of Neptune. It seemed strange to both him and, later, Kuiper that the distribution of mass in the Solar System, excluding the Sun, should be highest in the centre – at Jupiter and Saturn – then trail off at Uranus and Neptune and fade away to nothing beyond those outer giants. Moreover, the reappearance of 'long-period' comets like Hale-Bopp and Hyakutake in the late 1990s implies that they normally reside in the far-off Oort Cloud, extending a thousand times beyond Pluto.

True, some long-period comets do have the misfortune of venturing too close to Jupiter and their orbits are altered to shorter periods. However, even the biggest planet in our Solar System does not have the gravitational power to account for *all* the short-period comets – like Halley's Comet, for example – that frequent the inner heliosphere. This led some astronomers to speculate that, in addition to the Oort Cloud, there may be another, closer source of short-period comets, and a search beyond the orbit of Neptune revealed the first KBO. The largest-known KBO is currently Varuna, at 900 km in diameter – much larger than Ceres, the largest of the asteroids that orbit in the belt located between Mars and Jupiter. Due to their immense distance from the Sun, most KBOs have surface temperatures between 61 K and 33 K and are probably composed of extremely thick water-ice overlying a rocky core.

Nevertheless, every known KBO is very faint. Even the brightest is a hundred times fainter than Pluto. Some scientists even suspect that Pluto and Charon are not a planet-and-moon system, but are simply the 'new largest' KBOs. A decade after its Pluto–Charon rendezvous, New Horizons should encounter one of these mysterious objects and conduct the first-ever detailed chemical and mineralogical survey of it.

OTHER OUTER-PLANET VENTURES

It seems unlikely that any other agency besides NASA has the resources to stage missions to the outer planets in the near future. Jupiter and Saturn are more easily accessible than Uranus and Neptune, and scientists have more reason to go there, thanks to the Jovian moons and Titan. However, until two years ago 'Neptune Orbiter' was listed as a potential flagship mission for consideration beyond 2007. Then, in the summer of 2002, engineer Steve Oleson of NASA's Glenn Research Center in Cleveland, Ohio, told the author that he just completed a plan to reach Uranus within a decade ...

At present, the only mission headed for an outer planet is Cassini–Huygens, which, at the time of writing, was a year away from encountering Saturn. After that, it would appear that future ventures hinge on the successful demonstration and operational use of advanced nuclear electric propulsion and solar sail technologies. If JIMO and the Interstellar Probe reach fruition and prove that low-cost spacecraft *can* be sent to far-off destinations in a fraction of the time it took the Voyagers, then we may see a multitude of missions to Jupiter, Saturn, Uranus and Neptune in the next few decades.

Bibliography

Beatty, J.K., Petersen, C.C. and Chaikin, A. (eds) (1999) *The New Solar System* (4th edn). Cambridge, Massachusetts: Sky Publishing Corporation.

Benedetti, L.R., Nguyen, J.H., Caldwell W.A., Liu, H., Kruger, M. and Jeanloz, R. (1999) Dissociation of CH_4 at high pressures and temperatures: diamond formation in giant planet interiors? *Science* 286: 100–102.

Bergstralh, J., Miner, E.D. and Matthews, M.S. (1991) *Uranus*. Tucson: University of Arizona Press.

Bergstralh, J. and Miner, E.D. (1991) The Uranian system. In J. Bergstralh, E.D. Miner and M.S. Matthews, *Uranus*. Tucson: University of Arizona Press, pp. 3–28.

Burgess, E. (1988) *Uranus and Neptune: The Distant Giants*. New York: Columbia University Press.

Burgess, E. (1991) *Far Encounter: The Neptune System*. New York: Columbia University Press.

Burns, J.A. (1999) Planetary rings. In J.K. Beatty, C.C. Petersen and A. Chaikin (eds), *The New Solar System* (4th edn). Cambridge, Massachusetts: Sky Publishing Corporation, pp. 221–240.

Crocco, G.A. (1956) One-ear exploration trip Earth–Mars–Venus–Earth. *Proceedings of the Eighth International Astronomical Congress*. Rome, Italy.

Cruikshank, D.P. (ed.) (1995) *Neptune and Triton*. Tucson: University of Arizona Press.

Cruikshank, D.P. (1999) Triton, Pluto and Charon. In Beatty, J.K., C.C. Petersen and A. Chaikin (eds), *The New Solar System* (4th edn). Cambridge, Massachusetts: Sky Publishing Corporation, pp. 285–296.

Cruikshank, D.P., Brown, R.H. and Clark, R.N. (1984) Nitrogen on Triton. *Icarus* 58: 233–305.

Davies, J.K. (1981) A brief history of the Voyager project: the end of the beginning. *Spaceflight* 23: 35–41.

Ehricke, K.A. (1957) Instrumented comets – astronautics of solar and planetary probes. *Eighth International Astronautical Conference*. Barcelona, Spain.

Elliot, J.L., Dunham, E.W. and Mink, D.J. (1977) The rings of Uranus. *Nature* 267: 328–330.

Elliot, J.L. and Kerr, R. (1984) *Rings*. Cambridge, Massachusetts: MIT Press.

Esposito, L.W., Brahic, A., Burns, J.A. and Marouf, E.A. (1991) Particle properties and processes in Uranus' rings. In J.T. Bergstralh, E.D. Miner and M.S. Matthews (eds), *Uranus*. Tucson: University of Arizona Press, pp. 410–465.

Flandro, G.A. (1966) Fast reconnaissance missions to the outer Solar System utilising energy derived from the gravitational field of Jupiter. *Astronautica Acta* 12: 329–337.

Flandro, G.A. (2001) From instrumented comets to grand tours: on the history of gravity assist. *American Institute of Aeronautics and Astronautics,* Paper 2001–0176.

French, R.G., Nicholson, P.D., Porco, C.C. and Marouf, E.A. (1991) Dynamics and structure of the Uranian rings. In J.T. Bergstralh, E.D. Miner and M.S. Matthews (eds), *Uranus*. Tucson: University of Arizona Press, pp. 327–409.

Greeley, R. (1994) *Planetary Landscapes* (2nd edn). New York: Chapman & Hall.

Greeley, R. (1999) Europa. In J.K. Beatty, C.C. Petersen and A. Chaikin (eds), *The New Solar System* (4th edn). Cambridge, Massachusetts: Sky Publishing Corporation, pp. 253–262.

Hammel, H.B., Lockwood, G.W., Mills, J.R. and Barnet, C.D. (1995) Hubble Space Telescope imaging of Neptune's cloud structure in 1994. *Science* 268: 1740.

Hanlon. M. (2001) *The Worlds of Galileo*. London: Constable.

Harland, D.M. (2001) *Jupiter Odyssey*. Chichester: Springer–Praxis.

Harland, D.M. (2001) *The Earth in Context*. Chichester: Springer–Praxis.

Harland, D.M. (2002) *Mission to Saturn*. Chichester: Springer–Praxis.

Hubbard, W.B. (1984) *Planetary Interiors*. New York: Van Nostrand Reinhold.

Hubbard, W.B., Brahic, A., Sicardy, B., Elicer, L.R., Roques, F. and Vilas, F. (1986) Occultation detection of a Neptunian ring-like arc. *Nature* 319: 636–640.

Hubbard, W.B. (1999) Interiors of the giant planets. In J.K. Beatty, C.C. Petersen and A. Chaikin (eds), *The New Solar System* (4th edn). Cambridge, Massachusetts: Sky Publishing Corporation, pp. 193–200.

Hunt, G. and Moore, P. (1989) *Atlas of Uranus*. Cambridge: Cambridge University Press.

Ingersoll, A.P. (1999) Atmospheres of the giant planets. In J.K. Beatty, C.C. Petersen and A. Chaikin (eds), *The New Solar System* (4th edn). Cambridge, Massachusetts: Sky Publishing Corporation, pp. 201–220.

Johnson, T. (1999) Io. In J.K. Beatty, C.C. Petersen and A. Chaikin (eds), *The New Solar System* (4th edn). Cambridge, Massachusetts: Sky Publishing Corporation, pp. 241–252.

Keeler, J.E. (1895) Spectroscopic proof of the meteoritic constitution of Saturn's rings. *Astrophysical Journal* 1:416–427.

Kivelson, M.G., Khurana, K.K., Joy, S., Russell, C.T., Southwood, D.J., Walker, R.J and Polanskey, C. (1997) Europa's magnetic signature: report from Galileo's pass on 19 December 1996. *Science* 276: 1239–1241.

Lawden, D.F. (1954) Perturbation maneuvers. *Journal of the British Interplanetary Society* 13:6.

Lellouch, E., Paubert, G., Moses, J.I., Schneider, N.M. and Strobel, D.F. (2003) Volcanically emitted sodium chloride as a source for Io's neutral clouds and plasma torus. *Nature* 421: 45–47.

Leutwyler, K. (2000) Ganymede's hidden ocean. *Scientific American*, 19th December.

Lissauer, J.J. (1985) Shepherding model for Neptune's arc ring. *Nature* 318: 544–545.

Littmann, M. (1990) *Planets Beyond*. Ontario: John Wiley & Sons.

Lockwood, G.W., Thompson, D.T., Hammel, H.B., Birch, P. and Candy, M. (1991) Neptune's cloud structure in 1989. Photometric variations and correlation with ground-based images. *Icarus* 90: 299–307.

Lorenz, R. and Mitton, J. (2002) *Lifting Titan's Veil*. Cambridge: Cambridge University Press.

McKinnon, W.B. (1999) Midsize icy satellites. In J.K. Beatty, C.C. Petersen and A. Chaikin (eds), *The New Solar System* (4th edn). Cambridge, Massachusetts: Sky Publishing Corporation, pp. 297–310.

McNab, D. and Younger, J. (1999) *The Planets*. London: BBC Worldwide.

Macknight, N. (ed.) (1991) *Space Year 1991*. Osceola, Wisconsin: Motorbooks International Publishers and Wholesalers.

Miner, E.D. (1998) *Uranus: The Planet, Rings and Satellites*. Chichester: Springer–Praxis.

Miner, E.D. and Wessen, R.R. (2002) *Neptune: The Planet, Rings and Satellites*. Chichester: Springer–Praxis.

Minovich, M.A. (1961) A method for determining interplanetary free-fall reconnaissance trajectories. *JPL Technical Memoir* 312: 130.

Moore, P. and Hunt, G. (1983) *The Atlas of the Solar System*. London: Mitchell Beazley.

Moore, P. (1996) *The Planet Neptune* (2nd edn). Chichester: Springer–Praxis.

Morrison, D. and Owen, T. (1996) *The Planetary System* (2nd edn). Reading, Massachusetts: Addison-Wesley.

Moulton, F.R. (1905) Evolution of the Solar System. *Astrophysical Journal* 22: 165.

Muhleman, D.O., Grossman, A.W., Butler, B.J. and Slade, M.A. (1990) Radar reflectivity of Titan. *Science* 248: 975–980.

Murray, B.C. (1989) *Journey into Space*. London: W.W. Norton & Company.

Owen, T. (1999) Titan. In J.K. Beatty, C.C. Petersen and A. Chaikin (eds), *The New Solar System* (4th edn). Cambridge, Massachusetts: Sky Publishing Corporation, pp. 277–284.

Pappalardo, R.T. (1999) Ganymede and Callisto. In J.K. Beatty, C.C. Petersen and A. Chaikin (eds), *The New Solar System* (4th edn). Cambridge, Massachusetts: Sky Publishing Corporation, pp. 263–276.

Peale, S.J., Cassen, P. and Reynolds, R.T. (1979) Melting of Io by tidal dissipation. *Science* 203: 892–894.

Petersen, C.C. and Brandt, J.C. (1995) *Hubble Vision*. Cambridge: Cambridge University Press.

Porco, C.C., Nicholson, P.D., Cuzzi, J.N., Lissauer, J.J. and Esposito, L.W. (1995) Neptune's ring system. In D.P. Cruikshank (ed.), *Neptune and Triton*. Tucson: University of Arizona Press, pp. 703–804.

Rannou, P., Hourdin, F. and McKay, C.P. (2002) A wind origin for Titan's haze structure. *Nature* 418: 853–856.

Reitsema, H.J., Hubbard, W.B., Lebofsky, L.A. and Tholen, D.E. (1982) Occultation by a possible third satellite of Neptune. *Nature* 215: 289–291.

Rothery, D.A. (1992) *Satellites of the Outer Planets*. Oxford: Clarendon Press.

Schenk, P.M. and Bulmer, M.H. (1998) Origin of montains on Io by thrust faulting and large-scale mass movements. *Science* 279: 1514.

Smith, P.H., Lemmon, M.T., Lorenz, R.D., Sromovsky, L.A., Caldwell, J.J. and Allison, M.D. (1996) Titan's surface, revealed by HST imaging. *Icarus* 119: 336–349.

Sromovsky, L.A., Fry, P.M., Dowling, T.E., Baines, K.H. and Limaye, S.S. (2001) Coordinated 1996 HST and IRTF imaging of Neptune and Triton III: Neptune's atmospheric circulation and cloud structure. *Icarus* 149: 459–488.

Spencer, J.R., Calvin, W.M. and Person, M.J. (1995) CCD spectra of the Galilean satellites: molecular oxygen on Ganymede. *Journal of Geophysical Research* 100: 19049–19056.

Standage, T. (2000) *The Neptune File*. London: Penguin.

Stone, E.C. and Miner, E.D. (1986) The Voyager 2 encounter with the Uranian system. *Science* 233: 39–43.

Taylor, S.R. (1992) *Solar System Evolution*. Cambridge: Cambridge University Press.

Trefil, J. (1999) *Other Worlds*. Washington, DC: National Geographic Society.

Turtle, E.P. and Pierazzo, E. (2001) Thickness of a Europan ice shell from impact crater simulations. *Science* 294: 1326–1328.

Tyson, N.C. de Grasse and Irion, R. (2000) *One Universe*. Washington, DC: Joseph Henry Press.

Van Allen, J.A. and Bagenal, F. (1999) Planetary magnetospheres and the interplanetary medium. In J.K. Beatty, C.C. Petersen and A. Chaikin (eds), *The New Solar System* (4th edn). Cambridge, Massachusetts: Sky Publishing Corporation, pp. 39–58.

INTERNET REFERENCES

Alexander, A., Galileo smells volcano's strong breath on Io.
http://www.planetary.org/html/news/articlearchive/headlines/2001/iovolcanoes.html
Accessed December 2002.

Bridges, A., Ocean lurks deep in Ganymede, Galileo finds.
http://www.space.com/searchforlife/ganymede_ocean_001215.html.
Accessed January 2003.

Britt, R.R., Earth might have been a ringed planet, like Saturn.
http://www.space.com/scienceastronomy/planetearth/earth_rings_020917.html
Accessed February 2003.

Britt, R.R., Jupiter's moon count soars to 52 with four new discoveries.
http://www.space.com/scienceastronomy/jupiter_moons_030310.html
Accessed March 2003.

Florida Today Staff, Europa has scientists excited.
http://www.floridatoday.com/space/explore/stories/1997/032297a.htm
Accessed December 2002.

Florida Today Staff, New images hint at wet and wild history for Europa.

http://www.flatoday.com/space/explore/releases/1997/n97066.htm
Accessed December 2002.

Jet Propulsion Laboratory, Closest-ever picture of volcanic moon Io released.
http://galileo.jpl.nasa.gov/news/release/press991022.html
Accessed October 2002.

Jet Propulsion Laboratory, Europa's ice crust is deeper than 3km, UA scientists find.
http://galileo.jpl.nasa.gov/news/release/press011108.html
Accessed October 2002.

Jet Propulsion Laboratory, Cambridge: Cambridge Uni Farewell, Io; Galileo paying last visit to a restless moon.
http://galileo.jpl.nasa.gov/news/release/press020115.html
Accessed October 2002.

Jet Propulsion Laboratory, Final looks at Jupiter's moon Io aid big-picture view.
http://galileo.jpl.nasa.gov/news/release/press020528.html
Accessed October 2002.

Jet Propulsion Laboratory, Galileo evidence points to possible water world under Europa's icy crust.
http://galileo.jpl.nasa.gov/news/release/press000825.html
Accessed October 2002.

Jet Propulsion Laboratory, Galileo findings boost idea of other-worldly ocean.
http://galileo.jpl.nasa.gov/news/release/press000110.html
Accessed October 2002.

Jet Propulsion Laboratory, Galileo finds Arizona-sized volcanic deposit on Io.
http://www.jpl.nasa.gov/galileo/status971105.html
Accessed December 2002.

Jet Propulsion Laboratory, Galileo Millennium mission status report for 18 January 2002.
http://galileo.jpl.nasa.gov/news/release/press020118.html
Accessed October 2002.

Jet Propulsion Laboratory, Galileo returns new insights into Callisto and Europa.
http://www.solarviews.com/eng/galpr8.htm
Accessed December 2002.

Jet Propulsion Laboratory, Galileo sees dazzling lava fountain on Io.
http://galileo.jpl.nasa.gov/news/release/press991217.html
Accessed October 2002.

Jet Propulsion Laboratory, Galileo spacecraft finds Europa has atmosphere.
http://www.jpl.nasa.gov/galileo/status970718.html
Accessed December 2002.

Jet Propulsion Laboratory Interview with Rosaly Lopes-Gautier.
http://www.jpl.nasa.gov/solar_system/features/lopes_index.html
Accessed December 2002.

Jet Propulsion Laboratory, Io's volcanoes erase one dating method but may provide another.
http://galileo.jpl.nasa.gov/news/release/press010314.html
Accessed October 2002.

Jet Propulsion Laboratory, Jupiter orbiter nears first visit to small moon, dusty ring. http://galileo.jpl.nasa.gov/news/release/press021029.html Accessed November 2002.

Jet Propulsion Laboratory, Jupiter particles' escape route found. http://galileo.jpl.nasa.gov/news/release/press010531-1.html Accessed October 2002.

Jet Propulsion Laboratory, Jupiter's moon Callisto may hide salty ocean. http://www.jpl.nasa.gov/galileo/news32.html Accessed December 2002.

Jet Propulsion Laboratory, Jupiter's white ovals take scientists by storm. http://www.jpl.nasa.gov/galileo/news31.html Accessed December 2002.

Jet Propulsion Laboratory, New Galileo images reveal Hawaiian-style volcano on Io. http://galileo.jpl.nasa.gov/news/release/press991104.html Accessed October 2002.

Jet Propulsion Laboratory, Solar System's largest moon likely has a hidden ocean. http://galileo.jpl.nasa.gov/news/release/press001216.html Accessed October 2002.

Jet Propulsion Laboratory, Spacecraft at Io sees and sniffs tallest volcanic plume. http://galileo.jpl.nasa.gov/news/release/press011004.html Accessed January 2003.

Jet Propulsion Laboratory, Spacecraft to fly over source of recent polar eruption on Io. http://galileo.jpl.nasa.gov/news/release/press010802.html Accessed October 2002.

Jet Propulsion Laboratory, Two spacecraft see new plume activity on Jovian moon Io. http://galileo.jpl.nasa.gov/news/release/press010329.html Accessed October 2002.

Jet Propulsion Laboratory, Wandering plumes, seeing red and slip-sliding away on Io. http://galileo.jpl.nasa.gov/news/release/press000518.html Accessed October 2002.

Malik, T., New estimate for thickness of crust over Europa's ocean. http://www.space.com/scienceastronomy/solarsystem/europa_icecrust_011113.html. Accessed December 2002.

Space.com Staff, Cassini and Galileo spy massive plumes on Io. http://www.space.com/scienceastronomy/solarsystem/io_plumes_010329.html. Accessed December 2002.

Space.com Staff, Cassini spacecraft snaps first photo of Saturn. http://www.space.com/scienceastronomy/cassini_saturn_021101.html Accessed November 2002.

Space.com Staff, Galileo sees tallest volcanic plume on Jupiter's moon Io. http://www.space.com/scienceastronomy/solarsystem/io_plumes_011004.html. Accessed December 2002.

Space.com Staff, Hubble: moons dance in Jupiter's eerie aurora. http://www.space.com/scienceastronomy/solarsystem/jupiter_aurora_001214.html Accessed December 2002.

Space.com Staff, New images of Jupiter's moon Io reveal volcanic action.
http://www.space.com/scienceastronomy/solarsystem/tupan_io_011211.html
Accessed December 2002.

Space Telescope Science Institute, Comet fragment slams into Jupiter.
http://hubblesite.org/newscenter/archive/1995/15/
Accessed November 2002.

Space Telescope Science Institute, Hubble again views Saturn's rings edge-on.
http://hubblesite.org/newscenter/archive/1995/31/
Accessed January 2003.

Space Telescope Science Institute, Hubble captures detailed image of Uranus' atmosphere.
http://hubblesite.org/newscenter/archive/1996/15/
Accessed June 2002.

Space Telescope Science Institute, Hubble captures volcanic eruption plume from Io.
http://hubblesite.org/newscenter/archive/1997/21/
Accessed October 2002.

Space Telescope Science Institute, Hubble clicks images of Io sweeping across Jupiter.
http://hubblesite.org/newscenter/archive/1999/13/
Accessed December 2002.

Space Telescope Science Institute, Hubble discovers new bright spot on Io.
http://hubblesite.org/newscenter/archive/1995/37/
Accessed December 2002.

Space Telescope Science Institute, Hubble discovers new Dark Spot on Neptune.
http://hubblesite.org/newscenter/archive/1995/21/
Accessed July 2002.

Space Telescope Science Institute, Hubble finds many bright clouds on Uranus.
http://hubblesite.org/newscenter/archive/1998/35/
Accessed June 2002.

Space Telescope Science Institute, Hubble finds ozone on Jupiter's moon Ganymede.
http://hubblesite.org/newscenter/archive/1995/36/
Accessed October 2002.

Space Telescope Science Institute, Hubble finds oxygen atmosphere on Jupiter's moon, Europa.
http://hubblesite.org/newscenter/archive/1995/12/
Accessed October 2002.

Space Telescope Science Institute, Hubble follows rapid changes in Jupiter's aurora.
http://hubblesite.org/newscenter/archive/1996/32/
Accessed November 2002.

Space Telescope Science Institute, Hubble makes a movie of Neptune's rotation and weather.
http://hubblesite.org/newscenter/archive/1996/33/
Accessed July 2002.

Space Telescope Science Institute, Hubble observations shed new light on Jupiter collision.
http://hubblesite.org/newscenter/archive/1994/48/
Accessed January 2003.

Space Telescope Science Institute, Hubble observes a new Saturn storm.
http://hubblesite.org/newscenter/archive/1994/53/
Accessed February 2003.
Space Telescope Science Institute, Hubble provides the first images of Saturn's aurorae.
http://hubblesite.org/newscenter/archive/1995/39/
Accessed January 2003.
Space Telescope Science Institute, Hubble provides a moving look at Neptune's stormy disposition.
http://hubblesite.org/newscenter/archive/1998/34/
Accessed July 2002.
Space Telescope Science Institute, Hubble provides the first images of Saturn's aurorae.
http://hubblesite.org/newscenter/archive/1995/39/
Accessed January 2003.
Space Telescope Science Institute, Hubble tracks Jupiter storms.
http://hubblesite.org/newscenter/archive/1995/18/
Accessed November 2002.
Space Telescope Science Institute, Hubble views Saturn ring-plane crossing.
http://hubblesite.org/newscenter/archive/1995/25/
Accessed January 2003.
Space Telescope Science Institute, Hubble Space Telescope helps find evidence that Neptune's largest moon is warming up.
http://hubblesite.org/newscenter/archive/1998/23/
Accessed July 2002.
Space Telescope Science Institute, Hubble Space Telescope observations of Neptune.
http://hubblesite.org/newscenter/archive/1995/09/
Accessed July 2002.
Space Telescope Science Institute, Hubble watches Uranus.
http://hubblesite.org/newscenter/archive/1997/36/
Accessed June 2002.
Space Telescope Science Institute, Huge spring storms rouse Uranus from winter hibernation.
http://hubblesite.org/newscenter/archive/1999/11/
Accessed June 2002.
Space Telescope Science Institute, NASA's Hubble Space Telescope views major storm on Saturn.
http://hubblesite.org/newscenter/archive/1991/04/
Accessed February 2003.
Space Telescope Science Institute, Satellite footprints seen in Jupiter aurora.
http://hubblesite.org/newscenter/archive/2000/38/
Accessed October 2002.
Space Telescope Science Institute, Saturn ring-plane crossing, November 1995.
http://hubblesite.org/newscenter/archive/1996/18/
Accessed February 2003.

Space Telescope Science Institute, UA scientist and team discover surface features cover Titan.
http://hubblesite.org/newscenter/archive/1994/55/
Accessed March 2003.

Sparks, H., Move over Europa: an ocean on Jupiter's Callisto?
http://www.space.com/scienceastronomy/solarsystem/callisto_water_010726.html
Accessed December 2002.

The Planetary Society Interview with Linda Morabito-Kelly.
http://www.planetary.org/voyager25/linda-kelly.html
Accessed December 2002.

The Planetary Society, Oxygen detected on the surface of Jupiter's moon Callisto.
http://www.planetary.org/html/news/articlearchive/headlines/1997/headln-121097.html
Accessed December 2002.

Views of the Solar System: Europa, Ice volcanoes reshape Europa's chaotic surface.
http://www.solarviews.com/eng/eurpr2.htm
Accessed January 2003.

Zak, A., Spacecraft: Planetary projects and concepts.
http://www.russianspaceweb.com/spacecraft_planetary_plans.html
Accessed January 2003.

Index